Transboundary freshwater dispute resolution

The UNU Programme on Integrated Basin Management focuses on water management, approaching the complex problematique from three particular angles: governance, capacity-building, and management tools. The programme is carried out through field-based research encompassing both natural and social sciences. It utilizes extensive networks of scholars and institutions in both developing and industrialized countries. This work is intended to contribute to policy-making by the United Nations and the international community, as well as to capacity-building in developing countries.

The Water Resources Management and Policy series disseminates the results of research carried out under the Programme on Integrated Basin Management and related activities. The series focuses on policy-relevant topics of wide interest to scholars, practitioners, and policy-makers.

Earlier books in this series are: *Hydropolities Along the Jordan River: Scarce Water and its Impact on the Arab-Israeli Conflict* by Aaron T. Wolf; *Managing Water for Peace in the Middle East: Alternative Strategies* by Masahiro Murakami; *Freshwater Resources in Arid Lands* edited by Juha I. Uitto and Jutta Schneider; *Central Eurasian Water Crisis: Caspian, Aral and Dead Seas* edited by Iwao Kobori and Michael H. Glantz; *Latin American River Basins: Amazon, Plata, and São Francisco* edited by Asit K. Biswas, Newton V. Cordiero, Benedito P.F. Braga, and Cecilia Tortajada; *Water for Urban Areas: Challenges and Perspectives* edited by Juha I. Uitto and Asit K. Biswas.

Transboundary freshwater dispute resolution: Theory, practice, and annotated references

By Heather L. Beach, Jesse Hamner, J. Joseph Hewitt, Edy Kaufman, Anja Kurki, Joe A. Oppenheimer, and Aaron T. Wolf

United Nations
University Press

TOKYO · NEW YORK · PARIS

© The United Nations University, 2000

The views expressed in this publication are those of the authors and do not necessarily reflect the views of the United Nations University.

United Nations University Press
The United Nations University, 53-70, Jingumae 5-chome,
Shibuya-ku, Tokyo, 150-8925, Japan
Tel: +81-3-3499-2811   Fax: +81-3-3406-7345
E-mail: sales@hq.unu.edu
http://www.unu.edu

United Nations University Office in North America
2 United Nations Plaza, Room DC2-1462-70, New York, NY 10017, USA
Tel: +1-212-963-6387   Fax: +1-212-371-9454
E-mail: unuona@igc.apc.org

United Nations University Press is the publishing division of the United Nations University.

Cover design by Joyce C. Weston
Cover photograph by Iwao Kobori

Printed in the United States of America

UNUP-1038
ISBN 92-808-1038-3

Library of Congress Cataloging-in-Publication Data

Transboundary freshwater dispute resolution: theory, practice, and annotated references / by Heather L. Beach ... [et al.].
    p.   cm.
  Includes bibliographical references and index.
  ISBN 92-808-1038-3
  1. Water-Law and legislation. 2. Transboundary pollution-Law and legislation. 3. Water rights (International law) I. Beach, Heather L. II. Title.
  K3496.T73   2000
  341.7'6253—dc21   00-008341

Contents

Preface

This work attempts to provide a comprehensive review of the relevant literature on managing conflicts stemming from the quantity and quality problems of water around the world. Current trends and projections suggest that disputes based on water scarcity escalate when the issue is not addressed effectively and in a timely manner. Proactive efforts to prevent these conflicts have been overwhelmed by pessimistic forecasting. This situation negatively affects multilateral cooperative efforts and results in attempts to pursue unilateral short-term gains and in some cases increases in military power. So far, few comprehensive and interdisciplinary analyses of such international surface water conflicts have been produced. Only fragmented findings and scattered experimentational endeavours are available to the conflict resolution community.

In this report we introduce knowledge from the fields of transboundary water disputes and conflict resolution. The latter incorporates the studies of political geography, economy, and hydrology. In addition, it draws upon the expertise of specialists in the areas of formal modelling, conflict resolution, and environmental and natural resources. The data in the report include a variety of factual information about individual cases as well as bilateral and multilateral agreements and general principles.

The literature surveyed indicates that while in many areas there has been extensive research and analysis, there continues to be a need for more studies on the specific situations that lead to conflicts over water and other environmental resources. Lateral learning, an attempt to understand the similarities between all conflicts over natural resources, will lend itself to future applications in predicting and preventing these conflicts.

A survey of international watersheds provides some bibliographical and general data collected from over 200 transboundary watersheds. Some countries share one or two river systems while others share quite a few. The India-Bangladesh Joint Rivers Commission counted 140 tributary systems in common. The fortunate corollary is that because water is such a scarce and vital resource, it has sometimes brought hostile co-riparians together into a cooperative mode, often with external incentives. A subset of case studies of the exhaustive list of international watersheds is examined in greater detail. A related effort is a compilation and analysis of relevant water treaties, and the rationale for the partial implementation of these treaties.

Acknowledgements

This project simply would not have been possible without the initiation and leadership of Ariel Dinar of the Agriculture and Natural Resources Division of the World Bank. While the final product constitutes a team effort at its best, the project is essentially his. The authors would also like to acknowledge and thank Jerome Delli Priscoli, Sandra Postel, and James Wescoat for their valuable comments in reviewing earlier drafts of this document. We are also grateful to Lili Monk for her copy-editing expertise.

Part 1

1

Introduction

The purpose of this document is to review the literature on water disputes and related water treaties to gain an understanding of why previous and current disputes over water have occurred and to seek out lessons to be learned in preventing similar future disputes. This work focuses on transboundary freshwaters. We use the term "transboundary" to refer to water that crosses between, or is shared by, nations, sub-national political units, economic sectors, or interests. For water that crosses between nations we refer to more specifically as "international waters." In addition, the sections on environmental resources represent lateral learning, an attempt to understand the similarities among conflicts over all natural resources. In future work, we will apply these lessons to several specific water disputes.

This document is divided into chapters on Theory and on Practice. Chapter 2 (Organizational theory) examines the theoretical literature that shapes how transboundary freshwaters (TFWs) are managed through institutions and law. The section dealing with negotiation theory broadly examines the diagnosis of conflict, the prognosis or resolution, and presents some analysis on individual and comparative case studies. Chapter 3 (Economic theory) details the literature related to using techniques of optimization and game theoretic models in the allocation and use of TFWs. Chapter 4 (Water disputes) explores the general background of water disputes and then follows with an in-depth comparative analysis. This analysis examines various issues of specific conflicts that are individually detailed in Chapter 7 (Case studies). Also included is a discussion of the history of treaties related to international waters. This portion provides a detailed summary of a treaties database. Chapter 5 (Environmental disputes) offers a general overview of a relatively new concept in environment literature, environmental security. In addition, there is a section on resources other than water, where issues of sources and types of conflicts related to natural resources such as oil, fishing, and air are examined. Some conclusions are gathered in Part 4. The Appendixes (Part 5) provide supplemental materials to the literature review. Chapter 7 is a detailed examination of 14 specific water disputes. Chapter 8 is a hardcopy preview of a systematic computer compilation of international water treaties and currently includes 140 treaties. Chapter 9 is an annotated bibliography of some materials found in our reference list, and Chapter 10 is a reference list of over 1,100 citations of related literature.

This document does not include the numerous works offering technical solutions. While engineering, hydrologic, or organizational aspects are extremely important, it seems that the problems of freshwater dispute resolution are superseded by constraints related to negotiating process by the sovereign stakeholder. Hence the focus of this work is more on political and social aspects, and the skills for dispute resolution. As Delli Priscoli (1989) explains, engineers and scientists need to expand beyond analytic solutions to water resources by adding techniques taken from the social sciences that are designed to facilitate reaching agreement. This is our focus, because hydropolitics are considered the main obstacle, and in general, as Biswas (1993) indicates, the management of international water resources was not given adequate attention in the last three decades.

The production of this document was conducted under the auspices of the Transboundary Fresh Water Disputes Project (TFWDP), a comprehensive and interdisciplinary analysis of international surface water conflicts. The TFWDP was an effort to create a qualitative and quantitative analysis of the situation at hand and to develop procedural and strategic templates for early intervention so as to help contain and manage conflicts. This survey of literature on water dispute resolution and related water treaties was funded in part by the World Bank Research Committee under the grant preparation fund with a majority of the research conducted during 1996.

Part 2: Theory

2
Organizational theory

Institutions and law[1]

Just as the flow of water ignores political boundaries, so too does its management strain the capabilities of institutional boundaries. While water managers generally understand and advocate the inherent powers of the concept of a watershed as a unit of management, where surface and groundwater, quantity and quality, are all inexorably connected, the institutions we have developed to manage the resource follow these tenets only in the exception. In the sections that follow, we review the current status of international water institutions and water law.

Water negotiations and institutional capacity

Frederiksen (1992) describes principles and practice of water resources institutions from around the world. He argues that while, ideally, water institutions should provide for ongoing evaluation, comprehensive review, and consistency among actions, in practice this integrated foresight is rare. Rather, he finds rampant lack of consideration of quality consider-ations in quantity decisions, a lack of specificity in rights allocations, dis-proportionate political power-by-power companies, and a general neglect for environmental concerns in water resources decision-making. Buck et al. (1993) describe an "institutional imperative" in their comparison of transboundary water conflicts in the United States (US) and the former

Soviet Union. Feitelson and Haddad (1995) take up the particular institutional challenges of transboundary groundwater.

To address these deficiencies at the international level, some have argued that international agencies might take a greater institutional role. Lee and Dinar (1995) describe the importance of an integrated approach to river basin planning, development, and management. Young et al. (1994) provide guidelines for coordination between levels of management at the global, national, regional, and local levels. Delli Priscoli (1989) describes the importance of public involvement in water conflict management. In other work (1992), he makes a strong case for the potential of Alternative Dispute Resolution (ADR) in the World Bank's handling of water resources issues. Trolldalen (1992) likewise chronicles environmental conflict resolution at the United Nations, including a chapter on international rivers. Most recently, the creation of the World Water Council has seen among its four primary challenges a "global institutional framework for water" (*WWC Bulletin*, December 1995).

While remaining optimistic, it is worth explicitly noting the difficulties that may present themselves as dispute resolution principles begin to permeate the government and non-government agencies responsible for transboundary resource negotiations. The first barrier that may preclude total reliance on ADR in its current state is that between scientific and policy analysis. As Ozawa and Susskind point out,

Scientific advice is [sometimes] reduced to an instrument for legitimating political demands. Scientific analysis, in turn, can distort policy disputes by masking, beneath a veneer of technical rationality, underlying concerns over the distribution of costs and benefits. (1985: 23)

Exacerbating this problem of science's tenuous relationship with policy analysis is the fact that training of diplomats is often in political science or law, while those scientists most competent to evaluate resource conflicts are rarely skilled in either diplomacy or policy analysis.

The second, somewhat subtler, barrier that diminishes ADR's usefulness in international water disputes is that between ADR practitioners and analysts. Zartman (1992) discusses a common practitioner's approach to environmental disputes either as a case of "problem-solving," where the disputants can dissociate themselves emotionally from the problem which is considered to be a distinct entity, a "game against nature;" or as a case of information dispute, where resolution becomes apparent in the process of clarifying the data. Zartman suggests that these views are incomplete, that they "assume away conflict, rather than explaining and confronting it" (1992: 114). He suggests steps, based on the ADR analyst's experience, to recognize conflicts of nature also as conflicts of interest: "Inherent in the conflict with nature is conflict among different parties' interests; inherent in problem solving is a need for conflict management" (1992: 114).

These barriers – between science and policy, and between practice and analysis – can individually lead to a convoluted and incomplete process of conflict resolution and, together, can preclude arrival at the "best" (Pareto-optimal or win-win) solution to a given problem. By concentrating on the process of conflict resolution, rather than the outcome, one can take on a much-needed dynamic, and ideally predictive, component.

International water law

According to Cano (1989: 168), international water law did not begin to be substantially formulated until after World War I. Since that time, organs of international law have tried to provide a framework for increasingly intensive water use, focusing on general guidelines that could be applied to the world's watersheds. These general principles of customary law, codified and progressively developed by advisory bodies and private organizations, are termed "soft law," and are not intended to be legally binding, but can provide evidence of customary law and may help crystallize that law. While it is tempting to look to these principles for clear and binding rules, it is more accurate to think of them in terms of guidelines for the process of conflict resolution.

The concept of a "drainage basin," for example, was accepted by the International Law Association (ILA) in the Helsinki Rules of 1966, which also provides guidelines for "reasonable and equitable" sharing of a common waterway (Caponera, 1985). Article V lists no fewer than 11 factors that must be taken into account in defining what is "reasonable and equitable."[2] There is no hierarchy to these components of "reasonable use"; rather they are to be considered as a whole. One important shift in legal thinking in the Helsinki Rules is that they address the right to "beneficial use" of water, rather that to water per se (Housen-Couriel, 1994: 10). The Helsinki Rules have explicitly been used only once to help define water use – the Mekong Committee used the Helsinki Rules definition of "reasonable and equitable use" in formulation of their Declaration of Principles in 1975, although no specific allocations were determined.[3]

Some nations raised objections as to how inclusive the process of drafting had been when the United Nations (UN) considered the Helsinki Rules in 1970. In addition and, according to Biswas (1993), more importantly, possibly interpreted as an infringement on a nation's sovereignty, some states (Brazil, Belgium, China, and France, for instance) objected to the prominence of the drainage basin approach. Others, notably Finland

and the Netherlands, argued that the most "rational and scientific" unit of management was the watershed. Others argued that, given the complexities and uniqueness of each watershed, general codification should not even be attempted. On 8 December 1970, the General Assembly directed its own legal advisory body, the International Law Commission (ILC) to study the "Codification of the Law on Water Courses for Purposes other than Navigation."

It is testimony to the difficulty of merging legal and hydrologic intricacies that the ILC, despite an additional call for codification at the UN Water Conference at Mar de Plata in 1977, has only just completed its task. For example, it took until 1984 for the term "international watercourse" to be satisfactorily defined (Krishna, 1995: 37–39). Problems both political and hydrological slowed the definition: in a 1974 questionnaire submitted to Member States, about half the respondents (only 20 per cent responded after eight years) supported the concept of a drainage basin (e.g. Argentina, Finland, and the Netherlands), while half were strongly negative (e.g. Austria, Brazil, and Spain) or ambivalent (Biswas, 1993); "watercourse system" connoted a basin, which threatened sovereignty issues; and borderline cases, such as glaciers and confined aquifers, both now excluded, had to be determined. In 1994, more than two decades after receiving its charge, the ILC adopted a set of 32 draft articles which, with revisions, were adopted by the UN General Assembly on 21 May 1997 as the "Convention on the Law of the Non-Navigational Uses of the International Watercourses."

The convention articles include language very similar to the Helsinki Rules, requiring riparian states along an international watercourse in general to communicate and cooperate. Included are provisions for exchange of data and information, notification of possible adverse effects, protection of ecosystems, and emergency situations. Allocations are dealt with through equally vague but positive language. Balanced with an obligation not to cause significant harm is "reasonable and equitable use" within each watercourse state, "with a view to attaining optimal utilization thereof and benefits therefrom." Based on seven factors, reasonable and equitable is defined similar to Helsinki.[4] The text of the convention does not mention a hierarchy of these factors, although Article 10 says both that, "in the absence of agreement or custom to the contrary, no use ... enjoys inherent priority over other uses," and that, "in the event of a conflict between uses ... [it shall be resolved] with special regard being given to the requirements of vital human needs."

When attempting to apply this reasonable but vague language to specific water conflicts, problems arise. For example, riparian positions and consequent legal rights shift with changing borders, many of which are still not recognized by the world community. Furthermore, international law only concerns itself with the rights and responsibilities of nations. Some political entities who might claim water rights, therefore, would not be represented, such as the Palestinians along the Jordan river or the Kurds along the Euphrates river.

Hydrography vs chronology

Extreme principles

Customary international law has focused on providing general guidelines for the watersheds of the world. In the absence of such guidelines, some principles have been claimed regularly by riparians in negotiations, often depending on where along a watershed a riparian state is situated. Many of the common claims for water rights are based either on hydrography, i.e. from where a river or aquifer originates and how much of that territory falls within a certain state, or on chronology, i.e. who has been using the water the longest.

Initial positions are usually extreme (Housen-Couriel, 1994; Matthews, 1984). Often claimed initially by an upstream riparian is the "doctrine of absolute sovereignty." This principle, referred to as the Harmon Doctrine (named for the nineteenth century US attorney-general who suggested this stance regarding a dispute with Mexico over the Rio Grande river), argues that a state has absolute rights to water flowing through its territory. Considering this doctrine was eventually rejected by the United States (itself a downstream riparian of several rivers originating in Canada), never implemented in any water treaty, nor invoked as a source for judgment in any international water legal ruling, the Harmon Doctrine is wildly overemphasized in the literature as a principle of international law.

The downstream extreme claim often depends on climate. In a humid watershed, the extreme principle advanced is "the doctrine of absolute riverain integrity," which suggests that every riparian has entitlement to the natural flow of a river system crossing its borders. This principle has reached acceptance in the international setting as rarely as absolute sovereignty. In an arid or exotic (humid headwaters region with an arid downstream) watershed, the downstream riparian often has an older water infrastructure that is in its interest to defend. The principle that rights are acquired through older use is referred to as "historic rights" (or "prior appropriations" in the US), that is, "first in time, first in right."

These conflicting doctrines of hydrography and chronology clash along many transboundary rivers, with positions usually defined by relative riparian positions. Downstream riparians, such as Iraq and Egypt, often receive less rainfall than their upstream neighbours and therefore have depended on river water for much longer historically. As a consequence, modern "rights-based" disputes often take the form of upstream riparians such as Ethiopia and Turkey arguing in favour of the doctrine of absolute sovereignty, with downstream riparians taking the position of historic rights.[5]

Moderated principles

It quickly becomes clear in a negotiation that keeping to an extreme position leads to very little room for bargaining. Over time, rights become moderated with responsibility such that most states eventually accept some limitation to both their own sovereignty and to the river's absolute integrity. The "doctrine of limited territorial sovereignty" reflects rights to reasonably use the waters of an international waterway, yet with the acknowledgement that one should not cause harm to any other riparian state.

In fact, the relationship between "reasonable and equitable use," and the obligation not to cause "appreciable harm," is the more subtle manifestation of the argument between hydrography and chronology. As noted above, the convention includes provisions for both concepts, without setting a clear priority between the two. The relevant articles are:

Article 5: Equitable and reasonable utilization and participation
1. Watercourse States shall in their respective territories utilize an international watercourse in an equitable and reasonable manner. In particular, an international watercourse shall be used and developed by watercourse States with a view to attaining optimal and sustainable utilization thereof and benefits therefrom, taking into account the interests of the watercourse States concerned, consistent with adequate protection of the watercourse.
2. Watercourse States shall participate in the use, development and protection of an international watercourse in an equitable and reasonable manner. Such participation includes both the right to utilize the watercourse and the duty to cooperate in the protection and development thereof, as provided in the present Convention.

Article 7: Obligation not to cause significant harm
1. Watercourse States shall, in utilizing an international watercourse in their territories, take all appropriate measures to prevent the causing of significant harm to other watercourse States.
2. Where significant harm nevertheless is caused to another watercourse State, the State whose use causes such harm shall, in the absence of agreement to such use, take all appropriate measures, having due regard for the provisions of articles 5 and 6, in consultation with the affected State, to eliminate or mitigate such harm and, where appropriate, to discuss the question of compensation.

Article 10: Relationship between different kinds of use
1. In the absence of agreement or custom to the contrary, no use of an international watercourse enjoys inherent priority over other uses.

2. In the event of a conflict between uses of an international watercourse, it shall be resolved with reference to the principles and factors set out in articles 5 to 7, with special regard being given to the requirements of vital human needs.

Not surprisingly, upstream riparians have advocated that the emphasis between the two principles be on "equitable utilization," since that principle gives the needs of the present the same weight as those of the past. Likewise, downstream riparians have pushed for emphasis on "no significant harm," effectively the equivalent of the doctrine of historic rights in protecting pre-existing use.

According to Khassawneh (1995: 24), the Special Rapporteurs for the ILC project had come down on the side of "equitable utilization" until the incumbency of J. Evensen, the third rapporteur who, along with Stephen McCaffrey, the final rapporteur for the project, argued for the primacy of "no appreciable harm." Commentators have had the same problem reconciling the concepts as the rapporteurs: Khassawneh (1995: 24) suggests that the latter rapporteurs are correct that "no appreciable harm" should take priority, while, in the same volume, Dellapenna (1995: 66) argues for "equitable use." The World Bank, which must follow prevailing principles of international law in its funded projects, recognizes the importance of equitable use in theory but, for practical considerations, gives "no appreciable harm" priority – it is considered easier to define – and will not finance a project which causes harm without the approval of all affected riparians (see World Bank, 1993: 120; Krishna, 1995: 43–45).

Even as the principles for sharing scarce water resources evolve and become more moderate over time, the essential argument still emphasizes the rights of each state, and rests on the fundamental dispute between hydrography and chronology. Resulting agreements tend to be more rigid than is useful, precluding shifting demographics or anthropologic variables within a basin. In addition, many terms that are inherently vague both for reasons of legal interpretation and for political expediencey – "reasonable," "equitable," and "significant," for example – make precise definitions difficult during negotiations. Moreover, by excluding navigation and other non-consumptive uses, the convention might hinder negotiators from "enlarging the pie" to achieve an agreement.

Summary

Water not only ignores our political boundaries, it evades institutional classification and eludes legal generalizations. Interdisciplinary by nature, water's natural management unit, the watershed – where quantity, quality, surface and groundwater all interconnect – strains both institutional and legal capabilities often past capacity. Analyses of international water institutions find rampant lack of consideration of quality considerations in quantity decisions, a lack of specificity in rights allocations, disproportionate political power by special interest, and a general neglect for environmental concerns in water resources decision-making. Very recently, these weaknesses are beginning to be addressed by, for example, the World Bank, United Nations, and the new World Water Council.

Customary legal principles have been equally elusive. The 1997 convention reflects the difficulty of merging legal and hydrologic intricacies: while the articles provide many important principles for cooperation, including responsibility for cooperation and joint management, they also codify the inherent upstream/downstream conflict by calling for both equitable use and the obligation not to cause appreciable harm. They also provide few practical guidelines for allocations – the heart of most water conflicts. In contrast to general legal principles, site-specific treaties have shown great imagination and flexibility, moving from "rights-based" to "needs-based" agreements in order to circumvent the argument over use versus harm.

Negotiation theory

The structure of the section is broadly divided into (a) conflict (diagnosis) and focuses more extensively on (b) resolution (prognosis), covering more general references to natural resources and domestic transboundary freshwater dispute resolution (TFWDR). The related area of modelling and game theory is dealt with in this review below in Game theory. A second inquiry into the subject includes brief coverage of (c) individual case studies, comparative cases and generalizations. For the sake of simplicity, the term "conflict resolution" (CR) is used generically, although "dispute" has often been mentioned as a smaller level of "conflict," and "termination," "dissolution," "reduction," "management" and other adjectives have been advanced as different from "resolution."

Conflict

Often the causes for TFWDs are attributed to the tangible aspects of water as a natural resource. Grey (1994) provides alarming figures regarding the "carrying capacity" of the environment as the habitat and provider for human beings, with the increasing per-capita demand of a rapidly growing population and a declining renewable water supply. Postel provides a sober analysis of the available and renewable water supplies, and reminds us that "viewed globally, fresh water is still undeniably abundant" (1984: 7) and can sustain a moderate standard of living. However, the distribution is uneven and most troubled areas of scarcity lie in the regions of Asia and Africa, which feature a high rate of population growth. What makes the picture worrying is that mismanagement may result in as much as a fourth of the world's reliable water supply being rendered unsafe for use by the year 2000. This brings Falkenmark and Widstrand (1992) to the conclusion that more than 20 countries are already experiencing "water stress" (fewer than 1,000 cubic metres per capita of renewable water resources).

According to Mather (1989), understanding the constraints that impede development of African river and lake basin resources is a precondition for planning. Hence, one has to take into account both the physical and climatic obstacles, socio-cultural characteristics and current priorities of national economies.

General environmental changes are discussed as affecting the freshwater situation and are thoroughly studied in Gleick, *Water in Crisis* (1993), which also provides the reader with a wealth of data covering issues of quantity and quality affecting the ecosystem, health, and agriculture. In separate works by Gleick (1988, 1990, 1992) on future climatic changes, he dwells on the effects of trends such as the rise in sea level and changes in the timing and distribution of precipitation and runoff on renewable sources of fresh water. By adding the growing demand for fresh water to already growing populations, climate change may seriously jeopardize the relations among nations sharing a river or lake, a present security consideration. A new concept is being debated, environmental security, which is examined below in Chapter 5 of this document. Politics and unilateral economic development strategies exacerbate the already mentioned difficulties into a crisis situation that leads Gleick and many other contributors (Ohlsson, 1992 also quoting Ismael Seageldin on the cover) to the conclusion of imminent water wars in the coming twenty-first century (Gleick 1993: 108–110).

Through an inductive approach, Frederiksen's (1992) coverage of India's Sardar Sarovar Project *Water Crisis in the Developing World* considers that more attention to the analysis of crisis for a developing world should include the short time available to act, the limited measures available for securing essential water supplies, the competing demands for funds to provide adequate means, and the minimal ability to manage unpredicted droughts.

Other sources of conflict relate to different values, beliefs, and attitudes among stakeholders, individuals, and groups, as illustrated in Lynne et al.'s (1990) study of water management in Florida. Clearly, in unresolved conflicts such obstacles need to be taken into account while trying to set up institutional arrangements. While the Florida study represents a case of rather effective management, improvements are still suggested.

The deteriorating water situation as a potential cause of war is also dealt with by Anderson (1991), who discusses the likelihood of conflict among riparian states while describing the influence of geographical location, national interest, military, political and economic dimensions on hydropolitics.

It is interesting to note, however, that books dealing generically with the causes of war such as Brown (1987) and Cashman (1993) do not particularly focus on water conflicts. Furthermore, Wolf (1996) challenges the often-dramatized assumption of wars in the past as resulting from such conflicts. Gleick's (1993) study of historic conflicts over water shows how water was used and manipulated as an instrument of war, but not necessarily as the main cause for engaging in actual warfare for control of natural resources. Still, the importance of geopolitics as a determinant for the need to share water resources has produced serious crises in dyadic and regional relations. Yet, in relation to the future, the theme of water-driven violence recurs in common with other authors, and there is a flagging of the seriousness of forthcoming conflict over water. Clarke (1991) considers that while freshwater shortage, poverty, and overpopulation have contributed to the international water crisis; he contends that the possibilities for mitigating such conflict are related to traditional and technological solutions. Quigg (1977), in "Water Agenda to the Year 2000" presents a comprehensive summary of water problems and issues, discussing on the one hand the development of water resources (pure drinking water, efficient irrigation, recharge and water mining, industrial recycling, the protection of watersheds, wetlands and the problem of arid lands). On the other hand, he examines wastewater and treatment (discharge standards, urban and agricultural runoff, toxic wastes, groundwater, disposal of sewage and sanitation) arguing that water should be regarded as a vulnerable and finite resource, as are food and energy.

With a more political outlook, impediments for conflict resolution are attributed less to the large number of stakeholders in many transboundary waterways, but to the asymmetry in the power relations of those stakeholders. In an international regime such as the one under study, the absence of authoritative allocation standards makes individual states more resistant to compromise, although Krasner (1985) does not consider such asymmetries to be insuperable.

For some authors, in cases of violent and protracted conflicts, TFWDR is viewed as a "low politics issue" and is often subordinated to the "high politics" of the overall dispute (Lowi, 1993). The Middle East is often given as an example. De Silva (1994) makes a similar point in relation to the intractable political conflicts in South Asia and adds that potential future conflicts may become even more exacerbated over the sharing of scarce resources, especially water and irrigation works.

Resolution

Moving into the prognosis, namely, of how to resolve TFWDs, the writings can be grouped according to the main approaches to the subject.

General theory conflict resolution

1. General theory

To a large extent the general theory on CR did not highlight the cases of intra or inter-state TFWD, stressing general principles and different types of conflicts (dyadic relations among neighbouring countries, borders, domestic minorities strife, economic exploitation, etc.). In concrete illustrations, the tendency of Kelman (1990) and Azar (1990) was to focus primarily on ethnopolitcal disputes or territorial disputes. Furthermore, Diamond and MacDonald's (1992) work on Multitrack Diplomacy and Montville's (1987) relevant work on "Track Two Diplomacy" does not directly refer to water disputes. Kaufman (1996) brought this subject down to the actual exercises that allow participants to move forward from an initial adversarial stage to the search for common ground, introducing a framework that may be applicable to TFWDR cases.

On the other hand, the potential for utilizing problem-solving workshops as negotiating techniques are more broadly explored in the specific area of natural resources. Bingham (1986) investigated a decade of environmental disputes in the United States and the development of the use of dispute resolution techniques, defined as "voluntary processes that involve some form of consensus building, joint problem solving, or negotiation" – excluding litigation, administrative procedures or arbitration (p. xv). These techniques were involved in at least 160 cases. In about 132 cases, the parties' objective was to find a solution, 78 per cent of them were successful in reaching agreements. Within such an extensive list, about 10 per cent were specific cases involving water resources, including water supply, water quality, flood protection, and the thermal effects of water plants. In addition there were cases of watershed management, fishing rights, and whitewater recreation. Many interesting findings come up in reference to the factors contributing to success and observations relating to the variety of stakeholders and the duration of negotiations. Bingham and others emphasized that one of the first areas for the search for common ground is the joint identification of the key factors in data collection that normally deal with technical issues of great complexity.

More specific work on environmental dispute resolution such as Zartman's (1992) article on "International Environmental Negotiation" provides a good link between the generalists and specialists in this field by identifying the main challenge and providing an enlarged pie with the largest shares possible for each party in the negotiations. A significant amount of research undertaken by Druckman (1993) on "Situational Levers of Negotiating Flexibility," resulting from a simulation on international negotiation on the regulation of gases contributing to the depletion of the ozone layer, provides light on the transformation from the initial rigid positions of the parties to the search for new solutions touching upon a large number of factors clustered into broad categories (issues, background factors, context, structure of conferences and teams and immediate situation) and analysed at the different stages, from pre-negotiation planning up to the endgame.

One of the most systematic attempts to assess the impact of Alternative Dispute Resolution (ADR) techniques as alternatives to adjudication in natural resources cases is provided by MacDonnell's (1988) leading piece to an issue of the *Natural Resources Journal* featuring articles on international and domestic cases of disputes. Identifying many types of stakeholder, from private to multiple actors including government agencies, and cross referencing the cases to types of dispute on natural resources, the author introduces the different approaches to their resolution. Water resources, together with land use, public lands, energy, and air quality are classified. Subsequently, mediation and facilitation as auxiliaries for the process of negotiation are discussed, and special emphasis is rightly given to the potential of collaborative problem solving for generating new options. Another article by Hayton (1993) examines the current status of cooperative arrangements for the development of water resources shared by two or more countries. Such arrangements may range from the simple exchange of data to the implementation of major projects and formal resolution of disputes. According to the author, there is a growing concern with the management of shared water resources, but the use and protection of water resources is still a distant goal, and increased institutionalization is required. Environmental resources other than water are examined in more detail in Chapter 5, Other resources, of this document.

Susskind and Weinstein (1980) provide "nine steps" for early identification of the parties that have a stake in the outcome of the dispute through to the follow-up for holding the parties to their agreed commitments. In a related book, Susskind and Cruikshank (1987) suggest how to move from a "win-lose" decisional framework into "all-gain" solutions by systematically defining the assisted and unassisted consensus-building process through the pre-negotiation, negotiation, and post-negotiation phases. The book provides the potential users of facilitators with specific advice on how to move from the planning stages through to the completion of the process. They also provide the readers with abundant references to other works on specific issues of the theory and practice of conflict resolution. As mentioned by Susskind and Cruikshank, the under-utilization of such methods has to do with the concern of public officials that delegating the decision to a consensual forum means losing control over the decision process and abdicating their responsibilities (1987: 241). The fear of being pressurized into compromises concluded by a group in which the stakeholder is in a weak position fails to grasp the voluntary nature of the exercise. One comprehensive analysis of both the nature of environmental conflict and dispute resolution theory is provided by Bacow and Wheeler (1984) in a book that provides the lessons drawn from eight cases in the United States. The bulk of the book explores negotiation and bargaining from a decision-theory perspective and incorporates the main elements to be applied in dispute resolution of the issue. Flashing the obstacles and elaborating on the incentives to negotiate, the different types of setting and technique are discussed using each case as an illustration of a particular dimension.

"Negotiated Rulemaking" (Reg-Neg), a particular technique developed in the United States for disputes arising in the environmental area has received a great deal of attention in this country and has been established as an official instrument in a Congressional Act in 1990. Used by the Environmental Protection Agency (EPA), the stages of Reg-Neg are conceived as: "evaluation of issues, parties; convening (2 phases); the actual negotiations and rulemaking" (Pritzker and Dalton, 1990). The different publications of Delli Priscoli present a survey of broad ADR "people oriented" techniques (1989) or its application to case studies (1988). He stresses that whereas incrementally such ADR principles are being introduced in domestic cases, the survey's incorporation into the process of solving TFW is nearly absent.

Additionally, a negotiation strategy was developed in cases for intractable environmental conflicts. Understanding the distracting effects (confused interests, technical disagreements, misunderstandings, questions of procedural fairness, escalation, and polarization) can bring the parties to assess the costs of the confrontational alternative.

2. Legal aspects of conflict resolution

A large body of the literature covers the rules and principles that emerged from different gatherings and in particular of the work of the International Law Commission in the area of international watercourses (Bourne 1992). Caponera has stressed cooperation in the drafting of International Water Resources in several documents and articles in academic journals (i.e. 1985, 1993, 1994). The section in itself necessitates a separate review of the literature and general references include work such as the FAO publications in 1978 and subsequently in 1984.

The rather large body of general principles brings up the question of volume and the further need for norm creation when the most serious question is weak regime formation and the lack of international enforcement mechanisms.

The normative value of such a body of literature should not be disregarded, even if recognizing the gap between the slow but significant norm-creation process and the application of such principles to the resolution of concrete cases. The issue of equity is addressed by Goldie (1985), who suggests cooperative management in place of competitive management to create a shared criterion for such measures, and could serve as a common basis for partners in conflict. But the question of water rights remains elusive and controversial, as discussed by McCaffrey (1992–3). The prevailing different doctrines are clearly favouring riparian positions according to geographic location and power asymmetries.

But clearly, the learning from the successful cases is not universal. McCaffrey states, "while there are numerous treaties regulating the utilization of water resources shared by two or more countries, international agreements are either inadequate or lacking entirely in some parts of the world where water is in greatest demand" (1992–3: 4). Hence, the use of international water law has a marginal value unless there is a common sharing of mechanisms and structure that could jointly use such principles for the advantage of the basin at large. For a more detailed discussion of water law, see the section Institutions and law in Chapter 2 and the section Water treaties in Chapter 4 of this document.

3. The role of third parties

Given the sensitivity of TFWD particularly when there is a fear of scarcity, the possibility of submission to arbitration is less likely than mediation or facilitation. In the particular cases of developing countries and water scarcity, the role of third parties, such as the contribution of international agencies, has been mentioned (Fano, 1977). Relevant works include several in-house publications of the World Bank stressing the role of this institution in contributing towards solutions of international waterways disputes. Kirmani and Rangeley's (1994) "Concepts for a More Active World Bank Role" in international inland waters, illustrates from the Indus Water Treaty that so far the Bank has made only limited direct interventions, and recommends a more proactive role in assisting riparian countries' efforts to establish cooperative arrangements. Reference to such a role in South Africa is made by Kuffner (1993) and by Rogers (1993) in reference to the "development triangle in South Asia" and in the "dying Aral Sea" (Serageldin, 1995). In other cases, reference to third parties in CR means an active intervention in the process itself using facilitation and mediation (Kaufman et al. 1997).

Whereas the effectiveness of mediation and problem solving has been highlighted in the context of domestic environmental conflicts, Dryzek and Hunter (1987) elaborate on the necessary conditions of this method for the resolution of international problems in the field of water resources. Issues such as pollution in the Mediterranean are provided, yet there is only one specific reference to a successful case of TFWDR through mediation: the Skagit river case, considered by some experts to be "one particularly straightforward United States-Canada issue" (Dryzek and Hunter, 1987: 96).

4. Lateral learning and expanding the package

Rather than focusing on water allocation itself, several authors have introduced externalities that can be conducive to resolution of what has been perceived as a finite resource conflict of a zero-sum nature. Allan (1992) introduces the concept of "virtual water" which adds as an incentive to scarce water allocation the international commitment for food security. The provision of wheat or other food from other countries in reasonable quantities and attractive prices may be an incentive to compromise.

The literature on technical solutions such as water transfers (Golubev and Biswas, 1979: 115) provides important tools for addressing mutual gains, and as such is a necessary albeit not sufficient condition for the resolution of TWDR. Kuffner's (1993) "Water Transfer and Distribution Schemes" article suggests the pooling of financial resources and investments in reservoirs and treatment facilities to provide greater security against local supply failure. Increasingly, the zero-sum issue of water allocation is seen within a more comprehensive water planning structure that includes a wider spectrum of objectives.

More remote yet relevant cases are offered from other areas of environmental dispute. An analogy has been made between solutions to energy and water disputes in terms of vital needs, supply-demand and pricing, and environmental damage (Brooks, 1994).

According to Gardner et al., water issues share attributes with other "Common-pool resources" defined as "sufficiently large natural and manmade resources that it is costly (but not necessarily impossible) to exclude potential beneficiaries from obtaining benefits from their use" (1991: 335), and as such, learning can be drawn from resolutions in other fields. This and other works provide an important insight for the facilitation of negotiation process in relation to common pool resources. Given the slow development of successful cases of TFWDR, looking for clues in other areas and using the tools of different disciplines is greatly needed. One case in point is Maida's contribution to Blackburn and Bruce's (1995) book *Mediating Environmental Conflicts: Theory and Practice*, borrowing ideas from *Law and Economic Perspectives*.

We have separately mentioned reference to domestic river dispute resolution as a source for lateral learning that can be applied to TFWDR. A large body of literature relates to agreements reached within states and suggests techniques and mechanisms that could be applicable to TFWD, calling for a systematic review of this literature's lessons. An innovative effort to learn from successful examples of domestic FWDR and the possible adaptation of their outcomes to transboundary disputes is provided by Bingham et al. (1994).

Even if the added element of sovereignty requires adaptation rather than copying of such mechanisms, the ideas can be to a large extent transferable. On the broader environmental issues we find several of such cases developed in Blackburn and Bruce (1995) as well as in Dworkin and Jordan's (1995) "Midwest Energy Utilities," Baird et al.'s (1995) "Mediating the Idaho Wilderness Controversy" and Mangerich and Luton's (1995) "The Inland Northwest Field Burning Summit." Amy (1987) stresses the role of mediation mechanisms in the United States as a tool for ADR and Brown (1984) presents the Central Arizona Water Control Study as a case for multi-objective planning and public involvement in a specific area of storage and flood control.

5. Integrative and institutional approaches

As mentioned earlier, the literature covering integrative solutions often emphasizes water management from a technical and engineering perspective, while in the area of CR, the term "management" has been used to refer to less ambitious outcomes than "resolution." This point is particularly relevant for TFWDs. For instance, in relation to Hennessy and Widgery's (1995) article on an holistic approach to river basin development, the appropriate water management is defined "as the use of the right solution to make development needs in a particular environment sustainable" (with examples from the Komati river basin in Swaziland, the Lesotho Highlands water project in South Africa and the project on the Roseires dam in Sudan).

Glasbergen (1995) elaborates the concept of "Network Management" as an organizational framework for the development of consensual approaches common to the method of "collaborative problem solving." A similar idea of the formation is "epistemic communities" (Keohane et al., 1992; Hass and Hass, 1995), which defines the riparian basin as an interdependent community with an interest that transcends the narrow view of each party and stresses the search for common ground across national lines. They further consider the conditions under which such networks or epistemic communities could develop.

Well-covered examples of shared management are the Canada/United States International Joint Commission (IJC) and the Mexico/United States International Boundary and Water Commission (IBWC) as shown in the pages of the *Natural Resources Journal* (Spring 1993), which calls attention to the progress made in resolving the multiple management issues throughout the Great Lakes basin and along the United States-Mexico border. While the latter had a lower degree of public participation in the process, we are nevertheless reminded that the agreement reached there on transboundary groundwater is the source for the Bellagio Draft Treaty, which is expected to serve as a standard for aquifers disputes among countries elsewhere.

The lack of sufficient agreements on underground water and conjunctive planning with surface water provides little precedent for reaching integrative solutions as stressed by Frederiksen (1996) and others. In some recent work on the Joint Management of the Israeli-Palestinian shared aquifers, Feitelson and Haddad (1995) provide a comprehensive framework of institutional arrangements, incorporating in a typology of 19 types a large variety of functions and mechanisms. Often the selection of case studies stresses the success that has been achieved in the Northern and Western hemispheres, among affluent societies, generally with abundant water resources.

Hofius (1991) covers the case of hydrologic cooperation among the Rhine basin countries and deals with the administrative problems associated with implementing the cooperation of several states bordering a large river basin. He stresses that programmes should not be too comprehensive and should lead to results within reasonable time.

Another reference to affluent societies is made by Frederiksen (1992) focusing on the international treaties between Canada and the United States and Mexico, the Rhine Riparian and the less effective treaties on the Baltic, the North, and Mediterranean Seas. Interstate institutions in Canada, Japan, and Australia are also introduced. The author rightly raises the question of applicability to developing countries and suggests that principles of organization are also available in farmer-owned irrigation entities that have existed for several hundred to two thousand years in places such as Nepal, southern India, Sri Lanka and Bali and the issue is not necessarily a radical change but improved results through sound legislation and, more importantly, comprehensive institutions. In Bali, Bell (1988) points out that the many engineering projects undertaken several decades ago were not due to scarcity of water in the physical sense but rather the economic cost of containing and managing it, an issue of relevance to other developing countries.

Many references are made to the successful way the US and Canada have resolved outstanding issues through an evolving shared management mechanism (Dworsky et al., 1993). Generally, observations for developing countries are being drawn, pointing out the lack of readily available information among the obstacles but on the other hand stressing that longstanding native institutional principles are in existence for many of such countries. This latter point recognizes that comprehensive institutions are in place and that the problem is to translate existing institutions into modern legislation and operating rules.

Case studies – illustrations and generalizations

Single case studies

As mentioned before, successful case studies of conflict resolution focus mostly on affluent, developed countries. Such is the case of Petts's (1988) contribution on Japan's Lake Biwa case and a large number of cases covering the United States, as mentioned in the section above dealing with domestic conflict resolution.

Cases of the developing third world focus on the difficulties in resolving conflict, and are illustrated in Gautam (1976), or Howell and Allan's (1994) edited book on the Nile, covering geography, hydrology, and historic aspects as well as proposals for the future management of the Nile waters. Another relevant work on the Nile, Hultin (1992), shows the impact of civil wars within many of the riparian states on the lack of resolution to the urgent issues at stake in the Basin.

In Islam's (1992) work on the Indo-Bangladesh common rivers, attention is given to the environmental and legal problems that have soured the relationship between India and Bangladesh. The competing claims for land where meandering rivers have altered land structure have already taken the form of armed skirmishes, and failure to tackle the problem could have a disastrous impact on the environment. As another example, "[m]any case studies of water conflicts point to the Middle East as a region with dyadic or subregional settings with gloomy forecastings that in 30 years no water will be available for agriculture and industry unless it is reused from elsewhere in the water cycle or new, highly expensive sources are developed" (Wolf, 1993: 825). Often, in the Israeli/Arab conflict, findings come to support one or the other side of the conflict, as in the case of Stauffer (1996). Stressing that the Malthusian imperative is still alive in the Middle East, the author reaches the conclusion that water may prove to be the ultimate stumbling block to an Arab-Israeli peace, and Israel may have to give up about one-half of its current total water consumption. Replacement of such quantity is going to be too costly and the best answer is to cut the agricultural subsidies. A large number of books and articles such as Kally and Fishelson (1993) relate to engineering solutions from water transfers, with monumental ideas, without necessarily covering the political and psychological aspects that seem to be the major obstacles for their implementation.

Comparative case studies

Often several case studies are brought up without rigorous comparative analysis. This is the case of Murphy and Sabadell (1986) who suggest a policy model for conflict resolution stressing the intranational political process when focusing on the negotiated settlements of the Paraná (Brazil-Paraguay), the Nile (Egypt-Sudan), and the Colorado (United States-Mexico) rivers. Priest (1992) reviews the cases of six rivers in the South Asian continent, the Middle East and Africa, and stresses how decolonization has affected disputes regarding water allocation and the motivations for dispute resolution. Additionally, Salewicz (1991) takes the case studies of the Danube and Zambezi river basins and deals with institutional and organizational aspects. Housen-Couriel (1994) searching through lessons for the Jordan river basin agreement focuses systematically on four treaty regimes that are presently in force (Columbia, Plata, and Indus river basins and Lake Chad) and an additional 18 cases.

Putting the emphasis in the process of negotiations rather than on the outcome (treaty), Delli Priscoli (1988) compares two cases of water resources in the United States granting general permits for wetland fill on Sanibel Island (Florida) and hydrocarbon exploration drilling throughout Louisiana and Mississippi, arguing that facilitation, mediation, and collaborative problem-solving techniques contributed to durable agreements among seemingly irreconcilable adversaries.

Learning from incorporating 63 propositions of "grounded theory" and through interviews with 30 environmental mediators, Blackburn (1995) extracts the parts specifically relevant to the learning for environmental mediation (Chapter 18) and provides the reader with a clear ten-stage approach with practical recommendations. In the same book, Guy and Heidi Burgess's "Beyond the Limits: Dispute Resolution of Intractable Environmental Conflicts" addresses the crucial issue of power asymmetry as a deterring factor for engaging in mediation and the tendency to resort to other power driven methods. Some misperceptions are mentioned, leading to polarization and escalation, suggesting ways to tackle constructively the "bitter-end syndrome" of such disputes. Even in such difficult processes, it is possible to discover win-win trade-offs emanating from opportunities of the results of the often crude power driven contests.

Summary

From this preliminary review, it is clear that the literature provides ample examples of cases with solutions to transwater disputes (TWD). A great portion of the work on conflict resolution (CR) stresses institutional and technical arrangements. CR is perceived as mechanisms that need to be incorporated once the agreement is reached, but few relate the incorporation of CR as instrumental to the process of reaching such agreements.

In retrospect, one of the most serious obstacles for resolution is how to prompt parties to act and look for innovative solutions before water disappears, or matters between countries reach a crisis situation. In a more optimistic vein, Newson (1992) comments on the development of an international movement towards sustainability in the management of large river basins. His two examples, however, illustrate that there is still a gap in the impact of the "Freshwater Europe Campaign" on the European Community, and the effect of the Earth Summit coalition on the developing world.

Notes

1. Some of these arguments are drawn from Wolf (1999). Equitable Water Allocations: The Heart of Transboundary Water Conflicts. *Natural Resources Forum* (February, forthcoming).
2. The factors include a basin's geography, hydrology, climate, past and existing water utilization, economic and social needs of the riparians, population, comparative costs of alternative sources, availability of other sources, avoidance of waste, practicability of compensation as a means of adjusting conflicts, and the degree to which a state's needs may be satisfied without causing substantial injury to a co-basin state.
3. While this is the sole case of the Helsinki Rules definitions being used explicitly in treaty texts, the concept of "reasonable and equitable use" is quite common, as is described below.
4. These factors include: geographic, hydrographic, hydrological, climatic, ecological, and other natural factors; social and economic needs of each riparian state; population dependent on the watercourse; effects of use in one state on the uses of other states; existing and potential uses; conservation, protection, development and economy of use, and the costs of measures taken to that effect; and the availability of alternatives, of corresponding value, to a particular planned or existing use.
5. For examples of these respective positions, see the exchange between Jovanovic (1985, 1986) and Shahin (1986) in respective issues of *Water International* about the Nile; and the description of political claims along the Euphrates in Kolars and Mitchell (1991).

Brahamaputra basins, several cooperative solutions to the conflict of regional water sharing are introduced. The common nature of these solutions is that they are technically and economically feasible, they are individually and regionally rational, and they are Pareto-admissible in the sense that no other solution can be preferred by any of the parties.

LeMarquand (1989) suggests a framework for developing river basins that is economically and socially sustainable. At the core of the approach is a river basin authority to coordinate basin-wide planning and execution of basin-wide, multi-purpose projects (water and other regional development). In the case of developing countries, the approach also includes a component to coordinate donor activities. Especially in international rivers, LeMarquand suggests the following conditions for successful water-sharing agreements: (1) similar perceptions of the problem, (2) similar characteristics of the utility function of the parties, (3) similar water production functions, (4) existence of some level of dialogue, (5) a small number of parties involved, and (6) at least one party having a desire to resolve the conflict.

Kally (1989), while evaluating the potential for cooperation in water-resources development between Middle East countries, examines a particular approach that is based on individual water-related projects among two or more parties in the region. It should be noted that in Kally's approach, the water-related projects cut across basins and do not focus on a particular basin. The author suggests that it is possible to envisage different combinations of various projects of potential interest to the particular parties as well as to all parties in the region. However, political considerations other than those related directly to water are likely to determine the level of cooperation in the region and the particular subset of projects to be selected.

Inter and intraregional allocations

Sprinz (1995) investigates the relationships between local (state) production-pollution, and international pollution-related conflicts. Although very specific to international environmental pollution conflicts, there are some features in this work that can be adapted to international water conflicts. The move from closed economies to a situation that allows international trade, international pollution regulation, and global environmental problems, produces a more stable and acceptable solution.

Using the Coase Theorem, which indicates that assigning of property rights to water users will allow optimal allocation of the scarce resource among users, Barret (1994) proves that when there is international interdependence, there is no guarantee that the allocation of water resources will be efficient. Barret applies simple modifications of the Prisoner's Dilemma game to various international water dispute case studies (Columbia, Indus, Rhine), under various treaty and institutional arrangement scenarios. The main conclusion from this work is that the allocation of the joint benefits from an allocation scheme of the basin water among the riparians is the key to an acceptable agreement.

Just et al. (1994) suggest an economic framework to deal with transboundary water issues, and apply it to the Middle East. The core of the approach is joint (with international help) planning and finance of water-related projects that may expand the resource base among countries in the region. Better use of existing sources can occur because the political and economic costs of changing the existing water use patterns are reduced when supply is higher.

Markets

Dudley (1992) introduces the concept of common property and capacity (of a water system such as a reservoir or a river) sharing in the context of water markets and potential disputes over water in large water systems. It is argued that because capacity sharing minimizes the interdependencies of behaviour between users of a water system, it provides a good basis for dividing up the system among the system users.

Case studies

Dinar et al. (1995) review pollution of international lake and reservoir water. It is obvious from the evidence accumulated over time and across the world that the nature of the resolution arrangement to water pollution-related disputes depends on the nature of the pollutant (monitoring and enforcement ability, remediation difficulty). Several case studies are used for illustration (Great Lakes, Aral sea, Mono lake).

Guariso et al. (1981) address the question of efficient use of scarce Nile water resources. Although not directly related to resolution of international conflicts over scarce water resources, one could use the argumentation in the analysis for further discussion of a regional approach. The multi-objective model of using Nile river water for Sinai suggests a trade-off between economic and political objectives, and introduces new (to the region) efficiency concepts, such as irrigation technologies, new cropping patterns, water application scheduling, crop rotation, and more.

Whittington and McClelland (1992) address future possible cooperation strategies for the Nile river basin. The common denominators of international cooperation opportunities discussed are joint development, monitoring, and management of the resource. Suggested mechanisms include individual projects that will benefit each of the riparians, and projects that will benefit some of the riparians. Included are trade-offs between investment – either directly in water-related projects or indirectly in agriculture – and water savings which may benefit all riparian countries.

Okidi (1988) examines macro-policy issues of state involvement in international water basin management in Africa. The paper suggests several issues to be considered concerning effective management: (1) over-politicization of institutions and programmes, (2) proliferation of institutions, (3) mix-up of economic exigencies with political prestige in water projects, and (4) over-centralization of institutions.

Giannias and Lekakis (1994) develop an optimization model for allocation of the Nestos water between water-user sectors in Greece and Bulgaria. Although the general framework of the model is similar to many others that have been reviewed here and elsewhere, it also includes some policy mechanisms to be agreed upon between the countries. Such policies include output price policy, price policy, taxes and subsidies for water-related industries, and trading in water rights. The model also takes into account the outcome of the settlement of the water allocation between these countries and the quality of water discharged to the Mediterranean.

Summary

Economics is one discipline that is used independently or jointly with other disciplines in explaining scarce resource disputes and indicating a set of possible and agreeable arrangements between the parties.

Solutions offered by economic approaches may look promising, but it is always necessary to identify the set of assumptions leading to such solutions. Even with this identification in mind, one can still argue that economic principles are among the sufficient, but not the necessary, conditions for a dispute to be solved.

Using economic terms, for a solution to a dispute to be attractive to the participants and to be economically sustainable, it needs to fulfil requirements for individual and group rationality. This need signifies that the resolution of a dispute for each participant is preferable to the status quo outcome, and to participation in any partial arrangement that includes a subset of the regional participants. The regional arrangement also fulfils requirements that all costs or gains are allocated.

As we well know, economics and politics play interactive roles in the evaluation of dispute resolution. Just as political considerations can effectively veto a joint project with an otherwise favourable economic outcome, a project with potential regional welfare improvements might influence the political decision-making process to allow the necessary cooperation.

Therefore, both economic and political considerations should be incorporated into evaluations of dispute resolution arrangements.

Game theory

Game theory is a relatively new branch of mathematics and the social sciences that has been used successfully to engineer improvements in policy and understanding of many market and non-market events. It is used to clarify decision-making in contexts where one player's best choice in a particular interaction depends, to some extent, on the choice of another player. This "best" choice is termed the "strategic choice." By working out the logic behind the purposeful behaviour of actors involved in some strategic interaction, it is possible to determine how individuals ought to make choices in a particular interaction if they adhere to principles of rationality. The principles of rational choice require that the players' behaviour is motivated by their own goals and values, as modified by their updatable expectations, and as constrained by their resources and the rules of the institutional context in which they find themselves. In the jargon of the theory, we say that a game's outcome depends upon the set of feasible outcomes, participants' choices, and the rules of the game.

This theory is easily shown to be relevant to the engineering of social outcomes.[1] Trying to guide social policy involves two steps: (1) the specification of social goals and (2) the design of institutions, rules, or strategies to channel the social outcomes toward those goals. The theory of games coupled with experimentation is ideally suited for these goals. After all, the idea behind a game is that institutions and agreements determine the rights and powers of participants. They determine both the acts available to players and the consequences that result from any pattern of acts taken by the set of participants.[2] The acts of the participants, and hence the social choice from among the feasible alternatives, depend upon both the choices of the actors and the institutions which define the processes and their rules. These rules, which govern, or at least influence, the outcome of the overall game, form the basic context of the decisions.

As in all science, the theory cannot be "sufficiently closed" without empirical understanding of the exact details of the rules, the institutions. Thus, any useful policy applications require a continual interplay between theoretical formulations, manipulation of real world assumptions, and careful observation to monitor the status of the theoretical predictions.

For us, game theory allows for an analysis of the economic and political aspects of a shared regional water problem in a manner that promises some increased leverage. International water disputes typically involve a relatively small number of participants, each with different objectives and perspectives.

A quantitative and theoretical analysis can be performed to show how a number of players might react to a situation in order to identify the likely properties of the outcomes resulting from rational choice on the part of the participants. One can then examine the likely outcomes to see if they conform to such criteria or goals as "Pareto-optimality," or whether they have standard stability characteristics (e.g. no player can gain by unilaterally moving away from that point). Therefore, one can investigate, using the tools of game theory, the prospects for cooperation in environments of choice. Such analysis can then be used to help design different and preferable contexts of interaction and negotiation.

Game theory has been applied to issues as diverse as national security, social justice, and religion. But it has been applied to international water conflicts only sporadically. Rogers (1969) analyses conflicting interests along the Lower Ganges and suggests strategies for cooperation between India and Pakistan. Dufournaud (1982) applies game theory to both the Columbia and the Lower Mekong to show that "mutual benefit" is not always the most efficient criterion to measure cooperative river basins. Dinar and Wolf (1994) use cooperative game theory to explore the economic pay-offs that might be generated in a technology-for-water exchange between Israel and Egypt, and how those pay-offs might be distributed to induce cooperation.

Many specific games have become models for particular problems. Certainly the most famous of such games is the two-person Prisoner's Dilemma game (2-PD). In such games one can examine the relationship between cooperation, self-interested behaviour, and efficiency. Political scientist R. Axelrod, has argued, in two-person situations, involving specific sorts of games,

A player who in an opening move acts generously and on a responding move acts cooperatively, never initiating attack, will outscore any other strategy, given time and averaging.[3] (cited in Painter, 1988)

In practice, however, the games being played between competing nations are far more complicated and the ensuing relationship between cooperative stances and receipt of rewards may be far weaker:

A strong positive relationship exists between tendencies to initiate and to receive international conflict. The correlation between cooperative initiation and receptive tendencies, however, is much weaker. (Platter and Mayer, 1989)

Nevertheless, other games can and have been used successfully as models of international conflict. In these and other essays it can be seen that game theory offers a framework for some level of analysis that might shed light on international water conflicts. For example, when the demand for water of a population in a water basin begins to approach its supply, the inhabitants have three choices:

- They can work unilaterally within the basin (or state) to increase supply – through wastewater reclamation, desalination, or increasing catchment or storage – or decrease demand, through conservation or greater efficiency in agricultural practices.
- They can cooperate with the inhabitants of other basins for a more efficient distribution of water resources. This cooperation usually involves a transfer of water from the basin with greater resources.
- Or, they can make no changes in planning or infrastructure and face each cycle of drought with increasing hardship. This is the option most often chosen by countries that are less developed or are racked by military strife.

These options are equally applicable to the problems facing inhabitants of a single basin that includes two or more political entities. Each can be modelled (see Falkenmark, 1989a and LeMarquand, 1977) for related work.

Although the last alternative may seem unreasonable, game theoretic models can help to explain how nations may make choices based on their underlying interests and the strategic structure of the game itself.[4] The modeller can then try to make prescriptions in such cases to change the contexts so as to lead to more efficient and welfare enhancing outcomes.

To solve the problem of water allocations cooperatively within an international water basin, a number of problems can be analysed using game theory:

- In international contexts, each sovereign party is free to break any agreement at little cost. Hence any engineered solution must be sensitive to the stability aspects of the proposed outcomes.
- For cooperation to occur, the parties must have some incentive which can justify the cooperation.

This latter point implies that for a cooperative solution to be accepted by the parties involved, it is required that (a) the joint cost or benefit is partitioned such that each participant is better off compared to a non-cooperative outcome; (b) the partitioned cost or benefit to any subset of participants (in the cooperative solution) are preferred by the subset to any other possible outcome they can guarantee themselves. Of course, in the real world of international relations, it also must be that all the costs are allocated.

The economic literature dealing with application of game theory solutions does not provide many examples of regional-international water sharing problems. As indicated above, Rogers (1969) applied a game theory approach to the disputed Ganges-Brahmaputra sub-basin that involves different uses of the water by India and Pakistan. The results suggest a range of strategies for cooperation between the two riparian nations that will result in significant benefits to each. In a recent paper, Rogers (1991) further discusses cooperative game theory approaches applied to water sharing in the Columbia basin between the US and Canada, the Ganges-Brahmaputra basin between Nepal, India, and Bangladesh, and the Nile basin between Ethiopia, Sudan, and Egypt. In-depth analysis is conducted for the Ganges-Brahmaputra case where a joint solution improves each nation's welfare more than any non-cooperative solution (Rogers, 1993).

Application of metagame theory, which is a non-numeric method to analyse political conflicts, has been applied to water-resources problems by Hipel et al. (1976). The resulting outcome of a conflict is a set of strategies most likely to occur and their pay-offs to each participant.

Becker and Easter (1994) have analysed water management problems in the Great Lakes region between different US states and between the US and Canada. A central planning solution is compared to a game theory solution with the result being in favour of the game theory solution.

Dinar and Wolf (1994), using a game theory approach, evaluate the idea of trading hydrotechnology for interbasin water transfers among neighbouring nations. They attempt to develop a broader, more realistic approach that addresses both the economic and political problems of the process. A conceptual framework for efficient allocation of water and hydrotechnology between two potential cooperators provides the basis for trade of water against water-saving technology. A game theoretic model is then applied to a potential water trade in the western Middle East, involving Egypt, Israel, the West Bank, and the Gaza Strip. The model allocates potential benefits from trade between the cooperators. The main findings are that economic merit exists for water transfer in the region, but political considerations may harm the process, if not block it entirely. Part of the objection to regional water transfer might be due to unbalanced allocations of the regional gains and, in part, to regional considerations not directly related to water transfer.

As the amount of water surplus decreases over time, however, the impetus towards conflict or cooperation (pay-offs) might change, depending on such political factors as relative power, level of hostility, legal arrangements, and form and stability of government.

Notes

1. See Plott (1978: 207) for a similar analysis.
2. The principles of game theory are not discussed here in detail, but can be found elsewhere. (See, for example, Shubik, 1982, and the other entries under "texts.")
3. Actually, even in the limited domain of his inquiry, his claim was suggestive, but wrong. To see the errors, read Bendor and Swistak (1995: 3596–600).
4. Bueno de Mesquita and Lalman (1992), for example, show how choices leading to war can be rational, despite a mutual preference by disputants for peacefully negotiated agreements.

Part 3: Practice

4

Water disputes

Water is likely to be the most pressing environmental concern of the next century (American Academy of Arts and Sciences, 1994). As global populations continue to grow exponentially, and as environmental change threatens the quantity and quality of natural resources, the ability for nations to resolve conflicts peacefully over internationally distributed water resources will increasingly be at the heart of stable and secure international relations. There are more than 200 international rivers in the world, covering more than one-half of the total land surface. Water has been a cause of political tensions, and occasional exchanges of fire, between Arabs and Israelis; Indians and Bangladeshis; Americans and Mexicans; and all 10 riparian states of the Nile river. Water is one of the few scarce resources for which there is no substitute, and over which there is poorly developed international law. The demand for water is also overwhelming, constant, and immediate (Bingham et al., 1994).

Comparative analysis and case studies

Resource conflicts will gain in frequency and intensity as water resources become relatively scarcer, and their use within nations can no longer be insulated from impacting on one's neighbours. A clear understanding of the details of how water conflicts have been resolved historically will be vital to those responsible for bringing together the parties of future conflicts.

What are readily available to the scholar and policy-maker interested in international water conflicts are the *results* of a particular period of negotiations – usually a treaty or other agreement that allocates the resource. This information tells us little about the *process* by which the disputes were resolved. For example, what were each side's opening positions? What underlying interests informed those positions? What obstacles were encountered during negotiations and how were those obstacles overcome? What principles were finally agreed to for water allocations? What provisions were established for resolving future water conflicts and enforcement mechanisms? Finally, has the agreement been effective? The cases presented in Chapter 7 do provide for some patterns. Some of them are discussed below.[1]

Anticipating possible water conflict

Given that the international community has neither the resources nor the time to help establish a basin-wide institution for integrated watershed management on each of the world's international rivers and aquifers, patterns do emerge which may be useful in anticipating likely conflicts.

Generally, the chronology of a typical water conflict is as follows: riparians of an international basin implement water development projects unilaterally first on water within their own territory, in attempts to avoid the political intricacies of the shared resource. At some point, as water demand approaches supply, one of the riparians, generally the regional power,[2] will implement a project that impacts on at least one of its neighbours. Egypt's plans for a high dam on the Nile river, or Indian diversions of the Ganges river to protect the port of Calcutta, and Turkey's GAP project on the Euphrates river to meet the needs of a new agricultural policy are examples of how countries continue to meet existing uses of water resources in the face of decreasing water availability.

In the absence of relations or institutions conducive to conflict resolution, the project that impacts on one's neighbours can become a flashpoint, and conflict among various parties is imminent. Each of these projects is preceded by indicators of impending or likely water conflict, which might include those given below.

Water quantity issues

Often, simply extrapolating water supply and demand curves will give an indication of when a conflict may occur, as the two curves approach each other. Conflicts over water in the Jordan river basin were inevitable during the mid-1960s, when demand outstripped supply in both Israel and

Jordan. Also, major shifts in supply might indicate likely conflict, due to greater upstream use or, in the longer range, to global change. The former is currently the case both on the Mekong river and on the Ganges river. Likewise, shifts in demand, due to new agricultural policies or movements of refugees or immigrants can also lead to problems. Water systems with a high degree of natural fluctuation can cause greater problems than relatively predictable systems.

Water quality issues

Any new source of point or nonpoint pollution, or any new extensive agricultural development resulting in saline return flow to the system, can indicate water conflict. Return flows from the state of Arizona into the Colorado river was the issue over which Mexico sought to sue the US in the 1960s through the International Court of Justice. It is also a point of contention on the lower Jordan river among Israel, Jordan, and Palestinians living in the West Bank.

Management for multiple use

Water is managed for a particular use or a combination of uses. A dam might be managed for storage of irrigation water, power generation, recreation, or a combination of all these. When the needs of riparians conflict, disputes are likely. Many upstream riparians, for instance, would manage the river within their territory primarily for hydropower, whereas the primary needs of their downstream neighbours might be timely irrigation flows. Chinese plans for hydropower generation and/or Thai plans for irrigation diversions would have an impact on Vietnamese needs for both irrigation and better drainage in the Mekong river delta.

Political divisions

A common indicator of water conflict is shifting political divisions that reflect new riparian relations. Such is currently the case throughout Central Europe as national water bodies such as the Aral sea, the Amu Dar'ya, and the Syr Dar'ya become international. Many of the conflicts presented here, including those on the Ganges river, the Indus river, and the Nile river, took on international complications as the central authority of hegemony, in these cases the British Empire, was dissipated. The converse – territorial integration – such as the unification of the two Germanies, can also present problems.

Indicators for type and intensity of conflict

Along with clues useful in anticipating whether or not water conflicts might occur, patterns based on past disputes may provide lessons for determining both the type and intensity of impending conflicts. These indicators might include the following.

Geopolitical setting

As mentioned above, relative power relationships, including riparian position, determine how a conflict unfolds. A regional power which is also an upstream riparian is in a more advantageous position to implement projects. These projects, in turn, are then catalysts for regional conflict. Turkey and India have been in such positions on the Euphrates river and the Ganges river, respectively. In contrast, the development plans of an upstream riparian may be held in check by a downstream power. An example would be Ethiopia's plans for Nile river development and its effect on Egypt.

Unresolved non water-related issues with one's neighbours could also have a deleterious effect in water conflicts. Israel, Syria, and Turkey, each and respectively have difficult political issues outstanding, which makes discussions on the Jordan river and Euphrates river more intricate.

Level of national development

The relative level of development of a party can affect the nature of water disputes in a number of ways. For example, a more developed region may have better options to alternative sources of water, or to different water management schemes, than less developed regions. This situation results in more options once negotiations begin. In the Middle East multilateral working group on water, a variety of technical and management options, such as desalination, drip irrigation, and moving water from agriculture to industry, have all been presented. These options, in turn, supplement discussions over allocations of the international water resource.

Different levels of development within a river basin, however, can exacerbate the hydropolitical setting. As a country develops, personal and industrial water demand tends to rise, as does demand for previously marginal agricultural areas. While balanced somewhat by more access to water-saving technology, a developing country will often be the first to develop an international resource to meet its growing needs. Thailand has been making these needs clear with its greater emphasis on Mekong river development relative to the other riparians.

The hydropolitical issue at stake

In a survey of 14 river basin conflicts, Mandel (1991) offers interesting insights relating to the issues at stake in a water conflict. He suggests that issues which include a border dispute in conjunction with a water dispute, such as the Shatt al-Arab waterway between Iran and Iraq and the Rio Grande river between the US and Mexico, can induce more severe conflicts than issues of water quality, such as the Colorado, Danube, and

Plata rivers. Likewise, conflicts triggered by human-initiated technological disruptions – dams and diversions – such as the Euphrates, Ganges,
Indus, and Nile rivers, are more severe than those triggered by natural
flooding, such as the Columbia and Senegal rivers.

Mandel's study also finds that there is no correlation between the
number of disputants and the intensity of the conflict. He thus argues
with the notion that river disputes with fewer parties are easier to
resolve.

Institutional control of water resources

An important aspect of international water conflicts is how water is controlled *within* each of the countries involved. Whether control of the
resource is vested at the national level, as in the Middle East, the state
level, as in India, or at the sub-state level, as in the United States, informs
and complicates international dialogue. It is also important to know
where control is vested institutionally. In Israel, for example, the Water
Commissioner is under the authority of the Ministry of Agriculture,
whereas in Jordan, the Ministry of Water has control over water resources. These respective institutional settings can shape internal political
dynamics quite differently, despite the similarity of issues under concern.

National water ethos

This term incorporates several somewhat ambiguous parameters which
together determine how a nation "feels" about its water resources. This
ethos, in turn, can help determine how much it "cares" about a water
conflict. Some factors of a water ethos might include:
- "mythology" of water in national history, e.g. has water been the
  "lifeblood of the nation?" Was the country built up around the heroic
  fellah? Is "making the desert bloom" a national aspiration?
- importance of water/food security issues in political rhetoric;
- relative importance of agriculture versus industry in the national
  economy.

Obstacles to successful negotiations

In addition to helping anticipate water conflicts, the patterns that begin to
emerge from past conflicts also suggest that there are common barriers to
successful negotiations. Early identification of these barriers, included
below, can help to overcome them more easily.

Lack of willingness to recognize other parties with the ability to block implementation

One limiting aspect of the International Court of Justice is that only
states may be parties to cases. This structure excludes minority political

or ethnic groups, as well as a whole range of political, environmental, and special interest groups who may have a stake in an international agreement. When the Middle East peace talks began, Palestinians had representation only as part of a joint delegation with the Jordanians. Currently, Kurdish interests have no representation by any of the parties to the Euphrates river dispute. All of these exclusions result in the interests of at least one party being ignored as disputes are being resolved; parties who may be able block implementation of a final agreement.

Scientific uncertainty/disputes

Except for the Mekong river, all of the basins included within this study include data disputes as a major component of the overall water conflict. In many of the countries, at least some water data are secret. Used as a stalling tactic in some negotiations has been the call for more study of data before decisions can be made. The Mekong Committee, as the notable exception, used joint data gathering as the first major cooperative task, precluding these kinds of problems.

National vs international settings

It should be clear from the cases presented in this study that both similarities and distinct differences are inherent between national and international water conflicts. Stressed more often are the differences, but just how different the two settings are is open to debate. Assumptions that are common include the following.

Institutions and authority

National cases are often played out in relatively sophisticated institutional settings, particularly in the West, while international conflicts are hampered by the lack of an institutional capacity for conflict resolution.

It has been argued, though, that even sophisticated institutions have often not been amenable to relinquishing the traditional, usually legal, approaches to resolving water conflicts, effectively presenting the same challenges as at the international setting.

Law and enforcement

The US and other countries have, over the years, established intricate and elaborate legal structures to provide both guidance in cases of dispute, and a setting for clarifying conflicting interpretations of that guidance. International disputes, in contrast, rely on poorly defined water law, a court system in which the disputants themselves have to decide jurisdiction and frames of reference before the hearing of a case, and little in the way of enforcement mechanisms. One result is that international water conflicts are rarely heard in the International Court of Justice. Likewise, of the international cases presented in this document, only the Mekong Committee has used the legal definition of "reasonable and equitable" use in its agreement.

Presumption of equal power

"All are equal in the eyes of the law," is a common phrase describing national legal frameworks. No such presumption exists in international conflicts, where power inequities define regional relations. Each of the watersheds presented here includes a hegemonic power, which brings its strength to bear in regional negotiations, and which often sees agreements tilt in its favour as a consequence.

Here, too, it has been argued that unequal resources, usually financial or political, result in real-world inequities finding their way into the national settings of conflict resolution as well.

The BATNA (Best Alternative to a Negotiated Agreement)

A difference commonly pointed out between national and international disputes is that, in national water conflicts, war is not usually a realistic alternative to failed negotiations. While it may be true that intranational "water wars" are not likely, the same is increasingly accepted as being true of the international setting. While shots have been fired, both nationally and internationally, and there has been troop mobilization between countries, no all-out war has ever been caused by water resources alone. As one analyst familiar with both strategic issues and water resources has noted, "Why go to war over water? For the price of one week's fighting, you could build five desalination plants. No loss of life, no international pressure, and a reliable supply you don't have to defend in hostile territory" (Tamir, in Wolf, 1995: 76).

While real differences do exist between the national and international settings for water conflict resolution, these distinctions may not be as great as is often thought. The fortunate corollary to this situation is that many of the successes of ADR in the national realm may be more applicable to the international setting than is commonly argued.

Summary

A clear understanding of the details of how water conflicts have been resolved historically is vital in discerning patterns that may be useful in resolving or, better, precluding, future conflict. Our investigations of 14 disputes suggest that, generally, the following pattern tends to emerge: riparians of an international basin implement water development projects unilaterally first on water within their territory, in attempts to avoid the political intricacies of the shared resource. At some point, as water demand approaches supply, one of the riparians, generally the regional power, will implement a project that impacts on at least one of its neighbours. This project can, in the absence of relations or institutions conducive to conflict resolution between the riparians, become a flashpoint, resulting in conflict. The comparative analysis also suggests other indicators of impending or likely water conflict; obstacles to successful negotiations; and observations regarding the national versus international settings.

Water treaties[3]

With the single exception of air to breathe, the human physiology needs no resource greater than water. Humans use water for a variety of needs, ranging from absolutely critical (drinking) to luxurious (swimming pools, fountains, or golf courses in the desert). Population pressures have a stronger impact every year on the amount of water available per capita in various parts of the world. International agreements about water address a growing relative scarcity. Agriculture requires nearly salt-free water, and concerns about industrial effluent mean that water quality issues will also find their way into international agreements. But the problem of scarcity, though growing, is not new. Water has been a source of contention for at least 4,500 years, as historians have found a treaty to end a water war between the ancient Mesopotamian states of Lagash and Umma (Cooper, 1983). More recently, treaties dating from the nineteenth century have addressed all facets of consumptive and non-consumptive water uses.

Because the basic problems associated with scarce water and water quality remain unchanged, the way in which governments solved difficulties in the past may prove useful for negotiations in the present. With this aim in mind, the authors have undertaken to collect and summarize all treaties addressing freshwater needs of the signatories in a Transboundary Freshwater Dispute Database. Our main criterion for selecting a treaty depended on whether its "major" focus was water as a scarce, consumable resource. Therefore, we include those that touch on transportation, fishing, and boundary demarcations only as they are relevant to issues of allocation and use.

The treaty database provides some confirmation or insight into other authors' ideas, such as Gulhati's (1973) comment that outside negotiators with additional resources to bring to bear can smooth or eliminate difficulties. The treaty database also shows areas in need of improvement or examination: perhaps if the reasoning behind the language of water allocations were better known, the numbers and bargains could be reused for equitable treaty making elsewhere or later in the same basin. As more treaties enter the database, more theories of negotiation also come up for analysis; some theories could be made new by application in a different region.

Studies of conflict do not usually centre on the treaties; if two states have a conflict, either the treaty does not satisfy one or either side (was the agreement negotiated fairly at the outset?), or a new development has changed the circumstances of the original treaty (climatic change and/or population growth). In addition, a treaty depends on good relations and good faith. In the absence of both, a treaty has little hope of maintaining peace. Yet of the 145 treaties making up the database in late 1997, the authors know of few if any that have been broken.

Treaties offer great amounts of information: they can tell about regional hegemony, how and which water needs are met, the relative importance of water in the political climate, development issues, and whether earlier treaties have successfully guided or guaranteed state behaviour.

Literature review

Treaty studies occupy only a small fraction of the dispute resolution literature. Most recently, Wescoat (1996) has produced "Main Currents in Early Multilateral Water Treaties: A Historical-Geographic Perspective." Legal scholars have discussed international law and treaty making, notably Teclaff (1991) and McCaffrey (1993). McCaffrey also offers theories about trends in treaty making, specifically the move towards integrated management from cut-and-paste approaches, the move away from navigation as the primary use, and the trend towards "equitable utilization." Hayton (1988, 1991) has argued that international law should include hydrologic processes in its theory. Dellapenna (1995) describes the evolution of treaty practice dating back to the mid-1800s, and Wescoat assesses historic trends of water treaties dating from 1648–1948 in a global perspective. Gulhati (1973) and Michel (1967) provide the most comprehensive analysis of a single treaty and the events and people leading up to its approval. The two authors discuss the 1960 Indus Waters Treaty and provide history that explains the wording and spirit of the treaty. Deeper analysis of major treaties, like Gulhati and Michel's work, is necessary for better interpretation of any statistical data the database may produce.

Background of modern water treaties

Treaties acknowledging and addressing water at least in terms of fishing, regularization, and navigation became common in the nineteenth century, based on the frequency of these treaties occurring in the treaty sources (see Methodology, below, for a list of sources). Population pressures may be responsible for water's rise in importance. Population-induced water stress may also change the parameters under which a treaty was previously concluded, rendering the treaty less applicable to the situation.

Water allocations are not included in early negotiations of an economic nature, such as treaties regulating pilotage or trade. In fact, few allocative treaties were negotiated prior to water needs and water stress reaching critical levels. Regarding hydropower, these treaties have decreased in significance because of a decrease in new dam construction. One exception may be Nepal, with an estimated 2 per cent (83,000 megawatts) (Aryal, 1995) of the world hydropower potential, but at present, geologic, engineering, and financial problems have slowed construction even in this country (Ganguly, personal communication).

Methodology

Many sources contain information about water-related treaties. The FAO (1978, 1984) indices of water treaties provide the greatest number – more than 3,600 relating to water use dating between 805 AD and 1984 – from which to choose relevant treaties. In addition to the FAO indices, law texts, journal articles, foreign policy documents or collections, personal contacts, and departments of state all provided additions to the database.

The treaties are identified as relevant based on their inclusion and treatment of one or more of the following issues: water rights, allocations, pollution, or principles for equitably addressing water needs; hydropower and reservoir/flood control development; environmental issues/water "rights" for riverine ecological systems; and on occasion, navigation, fishing, or border demarcation, usually in combination with unique and innovative conflict resolution methods.

All treaties were read thoroughly for specific and non-specific information. Some sources only contain excerpts or annotated treaty summaries. The full texts of those documents will eventually complete the database records for those agreements. Condensed treaties (some with direct quotes from the text) reside in the database; the reader will find the entire collection in Chapter 8. The textual information exists in discrete records, and the authors have taken some data – which may be meaningful when expressed statistically – and compiled it into percentages. As answers to further questions fill gaps in the database (such as how well each treaty has prevented diplomatic friction or how smoothly the negotiations proceeded), more meaningful quantitative analysis can be conducted.

Each treaty summary contains the following information: the name of the basin, principal focus, number of signatories, non-water linkages (such as money, land, or concessions in exchange for water supply or access to water), provisions for monitoring, enforcement, and conflict resolution, method and amount of water division, if any, and the date signed. Treaties signed before the mid-twentieth century are often incomplete or contain standard answers in most categories. Not surprisingly, population pressures affected fewer water treaties earlier in the century (that is, explicit allocations were less frequent in earlier treaties). Many of these treaties therefore address competition and conflict over water quantity less completely than do more modern treaties.

The category "principal focus" elicited the most possible answers (seven). Defining the principal focus of a treaty often proves difficult. Other categories are less difficult to manage. For instance, the existence of a commission (or council, technical advisory body, etc.) is easily determined. Either a treaty provides for it or it does not. Less clear are the powers of a commission; technical commissions *could* address disputes, but often the treaty deals with conflict resolution through other channels. If the agreements list no other form of dispute resolution, the authors assume that any conflict falls first to the advisory council and later to the respective signatory governments.

The database and its contents[4]

Number of signatories

Some treaties show common characteristics when expressed in statistical terms. For instance, a large majority of agreements have only two signatories (124 out of 145, or 86 per cent). Yet international rivers often have more than two riparians. The development and implementation of multilateral treaties have taken much more time than bilateral treaties. Even in situations where more than two parties have interests over a given body of fresh water (for instance, the Danube), few treaties (only 21 out of 145, or 14 per cent) include three or more parties.

It is unclear whether so many treaties are bilateral because only two states share a majority of international watersheds or because, according to negotiation theory, the difficulty of negotiations increases as the number of parties increases (Zartman, 1978). In basins with more than two riparians, this preference towards bilateral agreements can preclude the comprehensive regional management long advocated by water-resource managers. In addition, as "Balkanization" continues, i.e the fragmentation of countries into smaller, more homogeneous units, named for the historic and ongoing difficulties in the former Yugoslavia, the number of riparians will increase as well.

Multilateral treaties are still at a developmental stage, accounting for only 21 of the treaties in the current database. They usually address only minor environmental and data-gathering issues, although efforts to change that situation are underway. Some have established advisory bodies. None have follow-up treaties to add specifics lacking in the original agreements.

Wescoat (1996) examined multilateral treaties because they "reflect the influence of broad (rather than local) geopolitical situations," but the obvious prominence of bilateral treaties also indicates that countries prefer to negotiate one-on-one. Some nations have a predisposition to bilateral agreements: India's longstanding policy of bilateral-only negotiations presents a problem when attempting to develop a basin-level approach to managing the Ganges-Brahmaputra or Indus river systems. The Murray River Agreement, although not included in this database, has undergone substantial analysis as a "multilateral agreement." Although this agreement is among three Australian territories and is not international per se, it is often used as a model for the management of river basins among neighbours with more tenuous relationships, such as India and Bangladesh. All but three of the multilateral agreements listed in our database lack definite water allotments, although a few establish advisory and governing bodies among states to address this issue.

Of the 21 multilateral treaties/agreements, developing nations account for 13. Only one multilateral treaty exists among industrialized nations for allocations to a water source, namely the treaty regarding water withdrawals from Lake Constance signed by Germany, Austria, and Switzerland in 1966. None of the preindustrial-nation multilateral agreements specified any water allocations; instead all involved hydropower or other industrial uses.

The states surrounding the Aral sea signed an agreement in 1993 that addresses several issues, but the text itself does not address the issue of water allocations nor does it provide a blueprint for future water use. Like the Aral sea, Lake Chad also suffers from intense, poorly managed water resources, and extensive water withdrawals (Rangeley et al, 1994). The Chad Basin Treaty (1964), among Cameroon, Niger, Nigeria, and Chad, covers issues such as the economic development inside the basin, the lake's tributaries, and industrial uses of the lake, but does not address allocations. The agreement does create a commission, which, among other things, arbitrates disputes concerning implementation of the treaty. The commission prepares general regulations, coordinates the research activities of the four states, examines their development schemes, makes recommendations, and maintains contact among the four states.

Principal focus

Most treaties focus on hydropower and water supplies: 57 (39 per cent) of the treaties discuss hydroelectric generation, and 53 (37 per cent) distribute water for consumption. Nine (6 per cent) mention industrial uses, six (4 per cent) navigation, and six (4 per cent) primarily discuss pollution. Thirteen of the 145 (9 per cent) focus on flood control. The database includes one treaty that primarily discusses fishing (less than 1 per cent) (included in the database for other elements).

Monitoring

Seventy-eight treaties (54 per cent) have provisions for monitoring, while 67 (46 per cent) do not. When monitoring is mentioned, it is addressed in detail, often including provisions for data sharing, surveying, and schedules for collecting data.

Information sharing generally engenders good will and can provide confidence-building measures between co-riparians. Unfortunately, some states classify river flows as secret and others use lack of mutually acceptable data as a stalling technique in their negotiations. Most monitoring clauses contain only the most rudimentary elements, perhaps due to the time and labour costs of gathering data.

However, data collected by signatories of the treaty can provide a solid base for later discussions. India and Bangladesh previously could not agree on the accuracy of each other's hydrologic records, but eventually agreed on Ganges flow data and based a workable agreement on those data in 1977. The cooperation between engineers or among council members can result in the formation of an epistemic community, another positive outcome of data gathering/sharing. Treaties do not yet include provisions to monitor compliance, but such additions may bolster trust and increase the strength of these epistemic bonds.

Method for water division

Few treaties allocate water: clearly defined allocations account for 54 (37 per cent) of the agreements. Of that number, 15 (28 per cent) specify equal portions, and 39 (72 per cent) provide a specific means of allocation. There are four general trends in those treaties that specify allocations:
1. A shift in position often occurs during negotiations from "rights-based" criteria (whether hydrographic or chronological) in favour of "needs-based" values, based on irrigable land or population.
2. In the inherent disputes between upstream and downstream riparians over existing and future uses, the needs of the downstream riparian are more often delineated (agreements mention upstream needs only in boundary waters accords in humid regions) and existing uses, when mentioned, are *always* protected.
3. Economic benefits are not explicitly used in allocating water, although economic principles have helped guide definitions of "beneficial" uses and have suggested "baskets" of benefits, including both water and non-water resources, for positive-sum solutions.

4. The uniqueness of each basin is repeatedly suggested, both implicitly and explicitly, in the treaty texts.

This last point is exemplified in three unique treaties devised by negotiators: the 1959 Nile Waters Treaty divides the average flow based on existing uses, then evenly divides any future supplies projected from the Aswan High Dam and the Jonglei canal project; the Johnston Negotiations led to allocations between Jordan river riparians based on the irrigable land within the watershed; each party could then do what it wished with its allocation, including divert it out-of-basin; and the Boundary Waters agreement, negotiated with a hydropower focus between Canada and the United States, which provides for a greater minimum flow limit of the Niagara river over the famous falls during summer daylight hours, when tourism is at its peak.

Hydropower

Fifty-seven of the treaties (39 per cent) focus on hydropower. Power-generating facilities bring development, and hydropower provides a cheap source of electricity to spur developing economies. Some, however, suggest that the age of building dams will soon end, because of lack of funding for large dams, a general lack of suitable new dam sites, and environmental concerns.

Not surprisingly, mountainous developing nations at the headwaters of the world's rivers are signatories to the bulk of the hydropower agreements. Nepal alone, with an estimated 2 per cent of the world's hydropower potential, has four treaties with India (the Kosi river agreements of 1954, 1966, and 1978, and the Gandak power project in 1959) to exploit the huge power potential of the region.

Groundwater

Only three agreements deal with groundwater supply: the 1910 Convention between Great Britain and the Sultan of Abdali, the 1994 Jordan-Israeli and 1995 Palestinian-Israeli agreements. Treaties that focus on pollution usually mention groundwater, but do not quantitatively address the issue.

The complexities of groundwater law are described elsewhere in this work. Overpumping can impair an aquifer as a source for human consumption, through salinity from natural sources, seawater intrusion or evaporation deposition. Therefore, allocating groundwater is an especially intricate exercise.

The Bellagio Draft Treaty, developed in 1989, attempts to provide a legal framework for groundwater negotiations. The treaty requires joint management of shared aquifers and describes principles based on mutual respect, good neighbourliness, and reciprocity. While the Draft recognizes that obtaining groundwater data can prove difficult and expensive, and that mutually acceptable information relies on cooperative and reciprocal negotiations, it nevertheless provides a useful framework for future groundwater diplomacy.

Non-water linkages

Negotiators may facilitate the success of treaty negotiations by enlarging the scope of water disputes to include non-water issues. If pollution causes trouble in a downstream country, an upstream neighbour may opt to pay for a treatment plant in lieu of reduced inputs or reduced withdrawals. In such a case, lesser amounts of high-quality water may improve relations more than a greater quantity of polluted or marginal-quality water. Such tactics "enlarge the pie" of available water and other resources in a basin. Non-water linkages include capital, 44 (30 per cent); land, 6 (4 per cent); political concessions, 2 (1 per cent). Other linkages account for 10 treaties (7 per cent) and there are no linkages for 83 treaties (57 per cent).

Examples of these linkages can be found in the 1929 Nile agreement, in which the British agreed to give technical support to both Sudan and Egypt. In lieu of payments, the Soviet Union agreed to compensate lost power generation to Finland in perpetuity (the 1972 Vuoksa agreement). Britain even established a ferry service across newly widened parts of the Hathmatee river in India, in compensation for the inaccessibility problems created by a dam project in the late 1800s.

Compensation for land flooded by dam projects is common. For example, British colonies usually agreed to pay for water delivery and reservoir upkeep, and the British government agreed to pay for damage to houses. However, capital can provide compensation for a greater array of treaty externalities and requirements, such as the construction of new water delivery facilities (the India-Nepal Kosi river project agreements, signed in 1954 and 1966 provide two examples).

Treaties that allocate water also include payments for water – 44 treaties (30 per cent) include monetary transfers or future payments. As early as 1925, Britain moved towards equitable use of the rivers in its colonies: Sudan agreed to pay a portion of the income generated by new irrigation projects to Eritrea, since the Gash river flowed through that state as well. Some treaties also recognize the need to compensate for hydropower losses and irrigation losses due to reservoir storage (the 1951 Finland/Norway treaty and the 1952 Egypt/Uganda treaty both include such clauses). Again, these agreements emphasize the monetary aspect of water: they do not describe water as a right.

Because of individual specialization in conflict management or regional studies, one person may have difficulties finding information about similar

Table 1 **Selected treaty clauses regarding non-water linkages and unique water sharing**

Agreement	Clause
Exchange of notes constituting an agreement between the UK/ Uganda and Egypt regarding the construction of the Owen Falls dam in Uganda	Egypt pays Uganda £980,000 (loss of hydroelectric power) and also flood compensation (upon a later flood)
Convention between the governments of Yugoslavia and Austria concerning water economy questions relating to the Drava	Yugoslavia receives at least 50m schillings in industrial products for 82,500 MWH over 4 years
Johnston Negotiations	Syria: 132 MCM (10.3%); Jordan: 720 MCM (56%); Israel: 400 MCM (31.0%); Lebanon: 35 MCM. Based on area of irrigable land within the watershed in each country
Exchange of notes constituting and agreement between the US and Mexico concerning the loan of waters of the Colorado river for irrigation of lands in the Mexicali valley	USA releases 40,535 acre-feet (50 MCM) of water from September to December 1966 and will retain the same amount over one or three years, depending on the weather conditions that follow
Agreement between Finland and the USSR concerning the production of electric power in the part of the Vuoksi river bounded by the Imatra	The loss of 19,900 MWH generating capacity will be compensated to Finland in perpetuity

treaties in other areas. A survey of the Database reveals some interesting means of solving disputes. At least a few are worthy of a brief discussion. In addition to the two clauses in the 1952 Egypt/Uganda and the 1951 Finland/Norway treaties that provide financial compensation, other treaties address compensation or allocation without money. The common denominator among these selections is the willingness to use, as stated above in Method for water division, "baskets of benefits" instead of water or simple monetary transfers. Table 1 lists several non-water linkage or unique methods of sharing water.

Enforcement

Treaties may handle disputes with technical commissions, basin commissions, or via government officials. Fifty-two (36 per cent) of the treaties provide for an advisory council or conflict-addressing body within the parties' governments. Fourteen (10 per cent) refer disputes to a third party or the United Nations. Thirty-two (22 per cent) make no provisions

3

Economic theory

Economic theory can be used either directly or indirectly to explain resource problems and to settle disputes over scarce resources, such as
water. Economic concepts are applicable in the case of resource conflicts
arising from market failure; to the design of institutions and organizational solutions in terms of rules and structures that are socially desirable;
and to identify solutions that are associated with gains to all parties involved in the dispute (Loehman and Dinar, 1995). The literature provides
several methods that can be adapted to the resolution of conflicts. This
section reviews various approaches and how they have been applied to
resolving water as well as other resource-related conflicts.

Optimization models

Optimization models provide solutions that, economically, are preferable
to all parties and to each of the parties involved. A class of optimization
models applied to resource allocation problems can be found in the literature. The optimization models in this review are classified by several
approaches: regional planning, social planner approach, inter and intraregional allocations, and markets.

Optimization models and regional planning

Chaube (1992) applies a multi-level hierarchical modelling approach
to international river basins, in order to evaluate possible resolution

arrangements of the India-Bangladesh-Nepal-Bhutan conflict over the water of the Ganges-Brahmaputra river basin. The modelling approach, as applied in this paper, allows the utilization of existing models and institutional frameworks for the analysis of large-scale real life problems. By breaking the overall problem into hierarchical stages, this modelling approach can analyse the physical, political, economic, and institutional systems.

In contrast to Chaube who uses a static framework, Deshan (1995) presents a large system hierarchical dynamic programming model, which is applied to the Yellow river in China. By incorporating intertemporal effects, this approach allows for the testing of the likely future impact of water availability scenarios on the urban, storage, and hydropower sectors that compete for the scarce water of the river.

North (1993) applies a multiple objective model (MOM) to water-resource planning and management. MOMs are particularly important in water-related conflicts because water conflicts may arise when each party has different objectives in using the scarce resource. MOMs can compare the results of various optimization problems in terms of incommensurate values for economic, environmental, and social indicators.

Fraser and Hipel (1984) modify and apply the hypergame framework (Bennett, 1987) to analyse and solve actual conflicts, such as the Poplar river water diversion conflict between Saskatchuan, Canada and several US states. Hypergames are conflicts in which one or more of the players are not fully aware of the nature of the conflict situation. Such unaware players may not be aware of all the relevant parties; they may have a false understanding of other players' preferences, or may have an incorrect understanding of the options available to the other players. The conditions under which a hypergame approach is needed are relevant to many other water-related conflicts.

Kassem (1992) developed a river basin model that is driven by water demand in each of the nodes (stakeholders, users, countries, etc.) in the river basin. This comprehensive approach takes into account both the available water resources and the characteristics of water use by each of the use sectors. In addition, the model also allows for policy interventions in each of the river basin parties, in order to affect water conservation. Pricing, storage, and administrative quota restrictions are among such interventions.

Social planner approach

Rogers (1993) employs an engineering-economic approach to demonstrate the economic value of cooperation in river basin disputes. Using game theory concepts and technical-engineering data from the Ganges-

for dispute resolution, and 47 (32 per cent) of the texts are either incomplete or uncertain as to the creation of dispute resolution mechanisms. Can a technical advisory body address disputes? Perhaps, but as noted in Methodology (above), the treaties do not *expressly* provide for such activity.

Historically, force or the threat of force can ensure that a water treaty will be followed – but power is less desirable and more expensive as a guarantor of compliance than mutual agreement. Britain, for example, could force its colonial holdings to follow water treaties because it had one of the most powerful administrative and military organizations in the world. Similarly but more subtly, allocative agreements tend to favour regional hegemons because of their respective power. In some cases, such as the 1959 Nile Waters Treaty, other riparians do not appear in the treaty at all, clearly showing the balance of power in this basin.

While the conflict resolution mechanisms in these treaties do not generally show tremendous sophistication, new enforcement possibilities exist with new monitoring technology. It is now possible to manage a watershed in real time, using a combination of remote sensing and radio-operated control systems. In fact, the next major step in treaty development may well be mutually enforceable provisions, based in part on this technology of objective and highly detailed images, better chemical testing and more accurate flow computations than previously available.

Summary

The study of treaties does not occupy a significant portion of published literature, and therefore the useful information contained in international agreements remains largely undiscovered. More information is necessary regarding the success of treaties and whether the advisory/arbitration councils provide useful services in maintaining just and peaceful relations. The study of successes in some states may yield new ideas for negotiation in other regions. Hopefully the people responsible for the successful treaties can also provide input into the discussions concerning less successful or less amicable agreements. The authors hope that as additional treaties make the database more robust, clearer trends will emerge and scholars will find tools to broaden further the range of dispute resolution.

Notes

1. Some of these arguments, and many of these case studies, are summarized from Wolf's work in Bingham et al. (1994).

2. "Power" in regional hydropolitics can include riparian position, with an upstream riparian having more relative strength *vis-à-vis* the water resources than its downstream riparian, in addition to the more conventional measures of military, political, and economic strength. Nevertheless, when a project is implemented which impacts on one's neighbours, it is generally undertaken by the regional power, as defined by traditional terms, *regardless* of its riparian position.
3. The authors are indebted to those agencies that have helped fund different aspects of the Database, including the US Institute of Peace, the World Bank, the US Agency for International Development, Pacific Northwest National Labs, the Alabama Water Resources Institute, the University of Alabama, and the Oregon State University Department of Geosciences.
4. Some of the following is drawn from Hamner and Wolf (1998).

5

Environmental disputes

Environmental security

The concept of environmental security (also referred to as ecological security) represents an alternative to the current accepted paradigms for addressing threats related to environmental degradation in the post-Cold War world (Dabelko and Dabelko, 1995). Environmental security can also help explain the security needs surrounding freshwater resources while not being limited to these resources.

The study of environmental security is hampered by a lack of consensus on its definition. The literature on environmental security reflects a debate between the security-oriented school of thought and the environment-oriented school of thought.

History

The issue of environmental security has been gaining recognition as a legitimate geopolitical problem since the end of the Cold War. This type of "security" is as significant to countries as defence security. El-Ashry stated, "Nations are discovering that no matter how powerful or rich they are, they are hostage to environmental trends far from their shores" (1991: 16).

As early as 1977 environmentalist Lester Brown argued for a redefinition of national security that would include the environment. Following suit, Richard Ullman sought a similar objective in a 1983 article entitled "Redefining Security." The concept termed environmental security was officially introduced at the 42nd session of the United Nations General Assembly in 1987. The concept gained greater recognition in 1988 when former Soviet Foreign Minister Eduard Shevardnadze told the General Assembly that global environmental threats are quickly "gaining an urgency equal to that of the nuclear and space threats" (El-Ashry, 1991).

Defining the concept

The concept of environmental security raises several intriguing questions: what is environmental security? To whom is it important? Why? Is a concept like environmental security essential for linking environmental degradation or resource scarcity to intra- and interstate conflict? If environmental change and degradation do pose a security threat, whose security does it threaten and how? Should "security" be redefined to incorporate environmental issues and threats, or should environmental issues be seen as a variable of conflicts?

Another area being questioned centres on defining what the threats are that environmental security is attempting to resolve. On one hand, these threats are in the form of violence and conflict. Conversely, others contend that the threat is degradation of environmental resources leading to a decline in the quality of life.

In 1994, Dalby described environmental security as "policies to protect the integrity of the environment from human threats, and simultaneously, to prevent political conflict and war as a result of environmental change and degradation" (in O'Loughlin, 1994: 72). In addition, the definition found in the *Dictionary of Geopolitics* notes that environmental security may also include threats that arise from political instabilities resulting from large numbers of people displaced due to environmental degradation. It can also refer to the environmental damage done by military preparations for the Cold War and damage in the aftermath of the Gulf War (1994: 72).

At the outset, this definition would appear to satisfy both the security view and the environmental view, it is, however, inherently contradictory. On one hand it suggests that we need to protect the environment from human threats, it then indicates we must protect humans from interacting in conflict and war, caused by changes and degradation of the environment.

Kumar (1995) suggests that to define the concept environmental security, one must begin with identifying the components of an ecological crisis that might threaten a state's security. The components of ecological crises start with some kind of human activity that impacts on the environment, may or may not cause a major environmental change, and might lead to large-scale social disruption. It is the social disruption that might cause the various kinds of conflict. Kumar clarifies that "it is only when conflict is created that we have a threat to security" (1995: 154). In his view, the key to analysing environmental threats to security is to determine the variables that allow or prevent the transition between component stages.

A debate has been taking place for the past several years between Homer-Dixon and Levy concerning the definitions and analysis of "environment" and "security" (see 1995, 1996). Levy (1995) proposes a definition of environmental security with "environment," emphasizing the connection with physical and biological systems, and "security," emphasizing protection of national values against foreign threats. He has come to the conclusion that the reason the analyses thus far have been unable to understand the role of the environment in sparking regional conflict is because we do not fully understand what causes conflict. The focus of future research should therefore focus on conflict and not the environment. In Levy's continuing work, the emphasis is on calling for analysis of the causes of regional military conflicts (several articles in 1995).

Homer-Dixon, as lead researcher for the Peace and Conflict Studies Program at the University of Toronto and the American Academy of Arts and Sciences, has attempted to identify links between environmental scarcity and acute conflict. He defines environmental security as:

A condition with two dimensions: The first dimension is safety from chronic threats caused by environmental problems, such as hunger and disease. The second dimension is protection from sudden and hurtful disruptions in patterns of daily life induced by environmental problems. Such threats can exist at all levels of income and development and can occur in homes, in jobs, or in communities. (1996: 56)

Homer-Dixon uses the term scarcity rather than security in a majority of his published work, focusing rather on the links between environmental stress and violence (1991, 1993, 1994, 1995, 1996). His definition of environmental scarcity examines the sources of renewable resource scarcity: supply-induced, demand-induced, and structural.

Due to the lack of consensus concerning the definition of environmental security, many sources include some discussion of the different viewpoints from the available literature (for example, see Dabelko and Dabelko, 1995).

The debate

Redefining security

It is apparent after examining the literature that both sides agree that the environment is a variable of conflict and security issues. This linkage does not mean that the environment is the primary trigger of the conflict or that it is a security issue in and of itself. The debate is over the details of where to place the environment and its problems into our established ways of thinking of conflict and security.

Many who are studying environmental security are arguing for a more holistic definition of security (Brown, 1977; Ullman, 1983; Mathews, 1989; Renner, 1989; Myers, 1993). This new definition would move security away from the traditional model of state behaviour. Mathews (1989) subscribes that the definition of national security should be broadened to include resource, environmental, and demographic issues. Dabelko and Dabelko offer that the "conception of security must instead be changed to reflect the new threats of environmental degradation" (1995: 8). Myers points to the idea of "one-world" living and thinking. In his opinion, growing environmental deficiencies generate conditions that yield and make conflict more likely. These deficiencies have and will serve to determine the source, aggravate core causes, and shape the nature of conflict (1993: 23).

Others oppose a redefinition of security but support the identification of environmental degradation as a major concern (Deudney, 1990, 1991; Dalby, 1992, 1994; Conca, 1994). Deudney questions the causality issue between environmental change and degradation and interstate conflict (Dabelko and Dabelko, 1995). Deudney's argument revolves around his belief that interstate violence, the traditional focus of national security, has "little in common with either environmental problems or solutions" and that "environmental degradation is not very likely to cause interstate wars" (1990: 461). He then claims that only when security from violence and environmental threats are similar can identifying environmental degradation as a threat to national security be useful. He then proceeds to argue that they are in fact different in nature, scope, origin, and degree of intentionality (1991: 23–4). Conversely, Gleick (1991), using very narrow definitions of environmental problems, argues that they are in fact related to violent conflict.

Westing (1989) perceives security as a legitimate quest for all but indicates one cannot look solely at the environmental component for success. His view is of a comprehensive human security (based on the Universal Declaration of Human Rights) that includes both environmental security and political security. Both of the sub-components must be satisfied to achieve security. Westing suggests that there are two prerequisites to achieve environmental security: (1) a protection requirement, the quality of the human environment; and (2) a utilization requirement, the sustainable use of renewable natural resources.

The military

Many authors are concerned about using a vocabulary that lends itself to military involvement with the overlap between environmental security and national security (Deudney, 1990; Dalby, 1992; Conca, 1993). Matthew suggests that one argument for this viewpoint is that "environmental security risks diluting the concept of national security which must be kept narrowly focused on military threats if it is to be usefully and effectively operationalized" (1995: 19). Another area of debate revolves around the military support effort for enforcement of this "security." Some suggest that military activities are major offenders, causing environmental degradation; according to this view, the military should be seen as part of the problem and not part of the solution. Still others are concerned "that this emphasis on environmental protection will hinder military readiness or war-fighting capabilities" (Matthew, 1995: 19).

Cause of conflict

Does environmental stress cause conflict? Many authors would respond with a resounding yes. In their argument for placing environmental change at the top of the priorities list of international politics, several authors have attempted to demonstrate the links through case studies (for example see, Westing, 1986; Myers, 1987; Gleick, 1993; Kumar, 1995; Homer-Dixon, 1991–1996).

Homer-Dixon's investigations (1993, 1994) found evidence of environmental scarcity serving as an underlying cause of intrastate conflict. This "subnational" conflict was primarily based on ethnic clashes created from environmentally induced population movements, and civil strife originating from economic productivity that was affected by environmental scarcities.

Concerning interstate conflict, Westing (1986) maintains that there were 12 conflicts in the twentieth century that contained distinct resource components. Homer-Dixon found that only a "few cases ... supported the interstate conflict hypothesis in terms of renewable resources as the source of conflict" (1994: 39).

Bringing it together

Several authors have seen their way through the debate and have made relevant reminders to those who study this concept of environmental security. Conca (1994) reminds that one must distinguish three ways that people or institutions might be viewing environmental security: (1) rhetorical endorsement; (2) institutional changes that reflect changing priorities; and (3) acceptance of fundamentally new conceptions of security.

Shaw (1996) advises that those who study environmental security must use the appropriate context. He indicates that there are three considerations for developing the relationship between environment and security:

First, it is important to recognize that both security and environmental issues are contextual; the extent and impact of a given problem is relative to its location and sensitivity of the system affected. Second, it is the security issue that provides the context for understanding the impacts of environmental issues and, third, the analysis of environmental issues must be compatible with the analyses of related security issues. (1996: 40)

The example Shaw uses to illustrate this point is that a water problem between Israel and Jordan would have vastly different implications if there were a similar problem between Canada and the US. He cautions that to establish a direct causal link between a "generic" environmental problem and the creation of violent conflict is problematic because of the uniqueness of the context in the different regions of the world. Dabelko and Dabelko support the idea of context when they suggest that "All issues of environmental degradation should not be forced to fit into the matrix of security and conflict" (1995: 8).

Water (transboundary resources)

Scholars are recognizing the importance of transboundary resources in the study of environmental security (Holst, 1989; Mathews, 1989; Lipschutz, 1992; Dabelko and Dabelko, 1995). An important fact is that the states primarily responsible for the problems are often not the ones who endure the majority of the damage. Dabelko and Dabelko point out that "What may be environmental hazards or resource shortages created entirely within one country, can dramatically affect neighboring states" (1995: 9). Mathews maintains "environmental strains that transcend national borders are already beginning to break down the sacred boundaries of national sovereignty" (1989: 162).

Water is becoming an extremely important environmental issue to many nations as demands upon it continue to increase. Although freshwater is a renewable resource, it is also a finite one. Each year, nature makes only so much water available in a given region. This supply can significantly drop below average in times of drought. In addition, as human need for water increases with population and industry, water demand can meet or, through mining groundwater, surpass, replenishable supplies (Postel, 1993: 10). This lack of water is a point of contention for many down-stream states that fear not only quantity but also quality issues (for example, the Euphrates, Ganges, Nile, and Rio Grande rivers).

Lipschutz suggests the perspective that if "people" believe that water rights are distributed inequitably or that the debate over these rights may be an issue in the future, this perception could lead to more conflict than the actual state of the water supply (1992: 5). Dabelko and Dabelko offer further that the nature of transboundary global environmental problems suggest that the best strategy for addressing these challenges between states is cooperation, not competition (1995: 5).

Summary

Even though environmental security is an evolving concept, there are already some common trends emerging. First, most researchers recognize the relevance of studying how environmental change and degradation impacts humans, whether or not it is caused by humans, and at all levels. Second, the general consensus appears to be that there is a need for understanding how to resolve natural resource disputes through both traditional and alternative dispute resolution techniques. Third, whether one wants to change the traditional paradigm of security or merely examine the environment as a contributor that can threaten one's security, there is an underlying theme that resource scarcity will probably lead to conflict in the future. Perhaps it is Matthew who has stated it best:

The environmental integrity of the planet and the welfare of humankind require tough choices between using resources and institutions that are at hand and forging new ones, reforming current practices and avoiding new stresses on the environment, and protecting the privileged position of industrial states and redistributing wealth and expertise. There is no clear path towards an environmentally secure future, but there are many routes likely to lead to conflict, violence, and misery. Avoiding these will demand innovation, pragmatism, and sacrifice. (Matthew 1995: 20)

Other resources

As pointed out by MacDonnell (1988), there are several sources and types of conflicts related to natural resources. The sources for disputes are rooted in the different values and importance people may associate with a particular resource (religious, production, etc.), their present and future use, and negative externalities associated with their use. Disputes may arise between private parties and government agencies, they may involve many private parties, and they also may involve different governments. The source and the type of the dispute may help in selecting the resolution technique (as is described in other parts of this review).

The overarching problem of resource scarcity

The availability of natural resources has decreased over time both in terms of quantity and quality as a result of industrial development, and the expansion of urbanization and agriculture. The inevitable consequences of increasing the demand for various natural resources represent an increase in the value for the use of these scarce resources and their services. A general review of the evidence of the growing scarcity of natural resources can be found in Young (1991). Antle and Heidebrink (1995) estimate the environment-development trade-offs faced by countries during the process of economic growth. What Young and Antle and Heidebrink do not discuss, which can easily be assessed from the growing volume of literature, is the increase in the number of disputes over these natural resources. A general discussion that also addresses other dimensions of sustainable development can be found in Redclift (1991). In addition to the economic dimension, Redclift introduces the political dimension and the epistemological dimension (which seeks ways of acquiring knowledge and incorporating it into the conceptual systems). A more focused review of settlement of public international disputes on shared resources can be found in Alheritiere (1985). There are several peaceful means of settling resources disputes, including direct negotiations, good offices, mediation, enquiry and conciliation, consultation, arbitration, and judicial settlements. Alheritiere also discusses the advantages and disadvantages associated with each option and with each resource.

The following sections address specific disputes associated with several resources as they were documented in the literature.

Oil

The case of oil disputes has several dimensions. On the one hand, disputes may arise on the basis of production rights, oilfield boundaries and extraction rights. On the other hand, disputes associated with oil pollution are becoming common. Disputes may arise also with regard to prices of oil within OPEC countries (Toman, 1982a, 1982b), but will not be discussed here.

Mitchell (1994) discusses two international oil pollution control alternatives: the discharge and the equipment subregimes. The discharge subregime – aimed at protecting coastlines by regulating oil discharge zones – is demonstrated using case studies of disputes between the United Kingdom and Germany, the Netherlands, and the US. The equipment subregime regulates types of vessels and safety equipment on vessels (e.g. segregated ballast tanks). Although opposed by several governments representing public shipping interests (France and Japan), and private shipping interests (Denmark, Germany, Greece, Norway, Sweden), an agreement was eventually approved. The compliance with the treaty is questioned in the paper. Several mechanisms are suggested that include enhancing transparency of the subregimes, facilitating potent but low-cost sanctions, and coercing compliance rather than deterring violation.

Devlin (1992) describes the relationship between the oil extraction policy of a political system and the style of governance in totalitarian regimes, using case studies of Iran and Iraq. Although not directly related to the problem of dispute resolution in natural resources that is reviewed here, the relevancy of domestic policies to the understanding of disputes is essential (LeMarquand, 1977) in the resolution process.

Valencia (1986) describes international conflicts over oil and mineral resources in overlapping claim areas across world regions. Similar to transboundary aquifers and fisheries, the problem here is a lack of sufficient definition of property rights. A preferred solution to such disputes, as demonstrated in the paper, is the agreement for joint development in areas of overlapping claims. Using numerous examples, common elements of such agreements include: definition of the extent of the area; the nature and functions of the joint management body; the contract type (duration and termination rules); financial arrangements; the process of selection of concessionaires or operators, procedures, and principles for conflict resolution, and transfer of technology.

Land

Similar to the case of oil, land disputes may arise on the basis of land rights, oilfield boundaries and extraction rights, or soil erosion problems. A study in Haiti demonstrates the cooperative approach towards the management of a common property resource – land in a watershed that is subject to transboundary erosion effects by upstream mismanagement. The analysis of this case study demonstrates that incentives associated with land conservation triggered a cooperative effort on the farmers' side. The flexible set of rules, consisting of small investments or labour inputs in the construction and maintenance of small dams, allowed all parties to cooperate. Factors contributing to better cooperation of the individual parties that may be relevant for other cases included: potential direct or indirect benefit; level of effort needed to maintain cooperation; land tenure; religious affiliation; initial wealth; and existing investment in soil conservation.

Claims by Alaskan natives over vast amounts of land date back to 1867 when Alaska was sold to the US. The natives protested the sale arguing that they owned the land. When Alaska became a state in 1958, it was important to settle the conflict with thought given to cultural and traditional considerations. Federal laws and policies have recognized the right of Alaskan natives to land and wildlife for subsistence. A process of hearings was established (Berger, 1988), which in turn established the Alaska Native Review Commission. This commission started a consultative process among Eskimo villages and NGOs. The commission prepared a document that includes all the testimonies and recommendations regarding the native ownership of land. Since the conflict has been handled in the form of a commission, it is not clear whether the parties involved have accepted its recommendations.

Irrigation schemes that are served by the same source of water can create land and ecological degradation disputes, as is the case in Chokwe, Mozambique (Tanner et al., 1993). The study identified several sources of disputes over the use of land and water resources, over damages to irrigation structures, and over yields. In the specific case of land, these disputes were resolved by authoritative powers such as local family leaders, the executive council of the project members, and by the representatives of the irrigation management company.

Roads

Sources of road disputes represent different values that residents and local or national governments assign to the land that is to be developed. Harashina (1988) reviews several disputes/case studies associated with the Tokyo Bay Area Artery Project. One important conclusion from the case studies is the need to start mediation at the same time as the planning of the project. This need means not only involving the stakeholders in the process of identifying and evaluating various options, but also assigning a mediator to resolve potential conflicts as they arise.

Fishing

Gramann and Burdge (1981) test the complementarity between recreation and fishing. Surveying visitors in Lake Shelbyville, Illinois, they found conflicting goals between skiers and fishermen who used the lake. The authors used discriminate analysis to describe the conflict between the two groups of visitors. Although the analysis showed only weak antagonism between the goals of skiers and fishermen, Gramann and Burdge's study allows us to understand the dynamics of conflict in recreation.

Air pollution

Acid rain

Acid rain pollution, caused by the burning of fossil fuels, is an international transboundary issue. Although acid rain has been recognized nationally and internationally as a polluting problem for some time, it was not until 1979 when the Convention on Long-Range Transboundary Pollution was signed, and then ratified in 1985 by 54 countries, that the issue gained international recognition. The convention establishes a data-gathering network and data-sharing system, and it provides regulations to deal with polluting substances (McCormick, 1985).

China is one of the primary coal users in the world, and produces the greatest amount of sulphur dioxide in Asia. In contrast, Japan uses several sources of energy, of which coal is only 16 per cent. It was found that cross-border pollution from China might contribute to acid rain in Japan (Matsuura, 1995). To manage this problem, Japan, through its Development Assistance programme, has developed a strategy to provide technical assistance to China so that it can improve its technological capability in reducing emissions that affect Japan.

Tahovnen et al. (1993) studied the acid rain dispute between Finland and the Soviet Union [sic]. Similar to the case of China and Japan, the asymmetry between the two countries introduced some interesting aspects to the dispute. The long border between these two countries introduced severe transboundary air pollution problems. Soviet sulphur emissions into Finland from industrial activity amounted to 651,000 tons per year. These emissions resulted in the acidification of forest soils. Although there are some transboundary Finnish air pollution effects on Soviet territory, these have not been properly documented. Solutions to the problem may include reduction of the emissions by reducing the industrial activity, improving the industrial processes to produce fewer emissions, or improving fuel standards. In reaching an agreement the paper employed both non-cooperative and cooperative approaches. A 50 per cent emission reduction agreement was the actual (real-life) resolution of the conflict. However, the agreement did not specify whether the 50 per cent reduction was in all pollution-contributing regions and this discrepancy could greatly affect the outcome of the agreement.

Additional readings on acid rain pollution can be found in Hawdon and Pearson (1993), Postel (1984), Kowalok (1993), Forster (1989), and Bhatti et al. (1992).

Global warming

The emission and accumulation of carbon dioxide in the atmosphere not only creates immediate air pollution problems, but also affects our global climate. Although it is very difficult to identify and measure the individual impact by one country on the climate of another, the cumulative effect over time can be observed. Cline (1992) believes that only an international cooperative strategy can be useful in managing this problem. He suggests that this international framework should draw upon economic, legal, institutional, and social factors. According to Cline, a cooperative approach that is unique to the global warming phenomena is for industrial countries in the northern hemisphere to invest in emission reduction technologies in developing countries in the southern hemisphere. By doing so the developed countries contribute to the reduction of global warming.

Summary

The literature reviewed in this section illustrates that other natural resource issues offer lessons that may be applied to the resolution of water conflicts. At the regional level, actors often have fared better at protecting their seas and fisheries and other common pool resources (i.e. deep seabed mining, grazing lands, fisheries, oilfields, rainforests, outer space, acid rain, and air pollution), than they do in protecting their transboundary waters. A systematic analysis of successful common pool resource institutions can lead to the development of design principles that may be applicable to other cases.

An analogy has been made between solutions to energy disputes and water disputes, suggesting that the "soft path" approach be used in terms of vital needs, supply-demand and pricing, and environmental damage. When dealing with aquifers, the comparison with underground oil is even more appealing, especially in border areas between two countries. However, one must acknowledge the many differences between the two resources.

Another field that offers lesson is that of domestic water agreements and national water law which, in the developed world and some developing countries, is much stronger and more resilient than international law. When asking ourselves what the main differences are between domestic and international watershed disputes, we are clearly extrapolating issues within the realm of the social sciences, including the dominant concept of sovereignty, the lack of enforcement mechanisms, the weakness of international law, the absence of the federal state as a built-in mediator, the disparities among the riparian in types of political regimes, the social structure, the chauvinistic attitudes of stakeholders, the cultural value system, and other issues.

This obvious but intriguing relationship requires more attention in our study. On the one hand, the relevance of the idea of national sovereignty clearly emerges as an obstacle, but working across national lines could generate other types of collective identity, overlapping interests of domestic groups of different riparians, and adherence to other regional configurations. The problem remains, so far, as to how the development of additional identities can be perceived not as threatening but as complementing the primordial national identity. On the other hand, it has been mentioned that domestic water management is "fragmented among sectors and institutions with little regard for conflicts or complemetaries among social, economic, and environmental objectives" (Serageldin, 1995). If that is the case at the domestic level, clearly its multiplication by the numbers of riparians and in different internal configurations makes the probability of a comprehensive but only technical approach rather slim.

A more thorough review of agreements signed among disputing provinces and states within federal governments raises some interesting points from the question of adaptation, to the guiding principles in international disputes. Nevertheless, many conclusions are significant in their work at the international level, both in determining the approach to conflict management as well as in drawing inferences from small to large-scale planning.

The learning of methods for solving territorial issues are of great importance. Granted that the intrinsic fluid and seasonal nature of waterways generates an interdependence that makes dispute-solving more difficult than resolving issues on set and fixed borders, let alone that rivers may also fulfil the added dimension of a national frontier. Perhaps because the differences are obvious, not enough sufficient lateral thinking about commonalties seems to have been done about this comparison.

Part 4

6

Conclusions and summary

This compilation of literature reviews was selected to show the direct link between the fields of dispute resolution and economics in the application of theory and techniques to conflicts over freshwater resources. The literature surveyed indicates that, in many areas, much study and analysis has taken place. However, even with the vast amount of literature available, there continues to be a need for more research on why conflicts over water and other environmental resources occur. A deeper understanding of the similarities among all conflicts over natural resources may provide applications for predicting and preventing these conflicts in the future.

In summary, Chapter 2 (Organizational theory) showed us that water not only ignores our political boundaries, it evades institutional classification and eludes legal generalizations. Interdisciplinary by nature, water's natural management unit, the watershed – where quantity, quality, surface, and groundwater all interconnect – strains both institutional and legal capabilities often past capacity. Analyses of international water institutions find rampant lack of consideration of quality considerations in quantity decisions, a lack of specificity in rights allocations, disproportionate political power by special interest, and a general neglect for environmental concerns in water-resources decision-making. The World Bank, United Nations, and the new World Water Council are now beginning to address these weaknesses.

Legal principles have been equally elusive. The 1997 Convention reflects the difficulty of merging legal and hydrologic intricacies. While the articles provide many important principles, including responsibility for cooperation and joint management, they also codify the inherent upstream/downstream conflict by calling for both equitable use and the obligation not to cause appreciable harm. They also provide little practical guidelines for allocations – the heart of most water conflicts. In contrast to general legal principles, site-specific treaties have shown great imagination and flexibility, moving from "rights-based" to "needs-based" agreements in order to circumvent the argument over use versus harm.

The section on negotiation theory indicated in this preliminary review shows a great effort to provide solutions to the TWD. A great part of the work on CR stresses institutional and technical arrangements; CR is perceived as mechanisms that need to be incorporated once the agreement is reached, but few relate to the incorporation of CR as instrumental to the process of reaching such agreements.

In retrospect, one of the most serious obstacles for resolution is insufficient information about a situation of imminent scarcity. Hence the problem is how to transform this knowledge and use it to look for shared innovative solutions.

Economics is one discipline that is used independently or jointly with other disciplines in explaining scarce resource disputes and indicating a set of possible and agreeable arrangements between the parties. Optimization models provide solutions offered by economic approaches that may look promising, but it is always necessary to identify the set of assumptions leading to such solutions. Even with this identification in mind, one can still argue that economic principles are among the sufficient, but not the necessary, conditions for a dispute to be solved.

Using economic terms, for a solution to a dispute to be attractive to the participants and to be economically sustainable, it needs to fulfil the requirements for individual and group rationality. In other words, the resolution of a dispute must be perceived for each participant as preferable to the status quo outcome, and to participation in any partial arrangement that includes a subset of the regional participants. The regional arrangement also fulfils requirements that all costs or gains are allocated.

As we well know, economics and politics play interactive roles in the evaluation of dispute resolution. Just as political considerations can effectively veto a joint project with an otherwise favourable economic outcome, a project with potential regional-welfare improvements might influence the political decision-making process to allow the necessary cooperation. Therefore, both economic and political considerations should be incorporated into evaluations of dispute resolution arrangements.

Game theory provides a vehicle through mathematics and the social sciences to engineer improvements in policy and understanding of many market and non-market events. In the jargon of the theory, we say that a

game's outcome depends upon the set of feasible outcomes, participants' choices, and the rules of the game. To solve the problem of water allocations within an international water basin cooperatively, a number of problems can be analysed using game theory:
- In international contexts, each sovereign party is free to break any agreement at little cost. Hence any engineered solution must be sensitive to the stability aspects of the proposed outcomes.
- For cooperation to occur, the parties must have some incentive that can justify the cooperation.

This latter point implies that for a cooperative solution to be accepted by the parties involved, it is required that the partitioned cost or benefit to any subset of participants is preferred by the subset to any other possible outcome they can guarantee themselves. Of course, in the real world of international relations, it must also be that all the costs are allocated.

As the amount of water surplus decreases over time, however, the impetus towards conflict or cooperation (pay-offs) might change, depending on such political factors as relative power, level of hostility, legal arrangements, and form and stability of government.

The study of treaties has not yet occupied a significant portion of published literature, and therefore the useful information contained in international agreements remains largely undiscovered. More information is necessary regarding the success of treaties and whether the advisory/ arbitration councils provide useful services in maintaining just and peaceful relations. The study of successes in some states may yield new ideas for negotiation in more tenuous regions. Hopefully the people responsible for the successful treaties can also provide input into the discussions for less successful or less amicable agreements. A hardcopy preview of a systematic computer compilation of international water treaties can be found in Chapter 8.

Environmental disputes (Chapter 5) serves to further our understanding of conflicts over resources other than water. A concept being utilized by some to examine the whole spectrum of natural resources conflict is environmental security. Even though environmental security is an evolving concept, there are already some common trends emerging. First, most recognize the relevance of studying how environmental change and degradation impacts upon humans, whether or not caused by humans, and at all levels, local, national, and global. Second, the general consensus appears to be that there is a need for understanding how to resolve natural resource disputes through both traditional and alternative dispute-resolution techniques. Third, whether one wants to change the traditional paradigm of security or merely examine the environment as a contributor that can threaten one's security, there is an underlying theme that resource scarcity will probably lead to conflict in the future.

A clear understanding of the details of how water conflicts have been resolved historically is vital in discerning patterns that may be useful in resolving or precluding future conflict. Our investigations of 14 disputes suggest that, generally, the following pattern tends to emerge: riparians of an international basin implement water development projects unilaterally first on water within their territory, in attempts to avoid the political intricacies of the shared resource. At some point, as water demand approaches supply, one of the riparians, generally the regional power, will implement a project that impacts on at least one of its neighbours. This project can, in the absence of relations or institutions conducive to conflict resolution, become a flashpoint, resulting in conflict. The comparative analysis also suggests indicators of impending or likely water conflict, obstacles to successful negotiations, and observations regarding national versus international settings. These 14 disputes are highlighted in detail in Chapter 7 (Case studies).

The production of this document was conducted under the auspices of the Transboundary Fresh Water Disputes Project (TFWDP), a comprehensive and interdisciplinary analysis of international surface water conflicts. The TFWDP is an effort to create a qualitative and quantitative analysis of transboundary water conflict and to develop procedural and strategic templates for early intervention so as to help contain and manage conflicts. The World Bank Research Committee under the grant preparation fund provided partial funding for this survey of literature on water dispute resolution and related water treaties.

Part 5: Appendixes

7
Case studies

Case studies of transboundary dispute resolution

List of cases

Watersheds

Danube
Euphrates
Jordan (including West Bank aquifers)
Ganges
Indus
Mekong
Nile
Plata
Salween

Aquifer systems

US-Mexico aquifers
West Bank aquifers (included with Jordan watershed)

Lakes

Aral sea
Great Lakes

Engineering works

Lesotho Highlands

Table 2 **Case studies**

NAME	RIPARIAN STATES (w/% of nat'l available water being utilized)#	RIPARIAN RELATIONS (w/dates of most-recent agreements)	AVERAGE ANNUAL FLOW (km³/yr)*	SIZE OF WATER-SHED (km²)*	CLIMATE	SPECIAL FEATURES
DANUBE	Albania (1.6) Austria (6.1) Bulgaria (7.1) Croatia (n/a) Czech Republic (n/a) Germany (43.8) Hungary (35.5) Italy (26.6) Moldavia (n/a) Poland (42.9) Romania (22.0) Slovakia (n/a) Slovenia (n/a) Switzerland (9.8) Ukraine (n/a) Yugoslavia (14.4)	Cold to warm (1994 Danube River Protection Convention)	206	810,000	Dry to humid	1994 Convention is first treaty developed through process of public participation
EUPHRATES	Iraq (86.3) Syria (102.0) Turkey (12.1)	Cool	46	1,050,000	Dry to Mediterranean	Ongoing tripartite dialogue but no international agreement

Basin	Riparian states	Treaty status		Area (km²)	Climate	Comments
JORDAN	Israel (95.6) Jordan (67.6) Lebanon (20.6) Palestine (100.0) Syria (102.0)	Cool to warm (1994 Treaty of Peace–Israel/Jordan; 1995 Interim Agreement – Israel/Palestine)	1.4	11,500	Dry to Mediterranean	Complex conflict and attempts at conflict resolution since 1919
GANGES-BRAHMA-PUTRA	China (19.3) Bangladesh (1.0) Bhutan (0.1) India (57.1) Nepal (14.8)	Cold to warm (1985 Agreement between India and Pakistan lapsed in 1988; new treaty in 1996)	971	1,480,000	Humid to tropical	Scheduled to be model/workshop case – limited riparians; on-going dispute
INDUS	Afghanistan (47.7) China (19.3) India (57.1) Pakistan (53.8)	Cool (1960 Indus Water Treaty between India and Pakistan)	238	970,000	Dry to humid subtropical	Scheduled as case to be "back-modelled"
MEKONG	Cambodia (0.1) China (19.3) Laos (0.8) Myanmar (0.4) Thailand (32.1) Vietnam (2.8)	Cool to warm (1957 Mekong Committee re-ratified as 1995 Mekong Commission)	470	790,000	Humid to tropical	Good example of resilience of agreement

Table 2 (cont.)

NAME	RIPARIAN STATES (w/% of nat'l available water being utilized)#	RIPARIAN RELATIONS (w/dates of most-recent agreements)	AVERAGE ANNUAL FLOW (km³/yr)*	SIZE OF WATERSHED (km²)*	CLIMATE	SPECIAL FEATURES
NILE	Burundi (3.1) Egypt (111.5) Eritrea (n/a) Ethiopia (7.5) Kenya (8.1) Rwanda (2.6) Sudan (37.3) Tanzania (1.3) Uganda (0.6) Zaire (0.2)	Cold to warm (1959 Nile Water Agreement only includes Egypt and Sudan)	84	2,960,000	Dry to tropical	Scheduled as complex model/workshop
PLATA	Argentina (3.5) Bolivia (0.7) Brazil (0.5) Paraguay (0.2) Uruguay (0.6)	Warm (1995 Mercosur (Southern Common Market) adds impetus to "hydrovia" canal project)	470	2,830,000	Tropical	Good example of inter-sectoral plus international dispute
SALWEEN	China (19.3) Myanmar (0.4) Thailand (32.1)	Cool to warm	122	270,000	Humid to tropical	Scheduled as conflict preclusion model/workshop

82

	Riparians (%)					
US-MEXICO aquifers (groundwater)	Mexico (22.3) United States (21.7)	Warm (1944 Water Treaty, modified in 1979)	n/a	n/a	Dry	Groundwater not included in original treaty, leading to uncertainty in relations
WEST BANK aquifers	Israel (95.6) Palestine (100.0)	Cool (1995 Interim Agreement)	n/a	n/a	Dry	Interim Agreement relegates groundwater allocations to future negotiations
ARAL	Afghanistan (47.7) Kazadhstan (n/a) Kyrgyzstan (n/a) Tajikistan (n/a) Turkmentistan (n/a) Uzbekistan (n/a)	Cool to warm (1993 and 1995 Agreements on Aral Action Plans)	1,020[1]	1,618,000	Dry to humid continental	Case of lake management exacerbated by internationalization of basin
GREAT LAKES	Canada (1.4) United States (21.7)	Warm	22,500[1]	509,200	Humid continental	Case of small number of riparians with good relations
LESOTHO HIGHLANDS	Lesotho (1.5) South Africa (28.4)	Warm	n/a	n/a	Humid marine	Interesting institutional arrangement exchanging water, financial considerations, and energy resources

Source: Kulshreshtha (1993)
* Sources: Gleick ed. (1993); UN Register of International Rivers (1978)
1 Values for lakes under "Annual Flow" are for storage volumes

Danube river

Case summary

River basin	Danube
Dates of negotiation	1985–1994
Relevant parties	All riparian states of the Danube
	Convention is the first designed through the process of public participation, including NGOs, journalists, and local authorities
Flashpoint	None – good example of "conflict preclusion"
Issues	
Stated objectives	To provide an integrated, basin-wide framework for protecting Danube water quality
Additional issues	
Water-related	Encourage communication between water-related agencies, NGOs, and individuals
Non-water	None
Excluded	Strong enforcement mechanism
Criteria for water allocations	None determined
Incentives/linkage	World Bank/donor help with quality control
Breakthroughs	No untoward barriers to overcome
Status	Convention signed in 1994. Too early to judge effectiveness of implementation

The problem

Prior to World War II, the European Commission of the Danube, with roots dating back to the 1856 Treaty of Paris and made up of representatives from each of the riparian countries, was responsible for administration of the Danube river. The primary consideration at the time was navigation, and the Commission was successful at establishing free navigation along the Danube for all European countries. By the mid-1980s, it became clear that issues other than navigation were gaining in importance within the Danube basin, notably problems with water quality. The Danube passes by numerous large cities, including four national capitals (Vienna, Bratislava, Budapest, and Belgrade), receiving the attendant waste of millions of individuals and their agriculture and industry. In addition, 30 significant tributaries have been identified as "highly polluted." The breakup of the USSR has also contributed to water quality deterioration, with nascent economies finding few resources for environmental problems, and national management issues being internationalized with redrawn borders. Recognizing the increasing degradation of water quality, in 1985 the (at the time) eight riparians of the Danube signed the "Declaration of the Danube Countries to Cooperate on Questions Concerning the Water Management of the Danube," commonly called the Bucharest Declaration. This Declaration led in turn to the 1994 Danube River Protection Convention.

Background

The Danube river basin lies at the heart of Central Europe and is Europe's second longest river, at a length of 2,857 km. The river's basin drains 817,000 km^2, including all of Hungary, most of Romania, Austria, Slovenia, Croatia, and Slovakia; and significant parts of Bulgaria, Germany, the Czech Republic, Moldova, and Ukraine. Territories of the Federal Republic of Yugoslavia – Bosnia and Herzegovina, and small parts of Italy, Switzerland, Albania, and Poland are also included in the basin. The Danube river discharges into the Black Sea through a delta that is the second largest wetland area in Europe. The river is shared by a large and ever-growing number of riparian states that for decades were allied with hostile political blocs; some of which are currently locked in intense national dispute. As a consequence, conflicts in the basin tended to be both frequent and intricate, and their resolution especially formidable.

Attempts at conflict management

World War II created new political alliances for the riparians, resulting in a new management approach. At a 1948 conference in Belgrade, the East Bloc riparians – a majority of the delegates – shifted navigation over to the exclusive control of each riparian. By the 1980s, though, quality considerations had led to the Bucharest Declaration of 1985, which reinforced the principle that the environmental quality of the river depends on the environment of the basin as a whole, and committed the riparians to a regional and integrated approach to water basin management, beginning with the establishment of a basin-wide unified monitoring network. Basin-wide coordination was strengthened at meetings in Sofia in September 1991, in which the riparians elaborated on a plan for protecting the water quality of the Danube. At that meeting, the countries and interested international institutions met to draw up an initiative to support and reinforce national actions for the restoration and protection of the Danube river. With this initiative, named the Environmental Program for the Danube River Basin, the participants agreed to create an interim task force to coordinate efforts while a convention to steer the programme was being negotiated.

Outcome

The principle of "participation" has been taken seriously in the work of the Environmental Program and the Coordination Unit. Initially, each riparian country was responsible for identifying two individuals to help coordinate activity within the basin. The first, a "country coordinator," usually a senior official, would act as liaison between the work of the programme and the country's political hierarchy. The second, a "country focal point," would coordinate the actual carrying out of the work plan.

In July 1992, the coordination unit held a workshop in Brussels to help facilitate communication between the coordinators, the focal points, and the donor institutions. Representatives from each of the (by then) 11 riparians and 15 donor and non-governmental organizations attended. An important outcome of the workshop was that the participants themselves designed a plan for each issue covered. One issue, for example, was an agreement to produce national reviews of data availability and priority issues within each country. The information would be used by prefeasibility teams funded by donors who were to identify priority investments in the basin. During the workshop, participants developed the criteria for the national reviews and agreed on a schedule for their completion.

The principle of participation was carried one level deeper at the third task force meeting in October 1993 in Bratislava. At that meeting, the task force agreed to prepare a "Strategic Action Plan" (SAP) for the Danube basin, with the provision that, "consultation procedures should be strengthened." This last point is particularly noteworthy because it is the first time public participation has been required during the development of an international management plan. This concept rejects the principle that internal politics within nations ought to be treated as a geopolitical "black box," whose workings are of little relevance to international agreements, and instead embraces the vital need for input at all levels in order to ensure that the plan has the support of the people who will affect, and be affected by, its implementation.

In principle, the individuals who participated in the workshops would form a nucleus that would not only have input in the drafting of a SAP, but would be involved in reviewing future activities that would be implemented as part of the Plan. By July 1994, two consultation meetings were held in each of the nine countries.

On 29 June 1994, in Sofia, the Danube river basin countries and the European Union signed the Convention on Cooperation for the Protection and Sustainable Use of the Danube River (the Danube River Protection Convention). The convention notes that the riparians of the Danube, "concerned over the occurrence and threats of adverse effects, in the short or long term, of changes in conditions of watercourses within the Danube River Basin on the environment, economies, and well-being of the Danubian States," agree to a series of actions, including:

– striving to achieve the goals of a sustainable and equitable water management, including the conservation, improvement and rational use of surface waters and groundwater in the catchment area as far as is possible;

– cooperating on fundamental water management issues and take all appropriate legal, administrative, and technical measures, to at least maintain and improve the current environmental and water quality conditions of the Danube river and of the waters in its catchment area and to prevent and reduce as far as possible adverse impacts and changes occurring or likely to occur;

– setting priorities as appropriate and strengthening, harmonizing, and coordinating measures taken and planned to be taken at the national and international level throughout the Danube basin aimed at sustainable development and environmental protection of the Danube river.

The Danube Convention is a vital legal continuation of a tradition of regional management along the Danube dating back 140 years. As a political document, it provides a legal framework for integrated watershed management and environmental protection along a waterway with tremendous potential for conflict.

In recent years, the riparian states of the Danube river have extended the principle of integrated management, and established a programme for the basin-wide control of water quality, which, if not the first such programme, has claims to being probably the most active and the most successful of its scale. The Environmental Program for the Danube River is also the first basin-wide international body that actively encourages public and NGO participation throughout the planning process, which, by diffusing the confrontational setting common in planning, may help preclude future conflicts both within countries and, as a consequence, internationally.

Euphrates basin

Case summary

River basin	Tigris-Euphrates
Dates of negotiation	Meetings since mid-1960s to present
Relevant parties	Iraq, Syria, Turkey
Flashpoint	Filling of two dams during low-flow period results in reduced flow to Iraq in 1975
Issues	
Stated objectives	Negotiate an equitable allocation of the flow of the Euphrates river and its tributaries between the riparian states
Additional issues	
Water-related	Water quality considerations Orontes river, which flows from Syria into Turkey
Non-water	Syrian support for PKK Kurdish rebels
Excluded	Possible connection between Tigris and Euphrates
Criteria for water allocations	None determined
Incentives/linkage	Financial: None Political: None
Breakthroughs	None
Status	Bilateral and tripartite negotiations continue with greater and lesser success – no agreement to date

The problem

In 1975, unilateral water developments came very close to provoking warfare along the Euphrates river. The three riparians to the river – Turkey, Syria, and Iraq – had been coexisting with varying degrees of hydropolitical tension through the 1960s. At that time, population pressures drove unilateral developments, particularly in southern Anatolia, with the Keban dam (1965–1973), and in Syria, with the Tabqa dam (1968–1973).

Background

Bilateral and tripartite meetings, occasionally with Soviet involvement, had been carried out between the three riparians since the mid-1960s, although no formal agreements had been reached by the time the Keban and Tabqa dams began to fill late in 1973, resulting in decreased flow downstream. In mid-1974, Syria agreed to an Iraqi request that Syria increase the flow from the Tabqa dam by 200 MCM/yr. The following year, however, the Iraqis claimed that the flow had been dropped from the normal 920 m3/sec to an "intolerable" 197 m3/sec, and asked that the Arab League intervene. The Syrians claimed that less than half the river's normal flow had reached its borders that year and, after a barrage of mutually hostile statements, pulled out of an Arab League technical committee formed to mediate the conflict. In May 1975, Syria closed its airspace to Iraqi flights and both Syria and Iraq reportedly transferred troops to their mutual border. Only mediation on the part of Saudi Arabia was able to break the increasing tension, and on 3 June, the parties arrived at an agreement that averted the impending violence. Although the terms of the agreement were not made public, Iraqi sources are cited as privately stating that the agreement called for Syria to keep 40 per cent of the flow of the Euphrates within it borders, and to allow the remaining 60 per cent through to Iraq.

Attempts at conflict management

The Southeast Anatolia Development Project (GAP is the Turkish acronym) has given a sense of urgency to resolving allocation issues on the Euphrates. GAP is a massive undertaking for energy and agricultural development that, when completed, will include the construction of 21 dams and 19 hydroelectric plants on both the Tigris and the Euphrates. 1.65 million hectares of land are to be irrigated and 26 billion kWh will be generated annually with an installed capacity of 7,500 MW. If completed as planned, GAP could significantly reduce downstream water quantity and quality.

A Protocol of the Joint Economic Committee was established between Turkey and Iraq in 1980, which allowed for Joint Technical Committee meetings relating to water resources. Syria began participating in 1983, but meetings have been intermittent at best.

A 1987 visit to Damascus by Turkish Prime Minister Turgut Ozal reportedly resulted in a signed agreement for the Turks to guarantee a minimum flow of 500 m3/s across the border to Syria. According to Kolars and Mitchell (1991), this total of 16 BCM/yr is in accordance with prior Syrian requests. However, according to Naff and Matson (1984), this is also the amount that Iraq insisted on in 1967, leaving a potential shortfall. A tripartite meeting between Turkish, Syrian, and Iraqi ministers was held in November 1986, but yielded few results.

Talks between the three countries were held again in January 1990, when Turkey closed the gates to fill the reservoir behind the Ataturk dam, the largest of the GAP dams, essentially shutting off the flow of the Euphrates for 30 days. At this meeting, Iraq again insisted that a flow of 500 m3/s cross the Syrian-Iraqi border. The Turkish representatives responded that this was a technical issue rather than one of politics and the meetings stalled. The Gulf War, which broke out later that month, precluded additional negotiations.

Outcome

In their first meeting after the war, Turkish, Syrian, and Iraqi water officials convened in Damascus in September 1992, but broke up after Turkey rejected an Iraqi request that flows crossing the Turkish border be increased from 500 m3/sec to 700 m3/sec. In bilateral talks in January 1993, however, Turkish Prime Minister Demirel and Syrian President Assad discussed a range of issues intended to improve relations between the two countries. Regarding the water conflict, the two agreed to resolve the issue of allocations by the end of 1993. Although an agreement has not, to date, been reached, Prime Minister Demirel declared at a press conference closing the summit that, "There is no need for Syria to be anxious about the water issue. The waters of the Euphrates will flow to that country whether there is an agreement or not" (Cited in Gruen, 1993). The issue remains unresolved.

Jordan river watershed

Case summary

River basin	Jordan river and tributaries (directly); Litani (indirectly)
Dates of negotiation	1953–1955; 1980s through the present
Relevant parties	United States (initially sponsoring); US and Russia (sponsoring multilateral negotiations) Riparian entities: Israel, Jordan, Lebanon, Palestine, Syria
Flashpoints	1951 and 1953 Syrian/Israeli exchanges of fire over water development in demilitarized zone; 1964–1966 water diversions
Issues	
Stated objectives	Negotiate an equitable allocation of the flow of the Jordan river and its tributaries between the riparian entities Develop a rational plan for integrated watershed development
Additional issues	
Water-related	Out-of-basin transfers Level of international control ("water master") Location and control of storage facilities Inclusion or exclusion of the Litani river
Non-water	Political recognition of adversaries
Excluded	Groundwater Palestinians as political entity (initially)
Criteria for water allocations	Amount of irrigable land within watershed for each state (in Johnston Negotiations); "needs-based" criteria developed in current peace talks
Incentives/linkage	Financial: US and donor communities have agreed to cost-share regional water projects Political: Multilateral talks work in conjunction with bilateral negotiations
Breakthroughs	Harza study of Jordan's water needs (in Johnston talks); question of water rights successfully relegated to bilateral talks; creation of a Palestinian Water Authority accepted by all parties

Status   Israel-Jordan Peace Treaty (1994); Israel-
                                  Palestine Interim Agreement (1993, 1995)
                                  each have major water components

The problem

The Jordan river flows between five particularly contentious riparians,
two of which rely on the river as their primary water supply. By the early
1950s, there was little room left for any unilateral development of the
river without impacting on other riparian states. The Johnston Negotiations, named after US special envoy Eric Johnston, attempted to mediate
the dispute over water rights among all the riparians in the mid-1950s.
Egypt was also included in the negotiations, because of its pre-eminence
in the Arab world. The initial issue was an equitable allocation of the
annual flow of the Jordan watershed among its riparian states – Israel,
Jordan, Lebanon, and Syria. Water is and continues to be a highly contentious issue among these countries, along with issues of land, refugees,
and political sovereignty. Until the current Arab-Israeli peace negotiations, which began in 1991, political problems were always handled separately from resource problems. Some experts have argued that by separating the two realms of "high" and "low" politics, each process was
doomed to fail. The initiatives that were addressed as strictly water-
resource issues, namely – the Johnston Negotiations of the mid-1950s,
attempts at "water-for-peace" through nuclear desalination in the late
1960s, negotiations over the Yarmuk river in the 1970s and 1980s, and
the Global Water Summit Initiative of 1991 – all failed to one degree or
another, because they were handled separately from overall political discussions. The resolution of water-resources issues then had to await the
Arab-Israeli peace talks to meet with any tangible progress.

Background

In 1951, several states announced unilateral plans for the Jordan watershed. Arab states began to discuss organized exploitation of two northern
sources of the Jordan – the Hasbani and the Banias. The Israelis made
public their "All Israel Plan," which included the draining of Huleh lake
and swamps, diversion of the northern Jordan river and construction of a
carrier to the coastal plain and Negev desert – the first out-of-basin
transfer for a watershed in the region.

In July 1953, Israel began construction on the intake of its National
Water Carrier at the Bridge of Jacob's Daughters, north of the Sea of

Galilee and in the demilitarized zone between Israel and Syria. Syria deployed its armed forces along the border and artillery units opened fire on the construction and engineering sites. Syria also protested to the UN and, though a 1954 resolution allowed Israel to resume work, the USSR vetoed the resolution. The Israelis then moved the intake to its current site at Eshed Kinrot on the northwestern shore of the Sea of Galilee. It was against this tense background that President Dwight Eisenhower sent his special envoy, Eric Johnston, to the Middle East in October 1953 to try to mediate a comprehensive settlement of the Jordan river system allocations, and design a plan for its regional development.

Attempts at conflict management

Johnston worked until the end of 1955 to reconcile US, Arab, and Israeli proposals in a Unified Plan amenable to all of the states involved. His dealings were bolstered by a US offer to fund two-thirds of the development costs. His plan addressed the objections of both Arabs and Israelis, and accomplished no small degree of compromise, although his neglect of groundwater issues would later prove a significant oversight. Though they had not met face-to-face for these negotiations, all states agreed on the need for a regional approach. Israel gave up integration of the Litani river, and the Arab states agreed to allow out-of-basin water transfers. The Arabs objected, but finally agreed, to storage at both the (unbuilt) Maqarin dam and the Sea of Galilee, so long as neither side would have physical control over the share available to the other. Israel objected, but finally agreed, to international supervision of withdrawals and construction. Allocations under the Unified Plan, later known as the Johnston Plan, were also delineated. Although the agreement was never ratified, both sides have generally adhered to the technical details and allocations, even while proceeding with unilateral development. Agreement was encouraged by the United States, which promised funding for future water development projects only as long as the Johnston Plan's allocations were adhered to. Since that time to the present, Israeli and Jordanian water officials have met several times a year, as often as every two weeks during the critical summer months, at so-called "Picnic Table Talks" at the confluence of the Jordan and Yarmuk rivers to discuss flow rates and allocations.

Outcome

By 1991, several events combined to shift the emphasis on the potential for "hydro-conflict" in the Middle East to the potential for "hydro-cooperation." The Gulf War in 1990 and the collapse of the Soviet Union caused a realignment of political alliances in the Middle East that finally made possible the first public face-to-face peace talks between Arabs and Israelis, in Madrid on 30 October 1991. During the bilateral negotiations between Israel and each of its neighbours, it was agreed that a second track be established for multilateral negotiations on five subjects deemed "regional," including water resources.

Since the opening session of the multilateral talks in Moscow in January 1992, the Working Group on Water Resources, with the United States as "gavel-holder," has been the venue by which problems of water supply, demand, and institutions have been raised among the parties to the bilateral talks, with the exception of Lebanon and Syria. The two tracks of the current negotiations, the bilateral and the multilateral, are designed explicitly not only to close the gap between issues of politics and issues of regional development, but to use progress on each to help catalyse the pace of the other, in a positive feedback loop towards "a just and lasting peace in the Middle East." The idea is that the multilateral working groups provide forums for relatively free dialogue on the future of the region and, in the process, allow for personal ice-breaking and confidence building to take place. Given the role of the Working Group on Water Resources in this context, the objectives have been more in the order of fact-finding and workshops, rather than tackling the difficult political issues of water rights and allocations, or the development of specific projects. Likewise, decisions are made through consensus only.

The pace of success of each round of talks has vacillated but, in general, has been increasing. By the third meeting in 1992, it became clear that regional water-sharing agreements, or any political agreements surrounding water resources, would not be dealt with in the multilaterals. Rather the role of these talks would be to deal with non-political issues of mutual concern, thereby strengthening the bilateral track. The goal in the Working Group on Water Resources became to plan for a future region at peace, and to leave the pace of implementation to the bilaterals. This distinction between "planning" and "implementation" became crucial, with progress only being made as the boundary between the two is continuously pushed and blurred by the mediators.

The multilateral activities have helped set the stage for agreements formalized in bilateral negotiations – the Israel-Jordan Treaty of Peace of 1994, and the Interim Agreements between Israel and the Palestinians (1993 and 1995). For the first time since these states came into being, the Israel-Jordan peace treaty legally spells out mutually recognized water allocations. Acknowledging that, "water issues along their entire boundary must be dealt with in their totality," the treaty spells out allocations for the Yarmuk and Jordan rivers, as well as Arava/Araba groundwater, and calls for joint efforts to prevent water pollution. In addition, "[recognizing] that their water resources are not sufficient to meet their needs," the treaty calls for ways of alleviating the water shortage through cooperative projects, both regional and international. The Interim Agreement also recognizes the water rights of both Israelis and Palestinians, but defers their quantification until the final round of negotiations.

Ganges river controversy

Case summary

River basin	Ganges river
Dates of negotiation	1960 to the present
Relevant parties	Pre-1971: India, Pakistan; Post-1971: India, Bangladesh
Flashpoint	India builds and operates Farakka barrage diversion of Ganges water without long-term agreement with downstream Bangladesh
Issues	
Stated objectives	Negotiate an equitable allocation of the flow of the Ganges river and its tributaries between the riparian states
	Develop a rational plan for integrated watershed development, including supplementing Ganges flow
Additional issues	
Water-related	Appropriate source for supplementing Ganges flow
	Amount of data necessary for decision-making
	Indian upstream water development
	Flood hazards mitigation
	Management of coastal ecosystems
Non-water	Appropriate diplomatic level for negotiations
Excluded	Other riparians, notably Nepal, until recently
Criteria for water allocations	Percentage of flow during dry season
Incentives/linkage	Financial: None
	Political: None
Breakthroughs	Minor agreements reached, but no long-term solution
Status	Short-term agreements reached in 1977, 1982, and 1985. Treaty signed in 1996

The problem

The problem over the Ganges is a typical example of the conflicting interests of up and downstream riparians. India, as the upper riparian, developed plans for water diversions for its own irrigation, navigability, and water supply interests. Initially Pakistan, and later Bangladesh, had interests in protecting the historic flow of the river for its own downstream uses. The potential clash between upstream development and downstream historic use sets the stage for attempts at conflict management.

Background

The headwaters of the Ganges and its tributaries lie primarily in Nepal and India, where snow and rainfall are heaviest. Flow increases downstream even as annual precipitation drops, as the river flows into Bangladesh – pre-1971 the eastern provinces of the Federation of Pakistan – and on to the Bay of Bengal.

On 29 October 1951, Pakistan officially called Indian attention to reports of Indian plans to build a barrage at Farakka, about 17 kilometres from the border. The barrage would reportedly divert 40,000 cubic feet per second (cusec)* out of a dry season average flow of 50,000 cusec from the Ganges into the Bhagirathi-Hooghly tributary, to provide silt-free flow into Calcutta bay, which would improve navigability for the city's port during dry months and keep saltwater from infiltrating the city's water supply. On 8 March 1952, the Indian government responded that the project was only under preliminary investigation, and that concern was "hypothetical."

Over the next years, Pakistan occasionally responded to reports of Indian plans for diversion projects of the Ganges, with little Indian response. In 1957, and again in 1958, Pakistan proposed that the services of the United Nations be secured to assist in planning for the cooperative development of the eastern river systems. India turned down these proposals, although it was agreed that water resources experts of the two countries should "exchange data on projects of mutual interests." These expert-level meetings commenced 28 June 1960.

Attempts at conflict management

The first round of expert-level meetings between India and Pakistan was held in New Delhi from 28 June–3 July 1960 reverse order, with three more rounds to follow by 1962. While the meetings were still in progress, India informed Pakistan on 30 January 1961 that construction had begun on the Farakka barrage. A series of attempts by Pakistan to arrange a

---

* Since all negotiations were in English units, that is what is reported here. Cusec = cubic feet per second = 0.0283 cubic metres per second.

meeting at the level of minister was rebuffed with the Indian claim that such a meeting would not be useful, "until full data are available." In 1963, the two sides agreed to have one more expert-level meeting to determine what data were relevant and necessary for the convening of a minister-level meeting.

The meeting at which data needs were to be determined, the fifth round at the level of expert, was not held until 13 May 1968. After that meeting, the Pakistanis concluded that agreement on data, and on the conclusions that could be drawn, was not possible, but that enough data were nevertheless available for substantive talks at the level of minister. India agreed only to a series of meetings at the level of secretary, in advance of a minister-level meeting.

These meetings, at the level of secretary, commenced on 9 December 1968 and a total of five were held in alternating capitals through July 1970. Throughout these meetings, the different strategies became apparent. As the lower riparian, the Pakistani sense of urgency was greater, and their goal was, "substantive talks on the framework for a settlement for equitable sharing of the Ganges waters between the two countries." India in contrast, whether actually or as a stalling tactic, professed concern at data accuracy and adequacy, arguing that a comprehensive agreement was not possible until the data available were complete and accurate.

These talks were of little practical value, and India completed construction of the Farakka barrage in 1970. Water was not diverted at the time, however, because the feeder canal to the Bhagirathi-Hooghly system was not yet completed.

Bangladesh came into being in 1971, and by March 1972, the governments of India and Bangladesh had agreed to establish the Indo-Bangladesh Joint Rivers Commission, "to develop the waters of the rivers common to the two countries on a cooperative basis." The question of the Ganges, however, was specifically excluded, and would be handled only between the two prime ministers.

At a minister-level meeting in Dhaka from 16–18 April 1975, India asked that, while discussions continue, the feeder canal at Farakka be run during the current period of low flow. The two sides agreed to a limited trial operation of the barrage, with discharges varying between 11,000 and 16,000 cusec in ten-day periods from 21 April to 31 May 1975, with the remainder of the flow guaranteed to reach Bangladesh. Without renewing or negotiating a new agreement with Bangladesh, India continued to divert the Ganges waters at Farakka after the trial run, throughout the 1975–1976 dry season, at the full capacity of the diversion – 40,000 cusec. There were serious consequences in Bangladesh resulting from these diversions, including desiccation of tributaries, salination along the coast, and setbacks to agriculture, fisheries, navigation, and industry.

Four more meetings were held between the two states between June 1975 and June 1976, with little result. In January 1976, Bangladesh lodged a formal protest against India with the General Assembly of the United Nations, which, on 26 November 1976, adopted a consensus statement encouraging the parties to meet urgently at the ministerial level for negotiations, "with a view to arriving at a fair and expeditious settlement." Spurred by international consensus, negotiations recommenced on 16 December 1976. At an 18 April 1977 meeting, an understanding was reached on fundamental issues, which culminated in the signing of the Ganges Waters Agreement on 5 November 1977.

Outcome

In principle, the Ganges Water Agreement covers:
1. sharing the waters of the Ganges at Farakka; and
2. finding a long-term solution for augmentation of the dry season flows of the Ganges.

The agreement would initially cover a period of five years. It could then be extended further by mutual agreement. The Joint Rivers Commission was again vested with the task of developing a feasibility study for a long-term solution to the problems of the basin, with both sides reintroducing plans along the lines described above. By the end of the five-year life of the agreement, no solution had been worked out.

In the years since, both sides and, more recently, Nepal, have had mixed success in reaching agreement. Since the 1977 accord:
- A joint communiqué was issued in October 1982, in which both sides agreed not to extend the 1977 agreement, but would rather initiate new attempts to achieve a solution within 18 months – a task not accomplished.
- An Indo-Bangladesh Memorandum of Understanding was signed on 22 November 1985, on the sharing of the Ganges dry season flow through 1988, and establishing a Joint Committee of Experts to help resolve development issues. India's proposals focused on linking the Brahmaputra with the Ganges, while Bangladesh's centred on a series of dams along the Ganges headwaters in Nepal.
- Although both the Joint Committee of Experts and the Joint Rivers Commission met regularly throughout 1986, and although Nepal was approached for possible cooperation, the work ended inconclusively.
- The prime ministers of Bangladesh and India discussed the issue of river water sharing on the Ganges and other rivers in May 1992, in New Delhi. Each directed their ministers to renew their efforts to achieve a long-term agreement on the Ganges, with particular atten-

tion to low flows during the dry season. Subsequent to that meeting, there has been one minister-level and one secretary-level meeting, at which little progress was reportedly made.

In December 1996, a new treaty was signed between the two riparians, based generally on the 1985 accord, which delineates a flow regime under varying conditions. While this agreement should help reduce regional tensions, issues such as extreme events and upstream uses are not covered in detail. Notably, Nepal, China, and Bhutan, the remaining riparians, but not party to the treaty, have their own development plans that could impact the agreement.

The very first season following signing of the treaty, in April 1997, India and Bangladesh were involved in their first dispute over cross-boundary flow. Water passing through the Farakka dam dropped below the minimum provided in the treaty, prompting Bangladesh to request a review of the state of the watershed.

Indus Water Treaty

Case summary

River basin	Indus river and tributaries
Dates of negotiation	1951–1960
Relevant parties	India, Pakistan
Flashpoint	Lack of water-sharing agreement leads India to stem flow of tributaries to Pakistan on 1 April 1948
Issues	
Stated objectives	Negotiate an equitable allocation of the flow of the Indus river and its tributaries between the riparian states
	Develop a rational plan for integrated watershed development
Additional issues	
Water-related	Financing for development plans
	Whether storage facilities are "replacement" or "development" (tied to who is financially responsible)
Non-water	General India-Pakistan relations
Excluded	Future opportunities for regional management
	Issues concerning drainage
Criteria for water allocations	Historic and planned use (for Pakistan) plus geographic allocations (western rivers vs eastern rivers)
Incentives/linkage	Financial: World Bank organized International Fund Agreement
	Political: None
Breakthroughs	Bank put own proposal forward after 1953 deadlock; international funding raised for final agreement
Status	Ratified in 1960, with provisions for on-going conflict resolution. Some suggest that recent meetings have been lukewarm. Physical separation of tributaries may preclude efficient integrated basin management

The problem

Even before the partition of India and Pakistan, the Indus posed problems between the states of British India. The problems became international only after partition, though, and the attendant increased hostility
and lack of supralegal authority exacerbated the issue. Pakistani territory,
which had relied on Indus water for centuries, now found the water
sources originating in another country, one with whom geopolitical relations were increasing in hostility.

Background

Irrigation in the Indus river basin dates back centuries. By the late 1940s
the irrigation works along the river were the most extensive in the world.
These irrigation projects had been developed over the years under one
political authority, that of British India, and any water conflict could be
resolved by executive order. The Government of India Act of 1935,
however, put water under provincial jurisdiction, and some disputes did
begin to crop up at the sites of the more extensive works, notably between the provinces of Punjab and Sind.

In 1942, a judicial commission was appointed by the British government to study Sind's concern over planned Punjabi development. The
Commission recognized the claims of Sind, and called for the integrated
management of the basin as a whole. The Commission's report was found
unacceptable by both sides, and the chief engineers of the two sides met
informally between 1943 and 1945 to try to reconcile their differences.
Although a draft agreement was produced, neither of the two provinces
accepted the terms and the dispute was referred to London for a final
decision in 1947.

Before a decision could be reached, however, the Indian Independence
Act of 15 August 1947 internationalized the dispute between the new
states of India and Pakistan. Partition was to be carried out in 73 days,
and the full implications of dividing the Indus basin seem not to have
been fully considered, although Sir Cyril Radcliffe, who was responsible
for the boundary delineation, did express his hope that, "some joint control and management of the irrigation system may be found" (Mehta,
1988: 4). Heightened political tensions, population displacements, and
unresolved territorial issues, all served to exacerbate hostilities over the
water dispute.

As the monsoon flows receded in the fall of 1947, the chief engineers of
Pakistan and India met and agreed to a "Standstill Agreement," which
froze water allocations at two points on the river until 31 March 1948,

allowing discharges from headworks in India to continue to flow into Pakistan.

On 1 April 1948, the day that the "Standstill Agreement" expired, in the absence of a new agreement, India discontinued the delivery of water to the Dipalpur canal and the main branches of the Upper Bari Daab canal. At an Inter-dominion conference held in Delhi on 3–4 May 1948, India agreed to the resumption of flow, but maintained that Pakistan could not claim any share of those waters as a matter of right (Caponera, 1987: 511). This position was reinforced by the Indian claim that, since Pakistan had agreed to pay for water under the Standstill Agreement of 1947, Pakistan had recognized India's water rights. Pakistan countered that they had historic rights, and that payments to India were only to cover operation and maintenance costs (Biswas, 1992: 204).

While these conflicting claims were not resolved, an agreement was signed, later referred to as the Delhi Agreement, in which India assured Pakistan that India would not withdraw water delivery without allowing time for Pakistan to develop alternate sources. Pakistan later expressed its displeasure with the agreement in a note dated 16 June 1949, calling for the "equitable apportionment of all common waters," and suggesting turning jurisdiction of the case over to the World Court. India suggested rather that a commission of judges from each side try to resolve their differences before turning the problem over to a third party. This stalemate lasted through 1950.

Attempts at conflict management

In 1951, Indian Prime Minister Nehru, whose interest in integrated river management along the lines of the Tennessee Valley Authority had been piqued, invited David Lilienthal, former chairperson of the TVA, to visit India. Lilienthal also visited Pakistan and, on his return to the US, wrote an article outlining his impressions and recommendations (the trip had been commissioned by *Collier's Magazine* – international water was not the initial aim of the visit). His article was read by Lilienthal's friend David Black, president of the World Bank, who contacted Lilienthal for recommendations on helping to resolve the dispute. As a result, Black contacted the prime ministers of Pakistan and India, inviting both countries to accept the Bank's good offices. In a subsequent letter, Black outlined "essential principles" that might be followed for conflict resolution. These principles included:

– the water resources of the Indus basin should be managed co-operatively;
– the problems of the basin should be solved on a functional and not on a political plane, without relation to past negotiations and past claims.

Black suggested that India and Pakistan each appoint a senior engineer to work on a plan for development of the Indus basin. A Bank engineer would be made available as an ongoing consultant.

Both sides accepted Black's initiative. The first meeting of the Working Party included Indian and Pakistani engineers, along with a team from the Bank, as envisioned by Black, and they met for the first time in Washington in May 1952.

When the two sides were unable to agree on a common development plan for the basin in subsequent meetings in Karachi, November 1952, and Delhi, January 1953, the Bank suggested that each side submit its own plan. Both sides did submit plans on 6 October 1953, each of which mostly agreed on the supplies available for irrigation, but varied extremely on how these supplies should be allocated.

The Bank concluded that not only was the stalemate likely to continue, but that the ideal goal of integrated watershed development for the benefit of both riparians was probably too elusive an arrangement at this stage of political relations. On 5 February 1954, the Bank issued its own proposal, abandoning the strategy of integrated development in favour of one of separation. The Bank proposal called for the entire flow of the eastern rivers to be allocated to India, and all of the western rivers, except for a small amount from the Jhelum, to be allocated to Pakistan. According to the proposal, the two sides would agree to a transition period while Pakistan would complete link canals dividing the watershed, during which India would continue to allow Pakistan's historic use to continue to flow from the eastern rivers.

The Bank proposal was given to both parties simultaneously. On 25 March 1954, India accepted the proposal as the basis for agreement. Pakistan viewed the proposal with more trepidation, and gave only qualified acceptance on 28 July 1954. The Pakistanis considered the flow of the western rivers to be insufficient to replace their existing supplies from the eastern rivers, particularly given limited available storage capacity. To help facilitate an agreement, the Bank issued an aide-mémoire, calling for more storage on the western rivers and suggesting India's financial liability for "replacement facilities" – increased storage facilities and enlarged link canals in Pakistan, which could be recognized as the cost replacement of pre-partition canals.

By 1959, the Bank evaluated the principal issue to be resolved as follows: which works would be considered "replacement" and which "development." Stated differently, for which works India would be financially responsible. To circumvent the question, Black suggested an alternate approach in a visit to India and Pakistan in May. Perhaps one might settle on a specific amount for which India was responsible, rather than arguing over individual works. The Bank might then help raise additional funds among the international community for watershed development. India was offered help with construction of its Beas dam, and Pakistan's plan, including both the proposed dams would be looked at favourably. With these conditions, both sides agreed to a fixed payment settlement, and to a 10-year transition period during which India would allow for Pakistan's historic flows to continue.

In August 1959, Black organized a consortium of donors to support development in the Indus basin and raised close to $900 million, in addition to India's commitment of $174 million. The Indus Water Treaty was signed in Karachi on 19 September 1960 and government ratifications were exchanged in Delhi in January 1961.

Outcome

The Indus Water Treaty addressed both the technical and financial concerns of each side, and included a timeline for transition. The main points of the treaty included:
– an agreement that Pakistan would receive unrestricted use of the western rivers, which, with minor exceptions, India would allow to flow unimpeded;
– provisions for three dams, eight link canals, three barrages, and 2,500 tube wells to be built in Pakistan;
– a 10-year transition period, from 1 April 1960 to 31 March 1970, during which water would continue to be supplied to Pakistan according to a detailed schedule;
– a schedule for India to provide its fixed financial contribution of $62 million, in 10 annual instalments during the transition period;
– additional provisions for data exchange and future cooperation.

The treaty also established the Permanent Indus Commission, made up of one Commissioner of Indus Waters from each country. The two Commissioners would meet annually in order to:
– establish and promote cooperative arrangements for the treaty implementation;
– promote cooperation between the parties in the development of the waters of the Indus system;
– examine and resolve by agreement any question that may arise between the parties concerning interpretation or implementation of the treaty;
– submit an annual report to the two governments.

In case of a dispute, provisions were made to appoint a "neutral expert." If the neutral expert fails to resolve the dispute, negotiators can be appointed by each side to meet with one or more mutually agreed-upon mediators. If either side (or the mediator) views mediated agreement as unlikely, provisions are included for the convening of a Court of Arbitration. In addition, the treaty calls for either party, if it undertakes any engineering works on any of the tributaries, to notify the other of its plans and to provide any data that may be requested.

Since 1960, no projects have been submitted under the provisions for "future cooperation," nor have any issues of water quality been submitted at all. Other disputes have arisen, and been handled in a variety of ways. The first issues arose from Indian non-delivery of some waters during 1965–1966, but became instead a question of procedure and the legality of commission decisions. Negotiators resolved that each commissioner acted as government representatives and that their decisions were legally binding.

One controversy surrounding the design and construction of the Salal dam was resolved through bilateral negotiations between the two governments. Other disputes, over new hydroelectric projects and the Wuller barrage on the Jhelum tributary, have yet to be resolved.

Mekong Committee

Case summary

River basin	Mekong river
Dates of negotiation	Committee formed 1957
Relevant parties	Cambodia, Laos, Thailand, Vietnam (directly), China, Myanmar (indirectly)
Flashpoint	None – studies by UN-ECAFE (1952, 1957) and US Bureau of Reclamation provide impetus for creation of Mekong Committee
Issues	
Stated objectives	Promote, coordinate, supervise, and control the planning and investigation of water resources development projects in the Lower Mekong basin
Additional issues	
Non-water	General political relations between riparians
Excluded	China and Myanmar were not included since inception; Cambodia not included between 1978 and 1991
Criteria for water allocations	Allocations have not been an issue; "reasonable and equitable use" for the basin defined in detail since 1975
Incentives/linkage	Financial: Extensive funding from international community
	Political: Facilitated relations between riparians, aid from both east and west despite political tensions
Breakthroughs	Studies by UN-ECAFE and US Bureau of Reclamation in 1950s
Status	Mekong Committee established in 1957 became the Interim Committee in 1978 with original members except for Cambodia. Early momentum has dropped off – extensive data networks and databases established, but few extensive projects implemented; none yet on the mainstream; Committee reratified as Mekong Commission in 1995

The problem

As is common in international river basins, integrated planning for efficient watershed management is hampered by the difficulties of coordinating between riparian states with diverse and often conflicting needs.
The Mekong, however, is noted mostly for the exceptions as compared
with other basins, rather than the similarities. For example, because the
region is so well watered, allocations per se are not a major issue. Also,
negotiations for joint management of the Mekong were not set off by a
flashpoint, but rather by creativity and foresight on the part of an authoritative third party – the United Nations – with the willing participation of the lower riparian states.

Background

The Mekong is the seventh largest river in the world in terms of discharge
(tenth in length). Rising in China, it then flows 4,200 kilometres through
Myanmar, Laos, Thailand, Cambodia, and finally through the extensive
delta in Vietnam into the South China Sea. The Mekong is also both the
first successful application of a comprehensive approach to planning
development of an international river and, at the same time, is one of
the least developed major rivers in the world, in part because of difficulties inherent in implementing joint management between its diverse
riparians.

A 1957 study performed by the United Nations Economic Commission
for Asia and the Far East (ECAFE) noted that harnessing the main stem
of the Mekong would allow hydropower production, expansion of irrigated land, a reduction of the threat of flooding in the delta region, and
the extension of navigability of the river as far as northern Laos. As had
earlier studies, the ECAFE report emphasized the need for comprehensive development of the river, and close cooperation between the riparians in coordinating efforts for projects and management. To facilitate
coordination, the report suggested the establishment of an international
body for exchanging information and development plans between the
riparian states. Ultimately, the report suggested, such a body might
become a permanent agency responsible for coordinating joint management of the Mekong basin. When the report was presented in the tenth-
anniversary meeting of ECAFE in Bangkok in March 1957, representatives from the four lower riparian states themselves adopted a resolution
calling for further study.

Attempts at conflict management

In mid-September 1957, after ECAFE's legal experts designed a draft charter for a "Coordination Committee," the lower riparians convened again in Bangkok as a "Preparatory Commission." The Commission studied, modified, and finally endorsed a statute that legally established the Committee for Coordination of Investigations of the Lower Mekong (Mekong Committee), made up of representatives of the four lower riparians, with input and support from the United Nations. The statute was signed on 17 September 1957.

The committee was composed of "plenipotentiary" representatives of the four countries, meaning that each representative had the authority to speak for their country. The committee was authorized to "promote, coordinate, supervise, and control the planning and investigation of water resources development projects in the Lower Mekong Basin."

The first committee session was on 31 October 1957, as was the first donation from the international community – 60 million francs (about $120,000) from France. With rapid agreement between the riparians came extensive international support for the work of the committee. By 1961, the committee's resources came to $14 million, more than enough to fund field surveys, which had been agreed to as priority projects. By the end of 1965, 20 countries, 11 international agencies, and several private organizations had pledged a total of more than $100 million. The secretariat itself was funded by a special $2.5 million grant made by UNDP. This group of international participants has been dubbed "the Mekong club," which has infused the international community with "the Mekong spirit."

Outcome

The early years were the most productive for the Mekong Committee. Networks of hydrologic and meteorologic stations were established and continued to function despite hostilities in the region, as were programmes for aerial mapping, surveying, and levelling. Navigation has been improved along the main stem of the river.

The work of the committee has also helped overcome political suspicion through increased integration. In 1965, Thailand and Laos signed an agreement on developing the power potential of the Nam Ngum river, a Mekong tributary inside Laos. Since most of the power demand was in Thailand, which was willing to buy power at a price based on savings in fuel costs, and since Laos did not have the resources to finance the project, an international effort was mobilized through the committee to help develop the project. As a sign of the committee's viability, the mutual flow of electricity for foreign capital between Laos and Thailand was never interrupted, despite hostilities between the two countries.

By the 1970s, the early momentum of the Mekong Committee began to subside, for several reasons. First, the political and financial obstacles necessary to move from data gathering and feasibility studies to concrete development projects have often been too great to overcome. A 1970 Indicative Basin Plan marked the potential shift between planning and large-scale implementation, including immense power, flood control, irrigation, and navigation projects, and set out a basin development framework for the following 30 years. In 1975, the riparians set out to refine the committee's objectives and principles for development in support of the Plan in a "Joint Declaration on Principles," including the first (and so far only) precise definition of "reasonable and equitable use" based on the 1966 Helsinki Rules ever used in an international agreement. The plan, which included three of the largest hydroelectric power projects in the world as part of a series of seven cascading dams, was received with scepticism by some in the international community (Kirmani, 1990: 203). At the current time, while many projects have been built along the tributaries of the Mekong within single countries, and despite the update of the Indicative Plan in 1987 and a subsequent "Action Plan" which includes only two low dams, no single structure has been built across the main stem.

Second, while the committee continued to meet despite political tensions, and even despite outright hostilities, political obstacles did take their toll on the committee's work. Notably, the committee became a three-member "interim committee" in 1978 with the lack of a representative government in Cambodia. Cambodia rejoined the committee as a full participant in 1991, although the latter still retained its "interim" status until 1995. Likewise, funding and involvement from the United States, which had been about 12 per cent of total aid to the committee, was cut off in June 1975 and has not been restored to significant levels.

Renewed activity came with the signing of the Paris Peace Agreement in 1991, after which Cambodia requested the reactivation of the Mekong Committee. The four lower riparians took up the call and spent the next four years determining a future direction for Mekong activities. The results of these meetings culminated in a new agreement, signed in April 1995, in which the Mekong Committee became the Mekong Commission. While it is too early yet to evaluate this renewed body, the fact that the riparians have made a new commitment to jointly manage the lower basin speaks for the resiliency of agreements put into place in advance of hot conflict. It should also be noted that Myanmar and China are still not party to the agreement, effectively precluding integrated basin management.

Nile Waters Agreement

Case summary

River basin	Nile river
Dates of negotiation	1920–1959 – treaties signed in 1929 and 1959
Relevant parties	Egypt, Sudan (directly); other Nile riparians (indirectly)
Flashpoint	Plans for a storage facility on the Nile
Issues	
Stated objectives	Negotiate an equitable allocation of the flow of the Nile river between Egypt and Sudan
	Develop a rational plan for integrated watershed development
Additional issues	
Water-related	Upstream vs downstream storage
Non-water	General Egypt-Sudan relations
Excluded	Water quality
	Other Nile riparians
Criteria for water allocations	Acquired rights plus even division of any additional water resulting from development projects
Incentives/linkage	Financial: Funding for Aswan High Dam
	Political: Fostered warm relations between Egypt and new government of Sudan
Breakthroughs	1958 coup in Sudan by pro-Egypt leaders made agreement possible
Status	Ratified in 1959. Allocations between Egypt and Sudan upheld till today. Other riparians, particularly Ethiopia, are planning development projects, which may necessitate renegotiating a more inclusive treaty

The problem

As the Nile riparians gained independence from colonial powers, riparian disputes became international and consequently more contentious, particularly between Egypt and Sudan. The core question of historic versus sovereign water rights is complicated by the technical question of where the river ought best be controlled – upstream or down.

Background

With the end of World War I, it became clear that any regional development plans for the Nile basin would have to be preceded by some sort of
formal agreement on water allocations. In 1920, the Nile Projects Commission was formed, with representatives from India, the United Kingdom, and the United States. The same year saw publication of the most
extensive scheme for comprehensive water development along the Nile,
now known as the Century Storage Scheme.

The plan worried some Egyptians, and was criticized by nationalists,
because all the major control structures would have been beyond Egyptian territory and authority. Some Egyptians saw the plan as a British
means of controlling Egypt in the event of Egyptian independence.

Attempts at conflict management

In 1925, a new water commission made recommendations, based on the
1920 estimates that would lead finally to the Nile Waters Agreement between Egypt and Sudan on 7 May 1929. Four billion cubic metres of water
per year (BCM/yr) were allocated to Sudan but the entire timely flow
(from January 20 to July 15) and a total annual amount of 48 BCM/yr was
reserved for Egypt. Egypt, as the downstream state, had its interests
guaranteed by:
- having a claim to the entire timely flow. This claim meant that any
  cotton cultivated in Sudan would have to be grown during the winter
  months;
- having rights to on-site inspectors at the Sennar dam, outside of
  Egyptian territory;
- being guaranteed that no works would be developed along the river or
  on any of its territory that would threaten Egyptian interests.

In accord with this agreement, one dam was built and one reservoir
raised, with Egyptian acquiescence.

The Aswan High Dam, with a projected storage capacity of 156 BCM/yr,
was proposed in 1952 by the new Egyptian government, however debate
over whether it was to be built as a unilateral Egyptian project or as a
cooperative project with Sudan kept Sudan out of negotiations until 1954.
The negotiations that ensued, and were carried out with Sudan's struggle
for independence as a backdrop, focused not only on what each country's
legitimate allocation would be, but whether the dam was even the most
efficient method of harnessing the waters of the Nile.

The first round of negotiations between Egypt and Sudan took place
between September and December 1954, even as Sudan was preparing

for its independence, scheduled for 1956. Negotiations broke off inconclusively, then briefly, and equally inconclusively, resumed in April 1955. Relations then threatened to degrade into military confrontation in 1958, when Egypt sent an unsuccessful expedition into territory in dispute between the two countries. In the summer of 1959, Sudan unilaterally raised the Sennar dam, effectively repudiating the 1929 agreement.

Sudan attained independence on 1 January 1956, but it was with the military regime that gained power in 1958 that Egypt adopted a more conciliatory tone in the negotiations that resumed in early 1959. Progress was speeded in part by the fact that any funding that would be forthcoming for the High Dam would depend on a riparian agreement. On 8 November 1959, the Agreement for the Full Utilization of the Nile Waters (Nile Waters Treaty) was signed.

Outcome

The Nile Waters Treaty had the following provisions:
- The average flow of the river is considered to be 84 BCM/yr. Evaporation and seepage were considered to be 10 BCM/yr, leaving 74 BCM/yr to be divided.
- Of this total, acquired rights have precedence, and are described as being 48 BCM for Egypt and 4 BCM for Sudan. The remaining benefits of approximately 22 BCM are divided by a ratio of $7^1/_2$ for Egypt (approx. 7.5 BCM/yr) and $14^1/_2$ for Sudan (approx. 14.5 BCM/yr). These allocations total 55.5 BCM/yr for Egypt and 18.5 BCM/yr for Sudan.
- If the average yield increases from these average figures, the increase would be divided equally. Significant decreases would be taken up by a technical committee, described below.
- Since Sudan could not absorb that much water at the time, the treaty also provided for a Sudanese water "loan" to Egypt of up to 1,500 MCM/yr through 1977.
- Funding for any project that increases Nile flow (after the High Dam) would be provided evenly, and the resulting additional water would be split evenly.
- A Permanent Joint Technical Committee to resolve disputes and jointly review claims by any other riparian would be established. The committee would also determine allocations in the event of exceptionally low flows.
- Egypt agreed to pay Sudan £E15 million in compensation for flooding and relocations.

Egypt and Sudan agreed that the combined needs of other riparians would not exceed 1,000–2,000 MCM/yr, and that any claims would be met with one unified Egyptian-Sudanese position. The allocations of the treaty have been held to until the present.

Ethiopia, which had not been a major player in Nile hydropolitics, served notice in 1957 that it would pursue unilateral development of the Nile water resources within its territory, estimated at 75–85 per cent of the annual flow, and suggestions were made recently that Ethiopia may eventually claim up to 40,000 MCM/yr for its irrigation needs both within and outside of the Nile watershed. No other state riparian to the Nile has ever exercised a legal claim to the waters allocated in the 1959 treaty.

Plata basin

Case summary

River basin	Plata
Dates of negotiation	Plata Basin Treaty signed 1969
Relevant parties	Argentina, Bolivia, Brazil, Paraguay, Uruguay
Flashpoint	None
Issues	
Stated objectives	Promote and coordinate joint development of the basin; "Hydrovia" proposed in 1989
Additional issues	
Water-related	Joint management
Non-water	None
Excluded	Treaty does not provide any supralegal authority
Criteria for water allocations	None
Incentives/linkage	Possibility of linking water projects with transportation infrastructure
Breakthroughs	None
Status	Intergovernmental Coordinating Committee functions; "Hydrovia" technical and environmental studies due in October 1996

The problem

A cooperative management body has been in place on the Plata basin since 1969. While generally successful and productive, the cooperative nature of basin management is being strained by the size and possible economic and environmental impacts of the proposed Hydrovia project, which is designed to improve barge transportation and represents the largest project for river development proposed to date.

Background

The Plata river basin drains more than 2 million km^2 of southeastern South America, including territory in Argentina, Bolivia, Brazil, Paraguay, and Uruguay. It encompasses some of the major rivers of the continent – the Paraná, the Paraguay, and the Uruguay – and the largest wetlands in the world – the Pantanal.

The states of the basin have traditionally been willing to cooperate with management of the watershed, and have stressed the river's binding them to each other. A 1969 umbrella treaty, to which all of the riparians are signatories, provides a framework for joint management of the basin.

This framework is being tested with a current river transportation proposal to dredge and straighten major portions of the Paraná and the Paraguay, including through the Pantanal wetlands. The initial backers of the proposal, which was dubbed "Hydrovia" ("waterway" in Spanish and Portuguese), were the governments of the Plata basin states. The project would allow year-round barge transportation – current conditions only allow for barges during the three dry months – and open up a major transportation thoroughfare for landlocked sections of the riparian states. Environmentalists and those whose livelihoods depend on traditional economies have expressed trepidation at the project.

Attempts at conflict management

The Plata Basin Treaty of 1969 provides an umbrella framework for several bilateral treaties between the riparians and a direction for joint development of the basin. The treaty requires open transportation and communication along the river and its tributaries, and prescribes cooperation in education, health, and management of 'non-water' resources (e.g. soil, forest, flora, and fauna). The foreign ministers of the riparian states provide the policy direction and a standing Inter-Governmental Coordinating Committee is responsible for ongoing administration.

Basin states agree to identify and prioritize cooperative projects, and to provide the technical and legal structure to see to their implementation. The treaty also has some limitations, notably the lack of a supra-legal body to manage the treaty's provisions. The necessity to go through each country's legal system for individual projects has resulted in some delays, or halts, in project implementation.

The treaty's success has been in the area of transportation, so it is not altogether surprising that the Hydrovia project has been put forward. The first meeting of the backers of the project was in April 1988, out of which the Intergovernmental Commission on the Paraná-Paraguay Hydrovia was formed.

Outcome

Positions between supporters and opponents of the project have sharpened, however, these positions are based on very little information. The Inter-American Development Bank has only recently helped to finance a technical and environmental feasibility study.

Salween river

Case summary

River basin	Salween river
Dates of negotiation	Joint working group established in 1989
Relevant parties	Myanmar, Thailand (directly); China (indirectly)
Flashpoint	None
Issues	
Stated objectives	Promote and coordinate joint development of hydropower projects within the Salween basin
Additional issues	
Water-related	Possibility of out-of-basin transfers to Thailand
Non-water	River flows through regions of ethnic unrest and drug trade
Excluded	China has not been included in any planning
Criteria for water allocations	None
Incentives/linkage	Possibility of linking water projects with transportation infrastructure
Breakthroughs	None
Status	Talks are in most preliminary stage; meetings continue although no plan for the basin, nor any main-stem project, has yet been established

The problem

The Salween basin is a good case of river planning in advance of conflict. Preliminary meetings are being held between Myanmar and Thailand, and some project feasibility studies are being implemented although, to date, no basin-wide plan, nor any main-stem project, has been implemented.

Background

The Salween originates in the Tibetan plateau and drains an area of 320,000 km^2 in China, Myanmar, and Thailand before it flows into the Gulf of Martaban. Despite the fact that studies since the 1950s have identified tremendous hydropower potential, the Salween is a relatively undeveloped basin – with only one major hydro-electric project at Baluchaung. The power companies of Thailand and Myanmar, as well as private Japanese concerns, have pursued individual feasibility studies but it is only since the 1970s that the potential of the basin as a whole has been investigated.

Attempts at conflict management

In June 1989, following the visit of a Thai government delegation to Rangoon, a joint technical committee was established between Thailand and Myanmar, made up primarily of representatives from the power companies of the two countries. Since that time, the committee has continued to meet and to pursue feasibility studies, but no project or management body has been developed. To date, China has not been included in discussions.

Outcome

As mentioned, the Salween is a basin in its earliest stages of development. What is noteworthy is that technical and management discussions have been proceeding in advance of major development projects, thus allowing for integrated management almost from the beginning.

Discussions have included issues outside of hydropower, and studies have suggested linkages between power, irrigation and drinking water diversions, barge transportation, and related surface infrastructure. Complicating management issues is the fact that sections of the watershed include regions of ethnic unrest and the tensions brought about by the international drug trade. Nevertheless, the basin offers the opportunity for integrated management to be implemented in advance of any flashpoint brought about by unilateral development.

US/Mexico shared aquifers

Case summary

River basin	Aquifers which straddle the US/Mexico boundary
Dates of negotiation	US-Mexico Water Treaty signed 1944; groundwater negotiations since 1973
Relevant parties	Mexico, United States
Flashpoint	Salinity crisis of 1961–1973 raised groundwater as important issue not detailed in 1944 treaty
Issues	
Stated objectives	Develop an equitable apportionment of shared aquifers
Additional issues	
Water-related	Pollution
Non-water	None
Excluded	None
Criteria for water allocations	None
Incentives/linkage	None
Breakthroughs	None
Status	Talks have been ongoing since 1973

The problem

The complications of groundwater are exemplified in the border region between the United States and Mexico where, despite the presence of an active supralegal authority since 1944, groundwater issues have yet to be resolved. Mentioned as vital in the 1944 treaty, and again in 1973, the difficulties in quantifying the ambiguities inherent in groundwater regimes have confounded the efforts of legal and management experts ever since.

Background

The border region between the United States and Mexico has fostered its share of surface-water conflict, from the Colorado to the Rio Grande/Rio Bravo. It has also been a model for peaceful conflict resolution, notably through the work of the International Boundary and Water Commission (IBWC), the supralegal body established to manage shared water resources as a consequence of the 1944 US-Mexico Water Treaty. Yet the difficulties encountered in managing shared surface-water pale in comparison to trying to allocate groundwater resources. Each aquifer system is generally so poorly understood that years of study may be necessary before one even knows what the bargaining parameters are.

Mumme (1988) has identified 23 sites in contention in six different hydrogeologic regions along the 3,300 kilometres of shared boundary. While the 1944 treaty mentions the importance of resolving the allocations of groundwater between the two states, it does not do so. In fact, shared surface-water resources were the focus of the IBWC until the early 1960s, when a US irrigation district began draining saline groundwater into the Colorado river and deducting the quantity of saline water from Mexico's share of freshwater. In response, Mexico began a "crash programme" of groundwater development in the border region, in order to make up the loss.

Attempts at conflict management

Ten years of negotiations resulted in a 1973 addendum to the 1944 treaty – Minute 242 of the IBWC, which limited groundwater withdrawals on both sides of the border, and committed each nation to consult the other regarding any future groundwater development. Allocations were not quantified and negotiations to do so have continued ever since.

A 1979 agreement – Minute 241 – grants the IBWC comprehensive authority to resolve conflicts arising from border water pollution. It has been suggested that this authority may be extended to encompass groundwater overpumping.

Outcome

It is testimony to the complexity of international groundwater regimes that despite the presence of an active authority for cooperative management, and despite relatively warm political relations and few riparians, negotiations have continued since 1973 without resolution.

Aral sea

Case summary

River basin	Aral sea and its tributaries, notably the Syr Darya and the Amu Darya
Dates of negotiation	Agreements signed in 1992 and 1993
Relevant parties	Kazakhstan, Kyrgystan, Tajikistan, Turkmenistan, and Uzbekistan (directly); Afghanistan, Iran, and China (indirectly); Russia has been active observer
Flashpoint	None – Soviet agricultural policies set off "creeping" crisis from 1960s
Issues	
Stated objectives	Stabilize and rehabilitate watershed, improve management, and build capacity of regional institutions
Additional issues	
Water-related	None
Non-water	General political relations between riparians
Excluded	Transboundary oil pipelines
Criteria for water allocations	Initially based on Soviet formula, now moving to "equitable use"
Incentives/linkage	Financial: Extensive funding from international community
	Political: Facilitated relations between riparians
Breakthroughs	Breakup of Soviet Union required coordination between new states
Status	Agreements reached in 1992, 1993. Initial programme implemented in 1995. Some concerns about funding, legal overlap, priorities.

The problem

The environmental problems of the Aral sea basin are among the worst in the world. Water diversions, agricultural practices, and industrial waste have resulted in a disappearing sea, salinization, and organic and inorganic pollution. The problems of the Aral, which previously had been an internal issue of the Soviet Union, became internationalized after its collapse in 1991. The five new major riparians – Kazakhstan, Kyrgyzstan, Tajikistan, Turkmenistan, and Uzbekistan – have been struggling since that time to help stabilize and, eventually to rehabilitate, the watershed.

Background

The Aral sea was, until comparatively recently, the fourth largest inland body of water in the world. Its basin covers 1.8 million km^2, primarily in what used to be the Soviet Union, and what is now the independent republics of Kazakhstan, Kyrgyzstan, Tajikistan, Turkmenistan, and Uzbekistan. Small portions of the basin headwaters are also located in Afghanistan, Iran, and China. The major sources of the sea, the Amu Darya and the Syr Darya, are fed by glacial meltwater from the high mountain ranges of the Pamir and Tien Shan in Tajikistan and Kyrgyzstan.

Irrigation in the fertile lands between the Amu Darya and the Syr Darya dates back millennia, although the sea itself remained in relative equilibrium until the early 1960s. At that time, the central planning authority of the Soviet Union devised the "Aral sea plan" to transform the region into the cotton belt of the USSR. Vast irrigation projects were undertaken in subsequent years, with the irrigated area expanding by over one-third from 1965 to 1988.

Such intensive cotton monoculture has resulted in extreme environmental degradation. Pesticide use and salinization, along with the region's industrial pollution, have decreased water quality, resulting in high rates of disease and infant mortality. Water diversions, sometimes totalling more than the natural flow of the rivers, have reduced the Amu Darya and the Syr Darya to relative trickles – the sea itself has lost 75 per cent of its volume, half its surface area, and salinity has tripled, all since 1960. The exposed seabeds are thick with salts and agricultural chemical residue, which are carried aloft by the winds as far as the Atlantic and Pacific oceans and further contribute to air pollution and health problems in the region.

Attempts at conflict management

The intensive problems of the Aral basin were internationalized with the breakup of the Soviet Union. Prior to 1988, both use and conservation of natural resources often fell under the jurisdiction of branches of the same Soviet agency, each of which acted as powerful independent entities. In January 1988, a state committee for the protection of nature was formed, which was elevated later to the Ministry for Natural Resources and Environmental Protection in 1990. The ministry, in collaboration with the republics, had authority over all aspects of the environment and the use of natural resources. This centralization came to an end with the collapse of the Soviet Union in 1991.

The five major riparians were initially regulated by ad hoc intergovernmental agreements based on Soviet quotas. In February 1992, the five republics negotiated an agreement to coordinate policies on their transboundary waters.

Outcome

The Agreement on Cooperation in the Management, Utilization and Protection of Interstate Water Resources was signed on 18 February 1992 by representatives from Kazakhstan, Kyrgyzstan, Tajikistan, Turkmenistan, and Uzbekistan. The agreement calls on the riparians, in general terms, to coordinate efforts to "solve the Aral Sea crisis," including exchanging information, carrying out joint research, and adhering to agreed-to regulations for water use and protection. The agreement also establishes the Interstate Commission for Water Management coordination to manage, monitor, and facilitate the agreement. Since its inception, the Commission has prepared annual plans for water allocations and use, and defined water use limits for each riparian state.

In a parallel development, an Agreement on Joint Actions for Addressing the Problems of the Aral Sea and its Coastal Area, Improving of the Environment and Ensuring the Social and Economic Development of the Aral Sea Region was signed by the same five riparians on 26 March 1993. This agreement also established a coordinating body, the Interstate Council for the Aral Sea, which was designated as the organization having primary responsibility for "formulating policies and preparing and implementing programs for addressing the crisis." Each state's minister of water management is a member of the Council. In order to mobilize and coordinate funding for the Council's activities, the International Fund for the Aral Sea was created in January 1993.

A long term "Concept" and a short-term "Program" for the Aral sea was adopted at a meeting of the Heads of Central Asian states in January 1994. The Concept describes a new approach to development of the Aral sea basin, including a strict policy of water conservation. The Aral sea itself was recognized as a legitimate water user for the first time. The Program has four major objectives:
- to stabilize the environment of the Aral sea;
- to rehabilitate the disaster zone around the sea;
- to improve the management of international waters of the basin; and

– to build the capacity of regional institutions to plan and implement these programmes.

Phase I of the Program, which will cost $260 million over three years, began implementation in 1995. These regional acitivies are supported and supplemented by a variety of governmental and non-governmental agencies, including the European Union, the World Bank, UNEP, and UNDP.

Despite this forward momentum, some concerns have been raised about the potential effectiveness of these plans and institutions. Some have noted that not all promised funding has been forthcoming. Others, Dante Caponera (1995), for example, have noted duplication and inconsistencies in the agreements, and warn that they seem to accept the concept of "maximum utilization" of the waters of the basin. Vinagradov (1996) has noted especially the legal problems inherent in these agreements, including some confusion between regulatory and development functions, especially between the Commission and the Council.

The International Joint Commission: Canada and the United States of America

Case summary

River basin	All transboundary waters along the US-Canada boundary
Dates of negotiation	1905–1909
Relevant parties	Canada (originally negotiating through UK), United States
Flashpoint	Water quality concerns of early twentieth century
Issues	
Stated objectives	To provide an institutional framework to deal with issues related to boundary waters
Additional issues	
Water-related	Water quality issues were re-emphasized in 1978
Non-water	1987 Protocol and 1991 Agreement added air pollution
Excluded	Tributaries to transboundary waters; some sovereignty issues
Criteria for water allocations	"Equal and similar rights"
Incentives/linkage	None
Breakthroughs	Canada accepted sovereignty argument; US accepted arbitration function
Status	Over 130 disputes have been averted or reconciled

The problem

Canada and the United States share one of the longest boundaries in the world. Industrial development in both countries, which in the humid eastern border region primarily has relied on water resources for waste disposal, had led to decreasing water quality along their shared border to the point where, by the early years of the twentieth century, it was in the interest of both countries to seriously address the matter. Prior to 1905, only ad hoc commissions had been established to deal, as they arose, with issues relating to shared water resources. Both states considered it within their interests to establish a more permanent body for the joint management of their shared water resources.

Background

Canada and the United States share a 6,400 km boundary between the main portions of their provinces and states, and an additional 2,400 km between the Canadian Northwest Territories and Alaska. Crossing these boundaries are some of the richest waterways in the world, not least of which are the vast water resources of the five Great Lakes. The ad hoc commissions which until then had been established to resolve water-related issues were not sufficient to handle the growing problem. Even the International Waterways Commission, established in 1905, only dealt with issues on a case-by-case basis.

Attempts at conflict management

As Canada and the United States entered into negotiations to establish a permanent body to replace the International Waterways Commission, the tone of the meeting was informed by the concerns of each state. For the United States, the overriding issue was sovereignty. While it was interested in the practical necessity of an agreement to manage transboundary waters, it did not want to relinquish political independence in the process. This concern was expressed by the United States position that absolute territorial sovereignty be retained by each state over the waters within its territory – tributaries should not be included in the Commission's authority. In addition, the new body might retain some of the ad hoc nature of prior bodies, so as not to acquire undue authority. Canada was interested in establishing an egalitarian relation with the United States. It was hampered not only because of the relative size and level of development of the two states at the time, but also because Canadian foreign policy was still the purview of the United Kingdom – negotiations had to be carried out between Ottawa, Washington, and London. Canada wanted a comprehensive agreement, which would include tributaries, and a commission with greater authority than the bodies of the past.

Outcome

The "Treaty Relating to Boundary Waters between the United States and Canada," signed between the United Kingdom and the United States in 1909, reflects the interest of each negotiating body. The treaty establishes the International Joint Commission with six commissioners, three appointed by the governments of each state. Canada accepted US sovereignty concerns to some extent – tributary waters are excluded. The

United States in turn accepted the arbitration function of the Commission and allowed it greater authority than the US would have liked.

The treaty calls for open and free navigation along boundary waters, allowing Canadian transportation also on Lake Michigan, the only one of the Great Lakes not defined as a boundary water. Although it allows each state unilateral control over all of the waters within its territory, the treaty does provide for redress by anyone affected downstream. Furthermore, the Commission has "quasi-judicial" authority: any project which would affect the "natural" flow of boundary waters has to be approved by both governments. Although the Commission has the mandate to arbitrate agreements, it has never been called to do so. The Commission also has investigative authority – it may have development projects submitted for approval, or be asked to investigate an issue by one or another of the governments. Commissioners act independently, not as representatives of their respective governments.

Water quality has been a focal concern of the Commission, particularly in the waterways of the Great Lakes. The Great Lakes-St Lawrence river system contains one-fifth of the world's surface fresh water and includes the industrial lifelines of each state. Perhaps as a consequence, the anti-pollution provisions of the treaty met little opposition on either side. A 1972 "Great Lakes Water Quality Agreement" calls for the states both to control pollution and to clean up waste waters from municipal and industrial sources. This led to the signing of a new agreement in 1978, and a comprehensive protocol in 1987, each of which expanded the Commission's authorities and activities with respect to water quality.

These agreements define specific water quality objectives – the 1987 Protocol called on the Commission to review "Remedial Action Plans," prepared by governments and communities, in 43 "Areas of Concern" – yet allow the appropriate level of government of each side to develop its own plan to meet objectives. The 1987 Protocol implemented an "eco-system" approach to pollution control, and called for the development of "lakewide management plans" to combat some critical pollutants. It also included new emphasis on non-point source pollution, groundwater contamination, contaminated sediment, and airborne toxics. In 1991, the two states signed an "Agreement of Air Quality" under which the Commission was given limited authority over joint air resources.

The International Joint Commission has met some criticism over the years; most recently some have questioned whether the limited authority of the Commission – politically necessary when the Commission was established – is really conducive to the "eco-system" approach called for in the 1987 Protocol or whether greater supralegal powers are necessary. Others have questioned the commitment of the Commission to the process of public participation. Nevertheless, given the vast amount of water resources under its authority, and the myriad layers of government to which it must be responsible, the Commission stands out as an institution which has effectively and peacefully managed the boundary waters of two nations for over some 90 years, reconciling or averting more than 130 disputes in the process.

Lesotho Highlands water project

Case summary

River basin	Senqu river
Dates of negotiation	1978–1986
Relevant parties	Lesotho, South Africa
Flashpoints	Water deficit in South African industrial hub
Issues	
Stated objectives	Negotiate technical and financial details of water transfer from Lesotho to South Africa
Additional issues	
Water-related	Hydropower for Lesotho internal consumption
Non-water	General development
Excluded	None
Criteria for water allocations	Amount for sale negotiated for treaty
Incentives/linkage	South Africa buys water from Lesotho and finances diversion; Lesotho uses payments and development aid for hydropower generation and general development
Breakthroughs	Financing arrangement negotiated which allowed for international funding
Status	Project completed in 1990; no complications despite significant shift in South African government

The problem

Lesotho, completely surrounded by South Africa, is a state poor in most natural resources, water being the exception. The industrial hub of South Africa, from Pretoria to Witwatersrand, has been exploiting local water resources for years and the South African government has been in search of alternate sources. The elaborate technical and financial arrangements that led to construction of the Lesotho Highlands project provide a good example of the possible gains of an integrative arrangement including a diverse "basket" of benefits.

Background

Development in Lesotho has been limited by its lack of natural resources and investment capital. Water is its only abundant resource, which is precisely what regions of neighbouring South Africa have been lacking. A project to transfer water from the Senqu river to South Africa was investigated in the 1950s, and again in the 1960s. The project was never implemented due to disagreement over appropriate payment for the water.

Attempts at conflict management

In 1978, the governments of Lesotho and South Africa appointed a joint technical team to investigate the possibility of a water transfer project. The first feasibility study suggested a project to transfer 35 m^3/sec, four dams, 100 km of transfer tunnel, and a hydropower component. Agreement was reached to study the project in more detail, the cost of the study to be borne by both governments.

The second feasibility study, completed in 1986, concluded that the project was feasible, and recommended that the amount of water to be transferred be doubled to 70 m^3/sec. A treaty between the two states was necessary to negotiate for this international project. Negotiations proceeded through 1986 and the "Treaty on the Lesotho Highlands Water Project between the Government of the Kingdom of Lesotho and the Government of the Republic of South Africa" was signed into law on 24 October 1986.

Outcome

The treaty spells out an elaborate arrangement of technical, economic, and political intricacy. A boycott of international aid for apartheid South Africa required that the project be financed, and managed, in sections. The water transfer component was entirely financed by South Africa, which would also make payments for the water that would be delivered. The hydropower and development components were undertaken by Lesotho, which received international aid from a variety of donor agencies, particularly the World Bank. Phase I of the Lesotho Highlands water project was completed in 1990, at a cost of $2.4 billion.

The Lesotho Highlands project provides lessons for the importance of a "basket" of resources being negotiated together. South Africa receives cost-effective water for its continued growth, while Lesotho receives revenue and hydropower for its own development. It is testimony to the resilience of these arrangements that no significant changes were made despite the recent dramatic political shifts in South Africa.

8

Treaty summaries

International water treaties

There follows a hard copy preview of a computerized compilation of international water treaties, listed first by date, and followed by a summary of each treaty according to district.

07/20/1874* Articles of agreement between the Edur Durbar and the British government
02/26/1885 Act of Berlin
08/10/1889 Agreement between Great Britain and France
04/15/1891 Protocol between Great Britain and Italy for the demarcation of their respective spheres ...**
09/16/1892 Amended terms of agreement between the British Government and the State of Jind, for regulating the supply of water for irrigation ...
08/29/1893 Agreement between the British government and the Patiala state regarding the Sirsa branch of the Western Jumna canal
02/04/1895 Exchange of letters between Great Britain and France
03/18/1902 Exchange of notes between Great Britain and Ethiopia
02/23/1904 Final working agreement relative to the Sirhind canal between Great Britain and Patiala, Jind and Nabha

* Date format: month/day/year.
** Some titles have been abbreviated, indicated by elision marks and/or square brackets.

05/09/1906	Agreement ... modifying the agreement signed at Brussels 12 May 1894
10/19/1906	Agreement between Great Britain and France
04/11/1910	Convention regarding the water supply of Aden between Great Britain and the Sultan of Abdali
05/05/1910	Treaty between Great Britain and the United States relating to boundary waters and boundary questions
09/04/1913	Exchange of notes constituting an agreement ... respecting the boundary between Sierra Leone and French Guinea
06/12/1915	Protocol ... for the delimitation of the frontier along the River Horgos
04/20/1921	Convention of Barcelona
10/28/1922	Convention between [Finland] and the [USSR] concerning the maintenance of river channels and the regulation of fishing on water courses ...
02/14/1925	Convention between [Norway] and [Finland] concerning the international legal regime of the waters of the Pasvik (Paatsjoki) and the Jakobselv ...
02/24/1925	Agreement between the United States of America and Canada to regulate the level of Lake of the Woods
06/15/1925	Notes exchanged ... respecting the regulation of the utilization of the waters of the River Gash
12/20/1925	Exchange of notes between Great Britain and Italy
07/01/1926	Agreement ... regulating the use of the water of the Cunene river
07/20/1927	Convention ... regarding various questions of economic interest ...
08/11/1927	Convention between Spain and Portugal to regulate the hydroelectric development of the international section of the River Douro
01/29/1928	Convention between the German Reich and the Lithuanian Republic regarding the maintenance and administration of the frontier waterways ...
05/07/1929	Exchange of notes ... in regard to the use of ... the River Nile for irrigation purposes
04/29/1931	Exchange of notes ... respecting the boundary between the mandated territory of South Africa and Angola
11/22/1934	Agreement ... regarding water rights on the boundary between Tanganyika and Ruanda-Urundi
05/11/1936	Exchange of notes ... regarding the boundary between Tanganyika Territory and Mozambique
11/07/1940	Exchange of notes between ... the United States of America

and ... Canada constituting an agreement regarding the development of certain portions ...

05/20/1941 Exchange of notes between the government of the United States and the Government of Canada ... concerning temporary diversion for power ...

11/27/1941 Exchange of notes constituting an agreement between the government of the United States and the Government of Canada relating to additional ...

05/22/1944 Declaration and exchange of notes concerning the termination of the process of demarcation of the Peruvian-Ecuadorean frontier

11/14/1944 Treaty between the United States of America and Mexico relating to the waters of the Colorado and Tijuana rivers, and of the Rio Grande ...

06/01/1945 Supplementary boundary treaty between [Argentina] and [Paraguay] on the river Pilcomayo

12/30/1946 Agreement concerning the utilization of the rapids of the Uruguay river in the Salto Grande area

02/03/1947 Treaty between the [USSR] and [Finland] on the transfer to the territory of the Soviet Union of part of the state territory of Finland in the region of ...

02/10/1947 Treaty of peace with Italy, signed at Paris, on 10 February 1947

05/04/1948 Inter-dominion agreement between the government of India and the government of Pakistan, on the canal water dispute between ...

05/31/1949 Exchanges of notes ... regarding the construction of the Owen Falls dam, Uganda

11/25/1949 Treaty concerning the regime of the Soviet-Romanian state frontier and final protocol

12/05/1949 Exchange of notes constituting an agreement between [Great Britain] ... and [Egypt] regarding the construction of the Owen Falls dam, Uganda

01/19/1950 Exchange of notes constituting an agreement between [Great Britain] (on behalf of ... Uganda) and [Egypt] regarding cooperation in meteorological ...

02/24/1950 Treaty between the [USSR] and [Hungary] concerning the regime of the Soviet-Hungarian state frontier and final protocol

02/27/1950 Treaty between the United States of America and Canada relating to the uses of the waters of the Niagara river

04/25/1950 State treaty concerning the construction of a hydroelectric power-plant on the Sauer at Rosport/Ralingen

06/09/1950 Convention between the [USSR] and [Hungary] concerning measures to prevent floods and to regulate the water regime in the area of the frontier ...

09/07/1950   Terms of reference of the Helmand River Delta Commission and an interpretive statement relative thereto, agreed by conferees of ...

10/16/1950   Agreement concerning the diversion of water in the Rissbach, Durrach and Walchen districts

10/16/1950   Agreement between [Austria] and [Germany] concerning the Österreichisch-Bayerische Kraftwerke AG

04/18/1951   Letters between the irrigation adviser and director of irrigation, Sudan government, and the controller of agriculture, Eritrea

04/25/1951   Agreement between [Finland] and [Norway] on the transfer from the course of the Näätämo (Neiden) river to the course of the Gandvik river ...

02/13/1952   Agreement concerning the Donaukraftwerk-Jochenstein Aktiengesellschaft

06/30/1952   Exchange of notes constituting an agreement between Canada and the United States of America relating to the St Lawrence Seaway project

07/16/1952   Exchange of notes constituting an agreement between the [UK/Uganda] and [Egypt] regarding the construction of the Owen Falls dam in Uganda

12/25/1952   Convention between the [USSR] and [Romania] concerning measures to prevent floods and to regulate the water regime of the river Prut

01/21/1953   Exchange of notes constituting an agreement between [Great Britain] and [Portugal] providing for the Portuguese participation in the Shiré valley ...

06/04/1953   Agreement between the Republic of Syria and the Hashemite Kingdom of Jordan concerning the utilization of the Yarmuk waters

11/12/1953   Exchange of notes constituting an agreement between the United States and Canada relating to the establishment of the St Lawrence River joint ...

04/16/1954   Agreement between [Czechoslovakia] and [Hungary] concerning the settlement of technical and economic questions relating to frontier water ...

04/25/1954   Agreement between the government of India and the government of Nepal on the Kosi project

05/25/1954   Convention between the governments of [Yugoslavia] and [Austria] concerning water economy questions relating to the Drava

11/18/1954   Agreement between [Great Britain/Rhodesia-Nyasaland]

with regard to certain ... natives living on the Kwando river

12/16/1954   Agreement between [Yugoslavia] and [Austria] concerning water economy questions in respect of the frontier sector of the Mura

04/07/1955   Agreement between [Yugoslavia] and [Romania] concerning questions of water control on water control systems and watercourses on or intersected ...

04/20/1955   Exchange of notes between Peru and Bolivia establishing a joint commission for study of ... joint use of the waters of Lake Titicaca

08/08/1955   Agreement between [Yugoslavia] and [Hungary] together with the statute of the Yugoslav-Hungarian Water Economy Commission

12/31/1955   Johnston Negotiations

01/20/1956   Agreement concerning cooperation between [Brazil] and [Paraguay] in a study on the utilization of the water power of the Acaray and Monday ...

04/09/1956   Treaty between the Hungarian People's Republic and the Republic of Austria concerning the regulation of water economy questions

08/18/1956   Agreement between the [USSR] and [China] on joint research operations to determine the natural resources of the Amur river basin and the prospects ...

10/13/1956   Treaty between [Czechoslovakia] and [Hungary] concerning the regime of state frontiers

12/05/1956   Agreement between [Yugoslavia] and [Albania] concerning water economy questions, together with the statute of the Yugoslav-Albanian Water ...

02/19/1957   Agreement between Bolivia and Peru concerning a preliminary economic study of the joint utilization of the waters of Lake Titicaca

05/14/1957   Treaty between the governments of the [USSR] and [Iran] concerning the regime of the Soviet-Iranian frontier and the procedure for the settlement ...

08/11/1957   Agreement between Iran and the Soviet Union for the joint utilization of the frontier parts of the rivers Aras and Atrak for irrigation and power ...

12/18/1957   Agreement between Norway and the Union of Soviet Socialist Republics on the utilization of water power on the Pasvik (Paatso) river

01/23/1958   Agreement between [Argentina] and [Paraguay] concerning a study of the utilization of the water power of the Apipe falls

03/21/1958   Agreement between [Czechoslovakia] and [Poland] concerning the use of water resources in frontier waters

04/04/1958   Agreement concerning water-economy questions between the government of [Yugoslavia] and [Bulgaria]

07/10/1958   State treaty between [Luxembourg] and [West Germany] concerning the construction of hydroelectric power installations on the Our river

07/12/1958   Agreement between the government of the French Republic and the Spanish government relating to Lake Lanoux

04/29/1959   Agreement between the [USSR], [Norway], and [Finland] concerning the regulation of Lake Inari by means of the Kaiakoski ... dam

10/23/1959   Indo-Pakistan agreement (with appendices) on East Pakistan border disputes

11/08/1959   Agreement between the government of the United Arab Republic and the government of Sudan

12/04/1959   Agreement between [Nepal] and [India] on the Gandak Irrigation and Power Project

01/11/1960   Agreement between Pakistan and India on West Pakistan-India border disputes

09/19/1960   Indus Waters Treaty

10/24/1960   Agreement relating to the construction of Amistad dam on the Rio Grande to form part of the system of international storage dams provided for by the ...

01/17/1961   Treaty relating to cooperative development of the water resources of the Columbia river basin (with annexes)

02/24/1961   Exchange of notes constituting an agreement concerning the treaty of 12 May 1863 to regulate the diversion of water from the River Meuse and the ...

04/26/1963   Exchange of notes constituting an agreement ... for the development of the Mirim lagoon

07/26/1963   Convention of Bamako

10/26/1963   Act ... states of the Niger basin

11/25/1963   Agreement ... relating to the Central African Power Corporation

11/30/1963   Convention between [Yugoslavia] and [Romania] concerning the operation of the Iron Gates water power and navigation ...

11/30/1963   Agreement between the Socialist Federal Republic of Yugoslavia and the Romanian People's Republic concerning the construction and operation ...

11/30/1963   Convention between the Socialist Federal Republic of Yugoslavia and the Romanian People's Republic concerning compensation for damage ...

01/22/1964   Exchange of notes constituting an agreement between Canada and the United States of America concerning the treaty relating to cooperative ...

01/22/1964   Exchange of notes constituting an agreement between Canada and the United States of America regarding sale of Canada's entitlement ...

02/11/1964   Agreement between Iraq and Kuwait concerning the supply of Kuwait with fresh water ...

05/22/1964   Convention and Statutes ... Lake Chad basin

07/16/1964   Convenio entre España y Portugal para Regular el Aprovechamiento hydroelectrico de los tramos internacionales de rio Duero y de sus afluentes

07/17/1964   Agreement between [Poland] and the [USSR] concerning the use of water resources in frontier waters

09/16/1964   Exchange of notes constituting an agreement between Canada and the United States of America authorizing the Canadian entitlement purchase ...

11/25/1964   Agreement concerning the river Niger commission and the navigation and transport on the river Niger

08/12/1965   Convention between Laos and Thailand for the supply of power

04/30/1966   Agreement between [West Germany], [Austria], and [Switzerland] relating to the withdrawal of water from Lake Constance

08/24/1966   Exchange of notes constituting an agreement concerning the loan of waters of the Colorado river for irrigation of lands in the Mexicali valley

12/19/1966   Revised Agreement between [Nepal] and [India] on the Kosi project

04/01/1967   Untitled: Agreement between South Africa and Portugal

09/28/1967   Franco-Italian convention concerning the supply of water to the Commune of Menton

12/07/1967   Treaty between [Austria] and [Czechoslovakia] concerning the regulation of water management questions relating to frontier waters

02/27/1968   Agreement between [Czechoslovakia] and [Hungary] concerning the establishment of a river administration in the Rajka-Gönyü sector ...

05/29/1968   Convenio y protocola adicional para regular el uso y aprovechamiento hidraulico de los tramos internacionales de los rios Miño, Limia, Tajo ...

10/23/1968   Agreement between the People's Republic of Bulgaria and the Republic of Turkey concerning cooperation ...

01/21/1969   Agreement between South Africa and Portugal

03/21/1969   Exchange of notes constituting an agreement for the construction of a temporary cofferdam at Niagara

03/21/1969   Exchange of Notes constituting an agreement between Canada and the United States of America for the temporary diversion for power purposes ...

07/04/1969   Convention concerning development of the Rhine between Strasbourg and Lauterbourg

01/30/1970   Convention of Dakar

12/16/1971   Agreement between [Romania] and the [USSR] on the joint construction of the Stinca-Costesti hydraulic engineering scheme

07/12/1972   Agreement between [Finland] and the [USSR] concerning the production of electric power in the part of the Vuoksi river bounded by the Imatra ...

11/24/1972   Statute of the Indo-Bangladesh Joint Rivers Commission

04/26/1973   Treaty between [Brazil] and [Paraguay] concerning the hydroelectric utilization of the water resources of the Paraná river ...

11/13/1973   Agreement between [Australia/Papua New Guinea] and [Indonesia] concerning administrative border arrangements

01/31/1975   Joint declaration of principles for utilization of the waters of the lower Mekong basin, signed by [Cambodia], [Laos], [Thailand], and [Vietnam]

03/06/1975   Agreement ... concerning the use of frontier watercourses

02/12/1976   Segundo protocolo

11/05/1977   Agreement between [Bangladesh] and [India] on sharing of the Ganges' waters at Farakka and on augmenting its flows

04/07/1978   Agreement between [Nepal] and [India] on the renovation and extension of Chandra canal, pumped canal, and distribution of the Western Kosi canal

06/30/1978   Convention relating to the creation of the Gambia River Basin Development Organization

07/03/1978   Treaty for Amazonian Cooperation

10/19/1979   Agreement on Paraná river projects

11/21/1980   Convention creating the Niger Basin Authority

07/20/1983   Meeting of the Joint Rivers Commission

10/01/1986   Treaty on the Lesotho Highlands water project between [Lesotho] and [South Africa]

10/08/1990   Convention ... on the international commission for the protection of the Elbe

03/26/1993   Agreement on joint activities in addressing the Aral sea ...

06/30/1994   Draft Convention on Cooperation for the Protection and Sustainable Use of the Danube River

10/26/1994   Treaty of peace between [Israel] and [Jordan], done at Arava/Araba crossing point ...

03/03/1995   Resolution of the Heads of States of the Central Asia [sic] on work of the EC of ICAS on implementation ...

04/05/1995   Agreement on the Cooperation for the Sustainable Development of the Mekong River Basin

09/28/1995   Israeli-Palestinian Interim Agreement on the West Bank and the Gaza Strip

12/12/1996   Treaty between [India] and [Bangladesh] on sharing of the Ganga/Ganges waters at Farakka

AMAZON BASIN

Treaty title	Declaration and exchange of notes concerning the termination of the process of demarcation of the Peruvian-Ecuadorean frontier
Basins involved	Amazon, Chira, Zarumilla, Tumbes
Main basin	Amazon
Date signed	05/22/1944
Signatories	Bilateral
Parties	Peru, Ecuador
Principal focus	Water supply
Non-water linkages	None
Comments on above	
Monitoring	No
Allocations	Peru agreed to supply water to Ecuadorian villages on the right bank of the "so-called old bed of the River Zarumilla." Obviously the specific needs of the villages are unclear, but Peru agreed to ensure an adequate water supply
Enforcement	None
Unequal power relationship	Not available
Information sharing	Not available
Conflict resolution	None
Method for water division	Unclear
Negotiations	
Additional comments	

Treaty title	Treaty for Amazonian Cooperation
Basins involved	Amazon
Main basin	Amazon
Date signed	07/03/1978
Signatories	Multilateral
Parties	Bolivia, Brazil, Colombia, Ecuador, Guyana, Peru, Surinam, Venezuela
Principal focus	Industrial uses
Non-water linkages	None
Comments on above	Treaty focuses most on development, not supply
Monitoring	No
Allocations	
Enforcement	None
Unequal power relationship	Yes
Information sharing	Yes
Conflict resolution	None
Method for water division	None
Negotiations	
Additional comments	They make mention of sustainable development and avoiding pollution

AMUR BASIN

Treaty title	Protocol ... for the delimitation of the frontier along the River Horgos
Basins involved	Horgos
Main basin	Amur
Date signed	06/12/1915
Signatories	Bilateral
Parties	China, Russia
Principal focus	Water supply
Non-water linkages	None
Comments on above	Primarily a border demarcation
Monitoring	No
Allocations	Existing canals were to remain in use; the remainder of water was to be divided equally
Enforcement	None
Unequal power relationship	Yes
Information sharing	None
Conflict resolution	None
Method for water division	Unclear
Negotiations	Tensions over the border continue to this day; Russia may have been trying to secure one front while fighting internal (revolution) and external (WWI) battles
Additional comments	Russia and China have the same border dispute as they have had for some time. The water supply was secondary and favoured the Russian position

Treaty title	Agreement between the [USSR] and [China] on joint research operations to determine the natural resources of the Amur river basin and the prospects ...
Basins involved	Amur
Main basin	Amur
Date signed	08/18/1956
Signatories	Bilateral
Parties	China, USSR
Principal focus	Hydropower
Non-water linkages	None
Comments on above	
Monitoring	Yes
Allocations	
Enforcement	Not available
Unequal power relationship	Not available
Information sharing	Yes
Conflict resolution	None

Method for water division	None
Negotiations	
Additional comments	Created a Joint Scientific Council on problems connected with the study

ARAKS, ATRAK BASIN

Treaty title	Agreement between Iran and the Soviet Union for the joint utilization of the frontier parts of the rivers Aras and Atrak for irrigation and power ...
Basins involved	Araks, Atrak
Main basin	Araks, Atrak
Date signed	08/11/1957
Signatories	Bilateral
Parties	Iran, USSR
Principal focus	Water supply
Non-water linkages	Land
Comments on above	Frontier modifications through the reservoirs
Monitoring	Not available
Allocations	Each party receives half of water for irrigation and hydropower generation
Enforcement	Not available
Unequal power relationship	Yes
Information sharing	Yes
Conflict resolution	Not available
Method for water division	Equal parts
Negotiations	
Additional comments	A joint commission was later added (although it had no conflict resolution capacity) for establishment of the boundary line, especially after alluvial deposits cause a shift in the thalweg

Treaty title	Treaty between the government of the [USSR] and [Iran] concerning the regime of the Soviet-Iranian frontier and the procedure for the settlement ...
Basins involved	Tedzen, Atrak, Araks, Harirud
Main basin	Atrak, Araks
Date signed	05/14/1957
Signatories	Bilateral
Parties	USSR, Iran
Principal focus	Pollution
Non-water linkages	None
Comments on above	
Monitoring	Yes
Allocations	Residents of either state are entitled to fish in frontier waters. They may water livestock, pro-

	vided that the livestock do not wander into the other's territory. Pollution control is mentioned
Enforcement	Not available
Unequal power relationship	Yes
Information sharing	Yes
Conflict resolution	Council
Method for water division	None
Negotiations	
Additional comments	

COLORADO BASIN

Treaty title	Treaty between the United States of America and Mexico relating to the waters of the Colorado and Tijuana rivers, and of the Rio Grande ...
Basins involved	Colorado, Rio Grande, Tijuana, Rio Bravo
Main basin	Colorado
Date signed	11/14/1944
Signatories	Bilateral
Parties	United States, Mexico
Principal focus	Water supply
Non-water linkages	Other linkages
Comments on above	Joint construction of dams and reservoirs for storage and flood controls and possible hydropower.
Monitoring	Yes
Allocations	US: All waters reaching the Rio Grande from the Pecos & Devils rivers, Goodenough spring, and Alamito, Terlingua, and Pinto creeks. Half of the flow below the lowest major international storage dam. One-third of the flow (431.721 MCM min.) reaching the Rio Grande from the Conchos, San Diego, San Rodrigo, Escondido and Salado rivers and the Las Vacas arroyo. (cont'd below)
Enforcement	None
Unequal power relationship	Yes
Information sharing	Yes
Conflict resolution	Not available
Method for water division	Complex but clear
Negotiations	Commission controls authorization of temporary diversions to one country, provided that "... the use of such water shall not establish any right to continue to divert it."
Additional comments	Half of all other flows reaching the Rio Grande, including "unmeasured tributaries," between Fort Quitman and the lowest international storage dam. Mexico: min. 1.850 BCM from Colorado (max 2.1

BCM) Mexico: all waters from the San Juan and Alamo rivers; half of Rio Grande below the lowest major international storage dam, and two-thirds of the Conchos, San Diego, San Rodrigo, Escondido, and Salado river, and the Las Vacas arroyo. Also one-half of all other flows reaching the Rio Grande, including "unmeasured tributaries" between Fort Quitman and the lowest international storage dam

Treaty title	Exchange of notes constituting and agreement concerning the loan of waters of the Colorado river for irrigation of lands in the Mexicali valley
Basins involved	Colorado
Main basin	Colorado
Date signed	08/24/1966
Signatories	Bilateral
Parties	USA, Mexico
Principal focus	Water supply
Non-water linkages	Money
Comments on above	Mexico will reimburse the US for any decrease in power generation at Hoover or Glen Canyon
Monitoring	Not available
Allocations	USA releases 40,535 acre-feet (50 MCM) of water from September to December 1966 and will retain the same amount over one or three years, depending on the weather conditions that follow
Enforcement	Not available
Unequal power relationship	Yes
Information sharing	Not available
Conflict resolution	None
Method for water division	Complex but clear
Negotiations	
Additional comments	

COLUMBIA BASIN

Treaty title	Treaty relating to cooperative development of the water resources of the Columbia River Basin (with annexes)
Basins involved	Columbia, Kootenai
Main basin	Columbia
Date signed	01/17/1961
Signatories	Bilateral
Parties	USA, Canada
Principal focus	Hydropower

Non-water linkages	Money
Comments on above	Canada will build reservoirs; US will pay $64.4 million for flood control, plus a fee for each flood
Monitoring	Yes
Allocations	Canada receives half the electricity generated by the plants. Either party may sell the electricity
Enforcement	None
Unequal power relationship	No
Information sharing	Yes
Conflict resolution	Council
Method for water division	Equal parts
Negotiations	
Additional comments	
Treaty title	Exchange of notes constituting an agreement between Canada and the United States of America concerning the treaty relating to cooperative ...
Basins involved	Columbia
Main basin	Columbia
Date signed	01/22/1964
Signatories	Bilateral
Parties	USA, Canada
Principal focus	Flood control
Non-water linkages	None
Comments on above	
Monitoring	Yes
Allocations	Canada and the US agree that the previous treaty "provides each of the a right to divert water for a consumptive use"
Enforcement	None
Unequal power relationship	No
Information sharing	Yes
Conflict resolution	None
Method for water division	Unclear
Negotiations	
Additional comments	
Treaty title	Exchange of notes constituting an agreement between Canada and the United States of America regarding sale of Canada's entitlement ...
Basins involved	Columbia
Main basin	Columbia
Date signed	01/22/1964
Signatories	Bilateral
Parties	USA, Canada
Principal focus	Hydropower
Non-water linkages	Money

Comments on above	This treaty refers to the 01/17/1961 treaty and actually sells the water for the price agreed upon in the earlier treaty (US $254,400,000)
Monitoring	Not available
Allocations	The sale of hydropower lasts for 30 years
Enforcement	Not available
Unequal power relationship	No
Information sharing	Not available
Conflict resolution	Not available
Method for water division	Complex but clear
Negotiations	
Additional comments	

Treaty title	Exchange of notes constituting an agreement between Canada and the United States of America authorizing the Canadian entitlement purchase ...
Basins involved	Columbia
Main basin	Columbia
Date signed	09/16/1964
Signatories	Bilateral
Parties	USA, Canada
Principal focus	Hydropower
Non-water linkages	Money
Comments on above	Sale of hydropower entitlement provided for; US $254,400,000 paid to Canada for treaty projects
Monitoring	Not available
Allocations	As before, Canada is entitled to receive one-half of the usable hydropower and one-half of the additional power resulting from additional stream flow created by channel alterations
Enforcement	Not available
Unequal power relationship	No
Information sharing	Yes
Conflict resolution	Other government agency
Method for water division	Complex/clear
Negotiations	The treaty also specifies exact numbers for compensation of lost downstream power: 2.70 mills per kilowatt-hour, and 46 cents per kilowatt of dependable (non-seasonal?) capacity for each month or fraction thereof
Additional comments	A permanent engineering board is also established

CONGO BASIN

Treaty title	Convention ... regarding various questions of economic interest ...

Basins involved	M'Pozo
Main basin	Congo
Date signed	07/20/1927
Signatories	Bilateral
Parties	Belgium, Portugal
Principal focus	Hydropower
Non-water linkages	Money
Comments on above	A portion of the electricity would go to Portugal (Angola), the downstream riparian
Monitoring	Not available
Allocations	
Enforcement	Not available
Unequal power relationship	Not available
Information sharing	Yes
Conflict resolution	UN/third party
Method for water division	Not available
Negotiations	Flooded areas would be compensated for any damages
Additional comments	15% of electric power generated by the dam would go to Angola

CUNENE BASIN

Treaty title	Agreement ... regulating the use of the water of the Cunene river
Basins involved	Cunene (Kunene)
Main basin	Cunene
Date signed	07/01/1926
Signatories	Bilateral
Parties	South Africa, Portugal
Principal focus	Hydropower
Non-water linkages	None
Comments on above	
Monitoring	Not available
Allocations	
Enforcement	Not available
Unequal power relationship	Not available
Information sharing	Not available
Conflict resolution	Not available
Method for water division	Equal parts
Negotiations	
Additional comments	Treaty concerned irrigation and hydropower. If the river diversion into South Africa resulted in a financial "gain," then payment was to be given for use of the waters

Treaty title	Exchange of notes ... respecting the boundary between the mandated territory of South Africa and Angola
Basins involved	Cunene
Main basin	Cunene
Date signed	04/29/1931
Signatories	Bilateral
Parties	South Africa, Portugal
Principal focus	Water supply
Non-water linkages	None
Comments on above	
Monitoring	Not available
Allocations	
Enforcement	Not available
Unequal power relationship	Not available
Information sharing	Not available
Conflict resolution	Not available
Method for water division	Unclear
Negotiations	
Additional comments	Supplied drinking water to the inhabitants of Ovamboland for drinking and cattle

Treaty title	Agreement between South Africa and Portugal
Basins involved	Cunene
Main basin	Cunene
Date signed	01/21/1969
Signatories	Bilateral
Parties	Portugal, South Africa
Principal focus	Water supply
Non-water linkages	Money
Comments on above	Financing provided by South Africa for the dam, compensation for flooded land and works area
Monitoring	Not available
Allocations	Portugal gets 50% of the flow (as measured at Ruacana). Portugal gets 50% of the flow for irrigation in Ovamboland, max 6m3/sec. No charge for the water.
Enforcement	Not available
Unequal power relationship	Not available
Information sharing	Not available
Conflict resolution	Not available
Method for water division	Equal parts
Negotiations	Aimed at regulating the flow, two hydroelectric plants, and water supply. Established a joint technical commission. Financial obligations of S. Africa limited to R 8,125,000 to be repaid @ 5% over 20 years

Additional comments	This treaty centres on money: S. Africa pays Portugal for kWh generated at the dam, with ratios established for payment versus % of flow

DANUBE BASIN

Treaty title	Treaty concerning the regime of the Soviet-Romanian state frontier and final protocol
Basins involved	Danube
Main basin	Danube
Date signed	11/25/1949
Signatories	Bilateral
Parties	USSR, Romania
Principal focus	Flood control
Non-water linkages	Money
Comments on above	Indirectly: any works undertaken will cost each side an equal amount
Monitoring	Not available
Allocations	
Enforcement	Not available
Unequal power relationship	Yes
Information sharing	Not available
Conflict resolution	Not available
Method for water division	None
Negotiations	
Additional comments	

Treaty title	Treaty between the [USSR] and [Hungary] concerning the regime of the Soviet-Hungarian state frontier and final protocol
Basins involved	Danube
Main basin	Danube
Date signed	02/24/1950
Signatories	Bilateral
Parties	USSR, Hungary
Principal focus	Flood control
Non-water linkages	None
Comments on above	
Monitoring	Yes
Allocations	
Enforcement	Not available
Unequal power relationship	Yes
Information sharing	Yes
Conflict resolution	Not available
Method for water division	None

Negotiations
Additional comments

Treaty title	Convention between the [USSR] and [Hungary] concerning measures to prevent floods and to regulate the water regime in the area of the frontier ...
Basins involved	Tisza
Main basin	Danube
Date signed	06/09/1950
Signatories	Bilateral
Parties	USSR, Hungary
Principal focus	Flood control
Non-water linkages	None
Comments on above	
Monitoring	Yes
Allocations	
Enforcement	Not available
Unequal power relationship	Yes
Information sharing	Yes
Conflict resolution	Not available
Method for water division	None
Negotiations	
Additional comments	

Treaty title	Agreement concerning the diversion of water in the Rissbach, Durrach, and Walchen districts
Basins involved	Isar, Rissbach
Main basin	Danube
Date signed	10/16/1950
Signatories	Bilateral
Parties	Austria, Germany (FRG)
Principal focus	Water supply
Non-water linkages	None
Comments on above	
Monitoring	Not available
Allocations	Austria "agrees to waive without compensation the right to lead off any waters of the Rissbach and its tributaries." Austria also agrees to "the leading-off without compensation of: "The Dürrach ... Kesselbach ... Blaserbach ... [and] the Dollmannbach" streams
Enforcement	Not available
Unequal power relationship	Not available
Information sharing	Not available
Conflict resolution	None
Method for water division	None
Negotiations	

Additional comments

Treaty title	Agreement between [Austria] and [Germany] concerning the Österreichisch-Bayerische Kraftwerke AG
Basins involved	Inn, Salzach
Main basin	Danube
Date signed	10/16/1950
Signatories	Bilateral
Parties	Austria, Germany (FRG)
Principal focus	Hydropower
Non-water linkages	Money
Comments on above	Creation of a joint-stock company which will pay some of the share capital in developing hydropower resources
Monitoring	Yes
Allocations	Water rights will be sold
Enforcement	Council
Unequal power relationship	No
Information sharing	Yes
Conflict resolution	Council
Method for water division	Unclear
Negotiations	
Additional comments	Waters of the Danube are excluded from this agreement; only those of the Inn and Salzach are under the effects of this treaty

Treaty title	Agreement concerning the Donaukraftwerk-Jochenstein Aktiengesellschaft
Basins involved	Danube
Main basin	Danube
Date signed	02/13/1952
Signatories	Multilateral
Parties	FRG (West Germany), Austria
Principal focus	Hydropower
Non-water linkages	None
Comments on above	
Monitoring	Not available
Allocations	Non-conflicting water right permits are to be issued simultaneously and are to be "as equal in scope as possible"
Enforcement	Council
Unequal power relationship	No
Information sharing	Yes
Conflict resolution	Council
Method for water division	Unclear
Negotiations	

Additional comments

Treaty title	Convention between the [USSR] and [Romania] concerning measures to prevent floods and to regulate the water regime of the river Prut
Basins involved	Danube, Prut
Main basin	Danube
Date signed	12/25/1952
Signatories	Bilateral
Parties	USSR, Romania
Principal focus	Flood control
Non-water linkages	None
Comments on above	
Monitoring	Yes
Allocations	
Enforcement	Not available
Unequal power relationship	Yes
Information sharing	Yes
Conflict resolution	Council
Method for water division	None
Negotiations	
Additional comments	

Treaty title	Agreement between [Czechoslovakia] and [Hungary] concerning the settlement of technical and economic questions relating to frontier water ...
Basins involved	Danube, Tisza
Main basin	Danube
Date signed	04/16/1954
Signatories	Bilateral
Parties	Hungary, Czechoslovakia
Principal focus	Flood control
Non-water linkages	None
Comments on above	Some services are provided by each side during the construction of works
Monitoring	Yes
Allocations	Each party "shall ... be free to use half the natural discharge ... exclusive of any increase brought about by artificial interference"
Enforcement	Not available
Unequal power relationship	Not available
Information sharing	Yes
Conflict resolution	Not available
Method for water division	Equal parts
Negotiations	
Additional comments	The states also agree to not grant any water use permit for the "execution on frontier watercourses

of hydraulic works which might adversely affect the discharge conditions or the bed"

Treaty title	Convention between the governments of [Yugoslavia] and [Austria] concerning water economy questions relating to the Drava
Basins involved	Drava
Main basin	Danube
Date signed	05/25/1954
Signatories	Bilateral
Parties	Austria, Yugoslavia
Principal focus	Water supply
Non-water linkages	Money
Comments on above	The Yugoslavs receive at least 50m schillings in industrial products for 82.5 GWh over 4 years
Monitoring	Yes
Allocations	A complex grid of daily flows, measured at Schwabeck, is to be maintained. For flows below 200 cms/above 300 cms, the difference in flow below Lavamünd can be augmented at Dravograd by withdrawing not more than 1 MCM from the reservoir.
Enforcement	Not available
Unequal power relationship	Not available
Information sharing	Yes
Conflict resolution	Council
Method for water division	Complex but clear
Negotiations	
Additional comments	

Treaty title	Agreement between [Yugoslavia] and [Austria] concerning water economy questions in respect of the frontier sector of the Mura
Basins involved	Mura
Main basin	Danube
Date signed	12/16/1954
Signatories	Bilateral
Parties	Austria, Yugoslavia
Principal focus	Flood control
Non-water linkages	None
Comments on above	
Monitoring	Yes
Allocations	
Enforcement	Not available
Unequal power relationship	Not available
Information sharing	Yes
Conflict resolution	Council
Method for water division	None

Negotiations

Additional comments   Created the Permanent Yugoslav-Austrian Commission for the Mura

Treaty title   Agreement between [Yugoslavia] and [Romania] concerning questions of water control on water control systems and watercourses on or intersected

...

Basins involved	Danube, Tisza
Main basin	Danube
Date signed	04/07/1955
Signatories	Bilateral
Parties	Romania, Yugoslavia
Principal focus	Flood control
Non-water linkages	None
Comments on above	
Monitoring	Yes
Allocations	
Enforcement	Council
Unequal power relationship	No
Information sharing	Yes
Conflict resolution	Council
Method for water division	Not available
Negotiations	

Additional comments   Created a Joint Technical Commission to record data and examine any projects that might affect the other party's water regime

Treaty title   Agreement between [Yugoslavia] and [Hungary] together with the statute of the Yugoslav-Hungarian water economy commission

Basins involved	Mura, Drava, Maros, Tisa, Danube
Main basin	Danube
Date signed	08/08/1955
Signatories	Bilateral
Parties	Hungary, Yugoslavia
Principal focus	Flood control
Non-water linkages	None
Comments on above	
Monitoring	Yes
Allocations	
Enforcement	Council
Unequal power relationship	Not available
Information sharing	Yes
Conflict resolution	Council
Method for water division	None
Negotiations	

Additional comments	Created a commission to oversee development and flood control on the frontier rivers
Treaty title	Treaty between the Hungarian People's Republic and the Republic of Austria concerning the regulation of water economy questions
Basins involved	Danube
Main basin	Danube
Date signed	04/09/1956
Signatories	Bilateral
Parties	Austria, Hungary
Principal focus	Water supply
Non-water linkages	None
Comments on above	
Monitoring	Yes
Allocations	
Enforcement	Council
Unequal power relationship	Not available
Information sharing	Yes
Conflict resolution	Council
Method for water division	Unclear
Negotiations	The treaty focuses heavily on money, expenditure sharing, and budgets
Additional comments	The contracting parties will discuss in Commission proceedings to grant water rights only after the measures for works to be taken have been discussed
Treaty title	Treaty between [Czechoslovakia] and [Hungary] concerning the regime of state frontiers
Basins involved	Danube
Main basin	Danube
Date signed	10/13/1956
Signatories	Bilateral
Parties	Czechoslovakia, Hungary
Principal focus	Flood control
Non-water linkages	None
Comments on above	
Monitoring	Not available
Allocations	
Enforcement	Not available
Unequal power relationship	No
Information sharing	Not available
Conflict resolution	Not available
Method for water division	None
Negotiations	
Additional comments	Each party agrees to not obstruct the flow of water in any manner unless both parties agree

Treaty title	Agreement between [Yugoslavia] and [Albania] concerning water economy questions, together with the statue of the Yugoslav-Albanian Water ...
Basins involved	Crni Drim, Beli Drim, Bojana, Lake Skadar
Main basin	Danube
Date signed	12/05/1956
Signatories	Bilateral
Parties	Albania, Yugoslavia
Principal focus	Hydropower
Non-water linkages	None
Comments on above	
Monitoring	Yes
Allocations	
Enforcement	Council
Unequal power relationship	Not available
Information sharing	Yes
Conflict resolution	Not available
Method for water division	None
Negotiations	
Additional comments	A Water Economy Commission was established

Treaty title	Agreement concerning water economy questions between the government of [Yugoslavia] and [Bulgaria]
Basins involved	Danube
Main basin	Danube
Date signed	04/04/1958
Signatories	Bilateral
Parties	Yugoslavia, Bulgaria
Principal focus	Industrial uses
Non-water linkages	None
Comments on above	The document does address customs issues, but not as a substitute for water
Monitoring	Yes
Allocations	
Enforcement	None
Unequal power relationship	Not available
Information sharing	Yes
Conflict resolution	None
Method for water division	None
Negotiations	
Additional comments	

Treaty title	Convention between [Yugoslavia] and [Romania] concerning the operation of the Iron Gates water power and navigation ...
Basins involved	Danube

Main basin	Danube
Date signed	11/30/1963
Signatories	Bilateral
Parties	Yugoslavia, Romania
Principal focus	Hydropower
Non-water linkages	None
Comments on above	
Monitoring	Yes
Allocations	
Enforcement	None
Unequal power relationship	Not available
Information sharing	Yes
Conflict resolution	Council
Method for water division	Equal parts
Negotiations	
Additional comments	

Treaty title	Agreement between the Socialist Federal Republic of Yugoslavia and the Romanian People's Republic concerning the construction and operation …
Basins involved	Danube
Main basin	Danube
Date signed	11/30/1963
Signatories	Bilateral
Parties	Yugoslavia, Romania
Principal focus	Hydropower
Non-water linkages	Land
Comments on above	Modification of international frontier to suit the dam
Monitoring	Yes
Allocations	Each state gets one-half the electric power (est. 2 million watts, or 10 billion kWh annually)
Enforcement	Not available
Unequal power relationship	Not available
Information sharing	Yes
Conflict resolution	Council
Method for water division	Equal parts
Negotiations	
Additional comments	

Treaty title	Convention between the Socialist Federal Republic of Yugoslavia and the Romanian People's Republic concerning compensation for damage …
Basins involved	Danube
Main basin	Danube
Date signed	11/30/1963
Signatories	Bilateral

Parties	Yugoslavia, Romania
Principal focus	Hydropower
Non-water linkages	Money
Comments on above	Compensation for damage caused by creation of the reservoir or construction of the dam
Monitoring	Not available
Allocations	
Enforcement	Not available
Unequal power relationship	Not available
Information sharing	Yes
Conflict resolution	Not available
Method for water division	None
Negotiations	
Additional comments	Another treaty signed this day dealt with "the determination of the value of investments and mutual accounting in connexion with the construction of the Iron Gates water power and navigation system on the River Danube"
Treaty title	Treaty between [Austria] and [Czechoslovakia] concerning the regulation of water management questions relating to frontier waters
Basins involved	Danube
Main basin	Danube
Date signed	12/07/1967
Signatories	Bilateral
Parties	Austria, Czechoslovakia
Principal focus	Navigation
Non-water linkages	Money
Comments on above	... mostly concerning maintenance and improvements
Monitoring	Yes
Allocations	"existing water rights in respect of frontier waters and the obligations connected therewith shall remain unaffected"
Enforcement	Council
Unequal power relationship	Not available
Information sharing	Yes
Conflict resolution	Council
Method for water division	Equal parts
Negotiations	
Additional comments	Scope of the treaty includes the following: "waters intersecting the state frontier and waters adjoining the state frontier where any ... measures applied to them in [either] state would have seriously adverse affects on water conditions in the territory of the other ..."

Treaty title	Agreement between [Czechoslovakia] and [Hungary] concerning the establishment of a river administration in the Rajka-Gönyü Sector ...
Basins involved	Danube
Main basin	Danube
Date signed	02/27/1968
Signatories	Bilateral
Parties	Czechoslovakia, Hungary
Principal focus	Navigation
Non-water linkages	None
Comments on above	
Monitoring	Not available
Allocations	
Enforcement	Not available
Unequal power relationship	No
Information sharing	Yes
Conflict resolution	Council
Method for water division	None
Negotiations	Created a joint river administration
Additional comments	

Treaty title	Agreement between [Romania] and the [USSR] on the joint construction of the Stinca-Costesti hydraulic engineering scheme
Basins involved	Prut
Main basin	Danube
Date signed	12/16/1971
Signatories	Bilateral
Parties	USSR and Romania
Principal focus	Hydropower
Non-water linkages	Money
Comments on above	Compensation for flooded land
Monitoring	Yes
Allocations	
Enforcement	Not available
Unequal power relationship	Yes
Information sharing	Yes
Conflict resolution	Council
Method for water division	Equal parts
Negotiations	Some border corrections made on an international boundary
Additional comments	

Treaty title	Draft Convention on Cooperation for the Protection and Sustainable Use of the Danube River
Basins involved	Danube
Main basin	Danube

Date signed	06/30/1994
Signatories	Multilateral (unsigned)
Parties	
Principal focus	Pollution
Non-water linkages	None
Comments on above	
Monitoring	Yes
Allocations	Contracting parties will "ensure efficient water quality protection and sustainable water use ..." "[That] means on the criteria of a stable, environmentally sound development ..."
Enforcement	None
Unequal power relationship	Not available
Information sharing	Yes
Conflict resolution	UN/third party
Method for water division	None
Negotiations	
Additional comments	Those having "considerable part" of the basin are those with greater than 2000 km^2 of basin area within the country's border. The treaty moves towards a comprehensive "water management" instead of focusing on a single aspect of the basin

DOURO BASIN

Treaty title	Convention between Spain and Portugal to regulate the hydroelectric development of the international section of the River Douro
Basins involved	Douro, Huebra, Esla, Tormes
Main basin	Douro
Date signed	08/11/1927
Signatories	Bilateral
Parties	Spain, Portugal
Principal focus	Hydropower
Non-water linkages	None
Comments on above	
Monitoring	Yes
Allocations	Each state receives the exclusive right to use the water that flows between set areas. Both states "undertake mutually to not reduce the volume of water which should reach the beginning of each zone of utilisation ..."
Enforcement	Council
Unequal power relationship	Not available
Information sharing	Yes

Conflict resolution	Council
Method for water division	Complex/clear
Negotiations	
Additional comments	

Treaty title	Convenio entre España y Portugal para regular el aprovechamiento hidroelectrico de los tramos internacionales de rio Duero y de sus afluentes
Basins involved	Douro
Main basin	Douro
Date signed	07/16/1964
Signatories	Bilateral
Parties	Spain, Portugal
Principal focus	Hydropower
Non-water linkages	None
Comments on above	
Monitoring	Yes
Allocations	
Enforcement	Council
Unequal power relationship	Not available
Information sharing	None
Conflict resolution	Council
Method for water division	Complex but clear
Negotiations	
Additional comments	ONLY hydroelectric uses. Nothing else discussed. The commission sets the distribution of water between the countries – namely, diversions that would reduce hydroelectric output. The "Limits Commission" also has a say. The International Consortium exists for ONLY industrial/economic collaboration on the rivers

DURANCE BASIN

Treaty title	Treaty of Peace with Italy, signed at Paris, on 10 February 1947
Basins involved	Lake of Mont Cenis
Main basin	Durance
Date signed	02/10/1947
Signatories	Multilateral
Parties	Italy, France (primarily), and the Allied Powers
Principal focus	Hydropower
Non-water linkages	Political concessions
Comments on above	Italy was to hand over archives regarding territory ceded in 1860; railway concessions; others

Monitoring	Not available
Allocations	Use of hydropower from district of Mont Cenis
Enforcement	Force/threat of Force
Unequal power relationship	Yes
Information sharing	Not available
Conflict resolution	None
Method for water division	None
Negotiations	
Additional comments	Naturally the "threat of force" is the enforcement here because Italy's involvement in World War II ended with this treaty

EBRO BASIN

Treaty title	Agreement between the government of the French Republic and the Spanish government relating to Lake Lanoux
Basins involved	Lake Lanoux, Carol, Font-Vive
Main basin	Ebro
Date signed	07/12/1958
Signatories	Bilateral
Parties	France, Spain
Principal focus	Hydropower
Non-water linkages	None
Comments on above	
Monitoring	Yes
Allocations	France agrees to return minimum 20 MCM to the Carol river annually
Enforcement	Not available
Unequal power relationship	Not available
Information sharing	Yes
Conflict resolution	UN/third party
Method for water division	Complex/clear
Negotiations	
Additional comments	Water deliveries based on a water year, not a calendar year

ELBE BASIN

Treaty title	Convention ... on the international commission for the protection of the Elbe
Basins involved	Elbe
Main basin	Elbe
Date signed	10/08/1990

Signatories	Bilateral
Parties	Germany, Slovak Federative Republic
Principal focus	Pollution
Non-water linkages	None
Comments on above	
Monitoring	Yes
Allocations	
Enforcement	Council
Unequal power relationship	Yes
Information sharing	Yes
Conflict resolution	Council
Method for water division	None
Negotiations	
Additional comments	

EUPHRATES BASIN

Treaty title	Agreement between Iraq and Kuwait concerning the supply of Kuwait with fresh water ...
Basins involved	Unspecified
Main basin	Euphrates
Date signed	02/11/1964
Signatories	Bilateral
Parties	Iraq, Kuwait
Principal focus	Water supply
Non-water linkages	None
Comments on above	
Monitoring	Not available
Allocations	Kuwait receives 120 million imperial gallons per day
Enforcement	Not available
Unequal power relationship	Yes
Information sharing	Not available
Conflict resolution	Not available
Method for water division	Complex/clear
Negotiations	Negotiations were to continue regarding larger water transfers
Additional comments	

Treaty title	Agreement ... concerning the use of frontier watercourses
Basins involved	Bnava Suta, Qurahtu, Gangir, Alvend, Kanjan
Main basin	Euphrates
Date signed	03/06/1975
Signatories	Bilateral

Parties	Iran, Iraq
Principal focus	Water supply
Non-water linkages	None
Comments on above	Part of a lengthy treaty in which borders were discussed, as well as mapping
Monitoring	Not available
Allocations	Flows of the Bnava Suta, Qurahtu, and Gangir rivers are divided equally. Flows of the Alvend, Kanjan Cham, Tib, and Duverij will be divided based on a 1914 commission report on the Ottoman/Iranian border "and in accordance with custom"
Enforcement	Council
Unequal power relationship	No
Information sharing	None
Conflict resolution	Other government agency
Method for water division	Equal parts
Negotiations	
Additional comments	

GAMBIA BASIN

Treaty title	Agreement between Great Britain and France
Basins involved	Gambia
Main basin	Gambia
Date signed	08/10/1889
Signatories	Bilateral
Parties	Great Britain, France
Principal focus	Navigation
Non-water linkages	None
Comments on above	
Monitoring	Not available
Allocations	
Enforcement	Not available
Unequal power relationship	No
Information sharing	Not available
Conflict resolution	Not available
Method for water division	Unknown
Negotiations	
Additional comments	The treaty established that France "had complete control of the Mellacoree River and Great Britain of the Scarcies"

Treaty title	Exchange of letters between Great Britain and France
Basins involved	Gambia
Main basin	Gambia

Date signed	02/04/1895
Signatories	Bilateral
Parties	Great Britain, France
Principal focus	Water supply
Non-water linkages	None
Comments on above	
Monitoring	Not available
Allocations	The amounts under previous use by local riverain inhabitants was to be continued (apparently not measured, however)
Enforcement	Not available
Unequal power relationship	No
Information sharing	Not available
Conflict resolution	Not available
Method for water division	Unknown
Negotiations	Negotiations were over state borders; the question of local inhabitants (who did not care about state borders) and their previous use of the river arose and was dealt with. Apparently water use was expected to be low. If not, no provisions are listed
Additional comments	Established that people dwelling near the river on "the right bank were given the right to use the river within previous limits, and subject to the laws of Sierra Leone concerning the navigation and use of water"
Treaty title	Agreement between Great Britain and France
Basins involved	Gambia
Main basin	Gambia
Date signed	10/19/1906
Signatories	Bilateral
Parties	Great Britain, France
Principal focus	Water supply
Non-water linkages	None
Comments on above	
Monitoring	Not available
Allocations	
Enforcement	Not available
Unequal power relationship	No
Information sharing	Not available
Conflict resolution	Not available
Method for water division	Not available
Negotiations	
Additional comments	Local inhabitants were given the right to use pasture and arable lands, springs and watering places in prior use even though a border separated them after the new borders were established

Treaty title	Exchange of notes constituting an agreement ... respecting the boundary between Sierra Leone and French Guinea
Basins involved	Gambia
Main basin	Gambia
Date signed	09/04/1913
Signatories	Bilateral
Parties	Great Britain, France
Principal focus	Hydropower
Non-water linkages	None
Comments on above	
Monitoring	Not available
Allocations	
Enforcement	Not available
Unequal power relationship	No
Information sharing	Not available
Conflict resolution	Not available
Method for water division	Not available
Negotiations	
Additional comments	Provided conditional approval for future hydropower exploits. Also provided water rights to inhabitants on either side of the river even in the parts of the river entirely controlled by France

Treaty title	Convention relating to the creation of the Gambia River Basin Development Organization
Basins involved	Gambia
Main basin	Gambia
Date signed	06/30/1978
Signatories	Multilateral
Parties	Gambia, Senegal, Guinea
Principal focus	Industrial uses
Non-water linkages	None
Comments on above	
Monitoring	Yes
Allocations	
Enforcement	Not available
Unequal power relationship	Not available
Information sharing	Yes
Conflict resolution	Council
Method for water division	Not available
Negotiations	
Additional comments	Created several levels of organizations relating to the Gambia river: Council of Ministers, The High Commission, and the Permanent Water Commission

GANGES BASIN

Treaty title	Agreement between the British government and the Patiala state regarding the Sirsa branch of the Western Jumna canal
Basins involved	Ganges
Main basin	Ganges
Date signed	08/29/1893
Signatories	Bilateral
Parties	Great Britain, India (Patiala)
Principal focus	Water supply
Non-water linkages	Money
Comments on above	Compensation for dam-flooded structures
Monitoring	Yes
Allocations	British government has sole control over the water supply in the canal "shares of water supply [are determined based on] the proportions of the areas commanded by the entire system in British and Patiala territory, respectively"
Enforcement	Not available
Unequal power relationship	Yes
Information sharing	Yes
Conflict resolution	None
Method for water division	Complex but clear
Negotiations	Negotiations with the British Empire again seem one-sided. Typical imperial-colonial relations
Additional comments	British had almost total control of the situation. Irrigation was provided by the British for the Patiala peoples
Treaty title	Agreement between the government of India and the government of Nepal on the Kosi project
Basins involved	Kosi
Main basin	Ganges
Date signed	04/25/1954
Signatories	Bilateral
Parties	India, Nepal
Principal focus	Hydropower
Non-water linkages	Money
Comments on above	Compensation for flooded lands, divided into 4 categories: Cultivated Lands, Forest, Village (immovable property), and Waste Land
Monitoring	Yes
Allocations	50% of hydropower generated goes to each party. India can regulate all the water supply, "without prejudice to the right of [Nepal] to withdraw for irrigation or any other purpose in Nepal such sup-

plies of water, as may be required from time to time ..."

Enforcement	None
Unequal power relationship	Yes
Information sharing	Yes
Conflict resolution	UN/third party
Method for water division	Complex/clear
Negotiations	
Additional comments	

Treaty title	Agreement between [Nepal] and [India] on the Gandak irrigation and power project
Basins involved	Gandak, Bagmati
Main basin	Ganges
Date signed	12/04/1959
Signatories	Bilateral
Parties	India, Nepal
Principal focus	Hydropower
Non-water linkages	Other linkages
Comments on above	Hydropower facilities, small irrigation canals (India will contribute to the cost of construction of canals smaller than 20 cusecs)
Monitoring	Yes
Allocations	Enough water to irrigate 40,000 acres (20 cusecs minimum) goes to Nepal. Similar amount goes to India, except the water will irrigate 103,500 acres. India has minimum 5,000 kW usage of the 15,000 kW plant, but India may charge an agreed-upon rate for the electricity.
Enforcement	Not available
Unequal power relationship	Yes
Information sharing	Yes
Conflict resolution	UN/third party
Method for water division	Complex/clear
Negotiations	
Additional comments	Nepal will continue to have the right to withdraw for irrigation "such supplies of water, as may be required from time to time ..." Note that this wording is verbatim from the 1954 agreement. Shortages will be pro-rated across both countries

Treaty title	Revised agreement between [Nepal] and [India] on the Kosi project
Basins involved	Kosi
Main basin	Ganges
Date signed	12/19/1966

Signatories	Bilateral
Parties	Nepal, India
Principal focus	Hydropower
Non-water linkages	Land
Comments on above	Land required for foreseen and unforeseen needs will be compensated by India. Foreseen requirements will be leased for 199 years
Monitoring	Yes
Allocations	Nepal "shall have every right to withdraw for irrigation and for any other purpose in Nepal water from the Kosi river and from [the basin] as may be required from time to time." India regulates the supplies in the river to produce power
Enforcement	Not available
Unequal power relationship	Yes
Information sharing	Yes
Conflict resolution	Other government agency
Method for water division	Unclear
Negotiations	
Additional comments	Stone, gravel, ballast, and timber compensated by India to Nepal. India pays compensation to Nepal for the loss of the land and immovable property flooded by the project

Treaty title	Statute of the Indo-Bangladesh Joint Rivers Commission
Basins involved	Ganges-Brahmaputra
Main basin	Ganges
Date signed	11/24/1972
Signatories	Bilateral
Parties	India, Bangladesh
Principal focus	Water supply
Non-water linkages	None
Comments on above	
Monitoring	Not available
Allocations	
Enforcement	Not available
Unequal power relationship	Yes
Information sharing	Not available
Conflict resolution	Not available
Method for water division	None
Negotiations	
Additional comments	Created a commission to "maintain liaison between the participating countries in order to ensure the most effective joint efforts in maximising the benefits from common river systems to both the countries," among other flood control responsibilities

Treaty title	Agreement between [Bangladesh] and [India] on sharing of the Ganges waters at Farakka and on augmenting its flows
Basins involved	Ganges
Main basin	Ganges
Date signed	11/05/1977
Signatories	Bilateral
Parties	India, Bangladesh
Principal focus	Water supply
Non-water linkages	None
Comments on above	
Monitoring	Yes
Allocations	75% of the average historic flows will be allocated; these Q-values are in 10-day increments from 1948–1973. Each 10-day period has its own allocations. India receives approx. 40% of that 75%
Enforcement	None
Unequal power relationship	Yes
Information sharing	Yes
Conflict resolution	Other government agency
Method for water division	Complex/clear
Negotiations	Treaty is to last for five years. No one has indicated if India held strictly to the agreement or not. There are provisions that essentially exclude third party negotiators.
Additional comments	If the flow dropped below 80% of the expected flow, Bangladesh will never receive less than 80% of that portion. If the flow exceeds the 75%, "the water shall be shared in proportion." India has the option to use 200 cusecs below the barrage but no more
Treaty title	Agreement between [Nepal] and [India] on the renovation and extension of Chandra canal, pumped canal, and distribution of the Western Kosi canal
Basins involved	Kosi
Main basin	Ganges
Date signed	04/07/1978
Signatories	Bilateral
Parties	Nepal, India
Principal focus	Water supply
Non-water linkages	Money
Comments on above	Money for the repairs and renovations provided by India in part. Nepal provides in-kind labour, surveying, and other efforts
Monitoring	Yes

Allocations	Nepal: 300 cusecs (in addition to the 64 already allocated). Nepal will acquire land beyond a certain point for use in the new distribution centre
Enforcement	Not available
Unequal power relationship	Yes
Information sharing	Yes
Conflict resolution	None
Method for water division	Complex/clear
Negotiations	
Additional comments	This treaty is for maintenance and new construction. The Chandra canal will be restored (by removing earth from the channel) to its previous 11 cumec capacity. Repairs to the headworks will also take place

Treaty title	Meeting of the Joint Rivers Commission
Basins involved	Ganges
Main basin	Ganges
Date signed	07/20/1983
Signatories	Bilateral
Parties	India, Bangladesh
Principal focus	Water supply
Non-water linkages	None
Comments on above	
Monitoring	Yes
Allocations	India: 39%. Bangladesh: 36%. Unallocated: 25% (and it was to remain unallocated)
Enforcement	None
Unequal power relationship	Yes
Information sharing	Yes
Conflict resolution	None
Method for water division	Complex but clear
Negotiations	The India-Bangladesh JRC is to investigate and study schemes to augment dry-season flow of the Ganges in an economic, feasible manner within three years
Additional comments	The treaty was to last only 18 months. Flood forecasting and warning arrangements were also discussed

Treaty title	Treaty between [India] and [Bangladesh] on sharing of the Ganga/Ganges waters at Farakka
Basins involved	Ganges
Main basin	Ganges
Date signed	12/12/1996
Signatories	Bilateral

Parties	India, Bangladesh
Principal focus	Water supply
Non-water linkages	None
Comments on above	
Monitoring	Yes
Allocations	
Enforcement	None
Unequal power relationship	Yes
Information sharing	Yes
Conflict resolution	Not available
Method for water division	Complex/clear
Negotiations	
Additional comments	Treaty signed on the 25th anniversary of the 1971 war in Pakistan that resulted in the creation of Bangladesh

GASH BASIN

Treaty title	Notes exchanged ... respecting the regulation of the utilization of the waters of the river Gash
Basins involved	Gash
Main basin	Gash
Date signed	06/15/1925
Signatories	Bilateral
Parties	Great Britain, Italy
Principal focus	Water supply
Non-water linkages	Money
Comments on above	Sudan would pay Eritrea a share of the income from lands irrigated by the Gash and in the Gash delta
Monitoring	Not available
Allocations	Eritrea would have use of 65 MCM. Eritrea could use half of the flow up to 17 CM/sec each, the excess going entirely to Kassala province
Enforcement	Not available
Unequal power relationship	Not available
Information sharing	Not available
Conflict resolution	Not available
Method for water division	Complex but clear
Negotiations	
Additional comments	Some water required to be left in the channel for downstream riparians. Eritrea was to receive 20% of all income over £50,000 from the Gash lands. Eritrea received a maximum of 65 MCM

Treaty title	Letters between the irrigation adviser and director of irrigation, Sudan government, and the controller of agriculture, Eritrea
Basins involved	Gash
Main basin	Gash
Date signed	04/18/1951
Signatories	Bilateral
Parties	Sudan, Eritrea
Principal focus	Water supply
Non-water linkages	None
Comments on above	Not known if this supersedes money due by previous agreement
Monitoring	Not available
Allocations	
Enforcement	Not available
Unequal power relationship	Not available
Information sharing	Not available
Conflict resolution	Not available
Method for water division	Complex but clear
Negotiations	
Additional comments	Reaffirmed previous irrigation amounts; Eritrea received a maximum of 65 MCM as before; now, the agreement was signed as independent nations

GREAT LAKES BASIN

Treaty title	Treaty between Great Britain and the United States relating to boundary waters and boundary questions
Basins involved	Great Lakes, Columbia, Niagara
Main basin	Great Lakes
Date signed	05/05/1910
Signatories	Bilateral
Parties	USA, Great Britain (Canada)
Principal focus	Water supply
Non-water linkages	None
Monitoring	Yes
Allocations	US may divert water above Niagara Falls (hydropower only) up to 20,000 cfs. UK (Canada) may divert (hydropower only) up to 36,000 cfs. Diversions set so the level of Lake Erie would not be affected. Both sides agree not to affect the natural flow of boundary rivers
Enforcement	None
Unequal power relationship	No

Information sharing	Yes
Conflict resolution	Council
Method for water division	Complex/clear
Negotiations	Disagreements that the Commission is unable to resolve are referred to an umpire, as prescribed by the Hague Convention (18 October 1907)
Additional comments	Smaller rivers also had water allocated for irrigation: St Mary and Milk rivers and their tributaries (in Montana, Alberta, and Saskatchewan) are treated as a single river. During the irrigation season (between 4/1 and 10/31), the USA gets prior appropriations of 500 cfs on the Milk River, or 75% of natural flow at that time, and Canada gets a prior appropriation of 500 cfs, or 75% of the natural flow at that time, from the St Mary river
Treaty title	Agreement between the United States of America and Canada to regulate the level of Lake of the Woods
Basins involved	Great Lakes, Rainy river
Main basin	Great Lakes
Date signed	02/24/1925
Signatories	Bilateral
Parties	Great Britain (Canada), United States
Principal focus	Flood control
Non-water linkages	Money
Comments on above	Canada paid the US $275,000 for protective works and measures necessary to regulate the lake levels. Additional costs will be split equally
Monitoring	Yes
Enforcement	Council
Unequal power relationship	No
Information sharing	Yes
Conflict resolution	Council
Method for water division	Not available
Negotiations	
Additional comments	
Treaty title	Exchange of notes between ... the United States of America and ... Canada constituting an Agreement regarding the development of certain portions ...
Basins involved	St Lawrence
Main basin	Great Lakes
Date signed	11/07/1940
Signatories	Bilateral
Parties	United States, Canada

Principal focus	Hydropower
Non-water linkages	Money
Comments on above	$1,000,000 for "preliminary engineering and other investigations" paid by the United States
Monitoring	Yes
Allocations	An additional 5,000 cfs allocated for hydropower to Canada
Enforcement	None
Unequal power relationship	No
Information sharing	Yes
Conflict resolution	None
Method for water division	Complex/clear
Negotiations	
Additional comments	
Treaty title	Exchange of notes between the government of the United States and the government of Canada ... concerning temporary diversion for power ...
Basins involved	Niagara
Main basin	Great Lakes
Date signed	05/20/1941
Signatories	Bilateral
Parties	USA, Canada
Principal focus	Hydropower
Non-water linkages	None
Comments on above	
Monitoring	Not available
Allocations	An additional 5,000 cfs for use in Canada to augment its war effort, plus 3,000 cfs offered by the US
Enforcement	Not available
Unequal power relationship	No
Information sharing	Not available
Conflict resolution	None
Method for water division	Complex/clear
Negotiations	
Additional comments	
Treaty title	Exchange of notes constituting an agreement between the government of the United States and the government of Canada relating to additional ...
Basins involved	Niagara
Main basin	Great Lakes
Date signed	11/27/1941
Signatories	Bilateral
Parties	USA, Canada
Principal focus	Hydropower
Non-water linkages	None

Comments on above	
Monitoring	Not available
Allocations	Canada receives an additional 6,000 cfs for hydropower generation and the US receives an additional 7,500 cfs for hydropower
Enforcement	Not available
Unequal power relationship	No
Information sharing	Not available
Conflict resolution	None
Method for water division	Complex but clear
Negotiations	
Additional comments	
Treaty title	Treaty between the United States of America and Canada relating to the uses of the waters of the Niagara river
Basins involved	Niagara
Main basin	Great Lakes
Date signed	02/27/1950
Signatories	Bilateral
Parties	United States, Canada
Principal focus	Hydropower
Non-water linkages	Money
Comments on above	Each state agrees to bear 50% of the costs of construction
Monitoring	Not available
Allocations	Between certain dates and times, Niagara Falls is allocated 100,000 cfs and at others, it is allocated 50,000 cfs. Power generated is split 50% apiece
Enforcement	None
Unequal power relationship	No
Information sharing	Yes
Conflict resolution	Council
Method for water division	None
Negotiations	
Additional comments	
Treaty title	Exchange of notes constituting an agreement between Canada and the United States of America relating to the St Lawrence Seaway project
Basins involved	St Lawrence
Main basin	Great Lakes
Date signed	06/30/1952
Signatories	Bilateral
Parties	USA, Canada
Principal focus	Hydropower
Non-water linkages	Money

Comments on above	Canada agrees to contribute $15 million towards the cost of channel enlargement
Monitoring	Not available
Allocations	
Enforcement	Not available
Unequal power relationship	No
Information sharing	Yes
Conflict resolution	None
Method for water division	None
Negotiations	
Additional comments	Both parties agree to equally split the costs of all development

Treaty title	Exchange of notes constituting an agreement between the United States and Canada relating to the establishment of the St Lawrence River joint ...
Basins involved	St Lawrence
Main basin	Great Lakes
Date signed	11/12/1953
Signatories	Bilateral
Parties	USA, Canada
Principal focus	Hydropower
Non-water linkages	None
Comments on above	
Monitoring	Not available
Allocations	
Enforcement	Not available
Unequal power relationship	No
Information sharing	Not available
Conflict resolution	Not available
Method for water division	None
Negotiations	Created a joint board of engineers to oversee/assist in the construction of works agreed to in the 29 October 1952 treaty
Additional comments	

Treaty title	Exchange of notes constituting an agreement for the construction of a temporary cofferdam at Niagara
Basins involved	Niagara
Main basin	Great Lakes
Date signed	03/21/1969
Signatories	Bilateral
Parties	USA, Canada
Principal focus	Hydropower
Non-water linkages	Other linkages

Comments on above	Costs of on-site data collection and dam construction will be borne based on previous agreements
Monitoring	Yes
Allocations	
Enforcement	None
Unequal power relationship	No
Information sharing	Yes
Conflict resolution	None
Method for water division	Complex but clear
Negotiations	
Additional comments	

Treaty title	Exchange of notes constituting an agreement between Canada and the United States of America for the temporary diversion for power purposes …
Basins involved	Niagara
Main basin	Great Lakes
Date signed	03/21/1969
Signatories	Bilateral
Parties	USA, Canada
Principal focus	Hydropower
Non-water linkages	Other linkages
Comments on above	Equal share in additional hydropower generation
Monitoring	Yes
Allocations	While the temporary cofferdam is in place, between 8,000 and 9,000 cfs will be diverted for additional power generation
Enforcement	None
Unequal power relationship	No
Information sharing	Yes
Conflict resolution	None
Method for water division	Equal parts
Negotiations	
Additional comments	Each power authority must contribute $385,500 in national currency for the construction of the cofferdam

GROUNDWATER BASIN

Treaty title	Convention regarding the water supply of Aden between Great Britain and the Sultan of Abdali
Basins involved	
Main basin	Groundwater
Date signed	04/11/1910
Signatories	Bilateral

Parties	Great Britain, Aden (Yemen)
Principal focus	Water supply
Non-water linkages	Money
Comments on above	3,000 rupees paid per month for upkeep and "rent" of the land
Monitoring	Not available
Allocations	
Enforcement	Not available
Unequal power relationship	Yes
Information sharing	Not available
Conflict resolution	None
Method for water division	None
Negotiations	
Additional comments	Wells to be dug at a suitable site and carried by canal for use in perpetuity by the British. One of the early groundwater agreements

GUADIANA BASIN

Treaty title	Convenio y Protocola Adicional Para Regular el Uso y aprovechamiento hidraulico de los tramos internacionales de los rios Mi-o, Limia, Tajo ...
Basins involved	Mi-o, Guadiana
Main basin	Guadiana
Date signed	05/29/1968
Signatories	Bilateral
Parties	Spain, Portugal
Principal focus	Hydropower
Non-water linkages	None
Comments on above	Construction authority is given to both parties on either river bank for cooperative efforts
Monitoring	Yes
Allocations	
Enforcement	Council
Unequal power relationship	
Information sharing	Yes
Conflict resolution	Council
Method for water division	Complex but clear
Negotiations	
Additional comments	A second monitoring body exists separate from the Commission

Treaty title	Segundo protocolo
Basins involved	Mi-o
Main basin	Guadina

Date signed	02/12/1976
Signatories	Bilateral
Parties	Spain and Portugal
Principal focus	Hydropower
Non-water linkages	None
Comments on above	Both countries will utilize companies from each country where possible for construction on the river
Monitoring	Yes
Allocations	
Enforcement	Council
Unequal power relationship	Not available
Information sharing	Yes
Conflict resolution	Council
Method for water division	None
Negotiations	
Additional comments	This treaty is an addendum to the 1968 treaty regarding only the Rio Mi-o, and not signed at the same time

HELMAND BASIN

Treaty title	Terms of reference of the Helmand River Delta Commission and an interpretive statement relative thereto, agreed by conferees of ...
Basins involved	Helmand
Main basin	Helmand
Date signed	09/07/1950
Signatories	Bilateral
Parties	Afghanistan, Iran
Principal focus	Water supply
Non-water linkages	None
Comments on above	
Monitoring	Yes
Allocations	To be determined by a third-party commission
Enforcement	None
Unequal power relationship	Not available
Information sharing	Yes
Conflict resolution	Not available
Method for water division	Unclear
Negotiations	
Additional comments	The Helmand River Delta Commission was created and given the task to measure and divide the river flows between the two signatories

INDUS BASIN

Treaty title	Articles of agreement between the Edur Durbar and the British government
Basins involved	Hathmatee
Main basin	Indus
Date signed	07/20/1874
Signatories	Bilateral
Parties	Great Britain, India (Edur)
Principal focus	Water supply
Non-water linkages	Other linkages
Comments on above	Double boat for passage through flooded areas provided by the British; some small bits of land
Monitoring	Yes
Allocations	Edur gets one half of the allocated water, and the British get one half for irrigation of their land as well
Unequal power relationship	Yes
Information sharing	Yes
Conflict resolution	Other government agency
Method for water division	Equal parts
Negotiations	Maharajah of Edur agreed to the construction of a weir. The British agreed to leave an outlet for water supply on one side. Maharajah suggested another site but it was found unsuitable
Additional comments	If any houses were flooded, the British would make up the costs according to an estimate by the project engineer
Treaty title	Amended terms of agreement between the British government and the State of Jind, for regulating the supply of water for irrigation ...
Basins involved	Indus
Main basin	Indus
Date signed	09/16/1892
Signatories	Bilateral
Parties	India (Jind), Great Britain
Principal focus	Water supply
Non-water linkages	Money
Comments on above	"Payment" for water/delivery
Monitoring	Not available
Allocations	Jind received water sufficient to irrigate 50,000 acres but the flow had the capacity to irrigate 60,000 acres. No storage capacity
Enforcement	Force/threat of force
Unequal power relationship	Yes

Information sharing	Not available
Conflict resolution	Not available
Method for water division	Complex but clear
Negotiations	Almost everything was done by executive order
Additional comments	The payments for the water were established by computed costs to irrigate similar states/areas in the British Empire. Cost was approximately Rs105,500 in 1892

Treaty title	Final working agreement relative to the Sirhind canal between Great Britain and Patiala, Jind and Nabha
Basins involved	Indus (Sirhind Canal)
Main basin	Indus
Date signed	02/23/1904
Signatories	Multilateral
Parties	Great Britain, India (Patiala, Jind, Nabha)
Principal focus	Water supply
Non-water linkages	Money
Comments on above	Operating expenses for the supply reservoir, compensation for accidental damage, if it occurs
Monitoring	Yes
Allocations	Patiala receives 82.6% of the water. Nabha receives 8.8%. Jind receives 7.6%. British villages receive water sufficient to irrigate the same proportion of its lands as of other villages nearby
Enforcement	None
Unequal power relationship	Yes
Information sharing	Yes
Conflict resolution	Other government agency
Method for water division	Complex but clear
Negotiations	
Additional comments	If the whole flow allocations cannot be met, the engineer may reduce flows proportionally, or he may deliver full proportion to one, then shut it off entirely while the others receive their full allotments

Treaty title	Inter-dominion agreement between the government of India and the government of Pakistan, on the canal water dispute between ...
Basins involved	Indus
Main basin	Indus
Date signed	05/04/1948
Signatories	Bilateral
Parties	India, Pakistan

Principal focus	Water supply
Non-water linkages	Money
Comments on above	Seignorage fees paid to India
Monitoring	Not available
Allocations	India was to reduce flow from upper Indus basin rivers progressively, to allow Pakistan to "develop areas where water is scarce and which were under-developed in relation to Parts of West Punjab"
Enforcement	None
Unequal power relationship	Yes
Information sharing	Yes
Conflict resolution	None
Method for water division	Unclear
Negotiations	
Additional comments	Also called the "Simla Agreement"
Treaty title	Indo-Pakistan agreement (with appendices) on East Pakistan border disputes
Basins involved	Indus
Main basin	Indus
Date signed	10/23/1959
Signatories	Bilateral
Parties	India, Pakistan
Principal focus	Water supply
Non-water linkages	Land
Comments on above	India grants some land for the Karnafuli dam
Monitoring	No
Allocations	India's land was flooded in exchange for "claims ... regarding the loss, if any, caused by the flooding ... should be settled
Enforcement	None
Unequal power relationship	Yes
Information sharing	None
Conflict resolution	Council
Method for water division	None
Negotiations	Each country agreed not to train border rivers to cut into the territory of the other. Outrageous reports to the press were also agreed to stop
Additional comments	The treaty created borders along rivers. Appendices set up "ground rules" for settlement of border disputes incurred in the field by military personnel. "Owing to a variety of reasons there have been occasional [gunshots] across the border." Recognizing the presence of hostility created a means with which to defuse border hostilities. Meetings between District Magistrates will take place in the second week of the month

Treaty title	Agreement between Pakistan and India on West Pakistan-India border disputes
Basins involved	Indus
Main basin	Indus
Date signed	01/11/1960
Signatories	Bilateral
Parties	Pakistan, India
Principal focus	Water supply
Non-water linkages	Land
Comments on above	Some border disputes were resolved by this document; each side relinquished claim to a section of disputed territory, although not all of it
Monitoring	Yes
Enforcement	Not available
Unequal power relationship	Yes
Information sharing	Yes
Conflict resolution	Other government agency
Method for water division	None
Negotiations	Pakistan relinquished claim to Chak Ladheke and India gave up their claim to three villages. The boundary at the Hussainiwala headworks was set firmly, and the boundary at the Suleimanke headworks was also agreed upon. A fifth boundary dispute was not resolved
Additional comments	The borders are ruled not to change in the event the river changes course. That is, in spots, the river could be entirely in one state or another. Regular meetings at the border are also provided for
Treaty title	Indus Waters Treaty
Basins involved	Indus
Main basin	Indus
Date signed	09/19/1960
Signatories	Bilateral
Parties	India, Pakistan
Principal focus	Water supply
Non-water linkages	Money
Comments on above	£62,060,000 as replacement costs of irrigation canals in regions formerly irrigating from Eastern rivers. Money paid to India if the 31 March 1970 expiration date is extended for up to three years.
Monitoring	Yes
Allocations	India: 100% of Eastern rivers, but some deliveries from those rivers will continue until 31 March 1970 or later if extended. Pakistan: 100% of Western rivers
Enforcement	Council

Unequal power relationship	Yes
Information sharing	Yes
Conflict resolution	Council, then a neutral third party
Method for water division	Complex but clear
Negotiations	Engineering plans were used first, then found lacking until political efforts could direct them. Third-party negotiators were necessary

JORDAN BASIN

Treaty title	Agreement between the Republic of Syria and the Hashemite Kingdom of Jordan concerning the utilization of the Yarmuk waters
Basins involved	Yarmuk
Main basin	Jordan
Date signed	06/04/1953
Signatories	Bilateral
Parties	Jordan, Syria
Principal focus	Water supply
Non-water linkages	Other linkages
Comments on above	Syria gets 75% of hydropower (not less than 3 MW mid-April to Mid-October)
Monitoring	Yes
Allocations	No less than 10 cm average shall flow from the dam, "for the irrigation of lands in Jordan and for other Jordanian schemes ..." "Syria shall retain [rights to use] all springs ... within its territory ... with the exception of waters welling up above the dam below the 250-metre level"
Enforcement	None
Unequal power relationship	Not available
Information sharing	Not available
Conflict resolution	Not available
Method for water division	Unclear
Negotiations	Syria bears 5% of the cost of the Maqarin installation and provides 20% of the workers
Additional comments	Jordan has the right to use the reservoir overflow and the generating station at Maqarin (and of course the Adasiya station, within Jordanian territory). Also, Jordan has "the right to use water superfluous to Syrian needs for its own purposes within Jordanian frontiers"
Treaty title	Johnston Negotiations
Basins involved	Jordan

Main basin	Jordan
Date signed	12/31/1955
Signatories	Multilateral (unsigned)
Parties	Israel, Jordan, Syria, Lebanon
Principal focus	Water supply
Non-water linkages	Money
Comments on above	US agreed to cost-share regional water projects if an agreement was reached
Monitoring	No
Allocations	Syria: 132 MC (10.3%). Jordan: 720 MCM (56%). Israel: 400 MCM (31.0%). Lebanon: 35 MCM. Based on area of irrigable land in each country
Enforcement	Economic
Unequal power relationship	Yes
Information sharing	None
Conflict resolution	UN/third party
Method for water division	Complex but clear
Negotiations	Johnston tried to separate resource issues from politics – and failed

Additional comments

Treaty title	Treaty of peace between [Israel] and [Jordan], done at Arava/Araba crossing point ...
Basins involved	Jordan, Yarmuk, Araba/Arava groundwater
Main basin	Jordan
Date signed	10/26/1994
Signatories	Bilateral
Parties	Israel, Jordan
Principal focus	Water supply
Non-water linkages	Other linkages
Comments on above	
Monitoring	Yes
Allocations	Yarmuk – Summer: Israel 12 MCM, Jordan gets the rest. Winter – Israel 13 MCM, Jordan gets the rest. Israel also takes 20 MCM, but will be returned later. Jordan – Summer: Israel maintains current use, equal to Jordan's. Winter: Jordan 20 MCM of the floods, both can pump flood excess to storage
Enforcement	Not available
Unequal power relationship	Not available
Information sharing	Yes
Conflict resolution	Council
Method for water division	Complex/clear
Negotiations	
Additional comments	Jordan also receives desalinated 10 MCM of approx. 20 MCM of saline springs. The two parties

will cooperate to find an additional 50 MCM of drinkable water. Israel can take 10 MCM over and above its current groundwater withdrawals, provided such withdrawals are hydrogeologically feasible and do not harm current Jordanian uses

Treaty title	Israeli-Palestinian interim agreement on the West Bank and the Gaza Strip
Basins involved	Jordan
Main basin	Jordan
Date signed	09/28/1995
Signatories	Bilateral
Parties	Israel, Palestine autonomy
Principal focus	Water supply
Non-water linkages	Money
Comments on above	Israel bears capital development costs for new water deliveries
Monitoring	Yes
Allocations	Israel recognizes Palestinian water rights
Enforcement	Council
Unequal power relationship	Not available
Information sharing	Yes
Conflict resolution	UN/third party
Method for water division	Complex/clear
Negotiations	From Israel, Palestinians get additional water: Hebron, 1 MCM; Ramallah, 0.5 MCM; Salfit, 0.6 MCM; Nablus, 1 MCM; Jenin, 1.4 MCM; Gaza, 5 MCM. Palestinians provide themselves with 2.1 MCM to Nablus; 17 MCM (Eastern aquifer) to Hebron, Bethlehem, Ramallah
Additional comments	Unsettling that they are looking for more engineering solutions to a badly overdrawn water budget: that 17 MCM cannot be sustainable

LAKE CHAD BASIN

Treaty title	Convention and statutes … Lake Chad Basin
Basins involved	Lake Chad
Main basin	Lake Chad
Date signed	05/22/1964
Signatories	Multilateral
Parties	Cameroon, Chad, Niger, Nigeria
Principal focus	Industrial uses
Non-water linkages	None

Comments on above	
Monitoring	Not available
Allocations	
Enforcement	Council
Unequal power relationship	No
Information sharing	Yes
Conflict resolution	Council
Method for water division	None
Negotiations	
Additional comments	It is a treaty primarily concerned with economic development inside the basin. The Commission prepares general regulations, coordinates the research activities of the four states, examines their development schemes, makes recommendations and maintains contact among the four states

LAKE TITICACA BASIN

Treaty title	Exchange of notes between Peru and Bolivia establishing a joint commission for study of ... joint use of the waters of Lake Titicaca
Basins involved	Lake Titicaca
Main basin	Lake Titicaca
Date signed	04/20/1955
Signatories	Bilateral
Parties	Peru, Bolivia
Principal focus	Hydropower
Non-water linkages	None
Comments on above	
Monitoring	Not available
Allocations	
Enforcement	Not available
Unequal power relationship	Not available
Information sharing	Yes
Conflict resolution	Not available
Method for water division	None
Negotiations	
Additional comments	Created a commission to examine economic opportunities
Treaty title	Agreement between Bolivia and Peru concerning a preliminary economic study of the joint utilization of the waters of Lake Titicaca
Basins involved	Lake Titicaca

Main basin	Lake Titicaca
Date signed	02/19/1957
Signatories	Bilateral
Parties	Peru, Bolivia
Principal focus	Hydropower
Non-water linkages	None
Comments on above	
Monitoring	No
Allocations	
Enforcement	Not available
Unequal power relationship	Not available
Information sharing	Yes
Conflict resolution	Not available
Method for water division	None
Negotiations	
Additional comments	"An estimate of the electricity consumption in both countries so that the construction of one or more hydroelectric stations capable of meeting the demand efficiently and equitably can be considered ..."

MARICA BASIN

Treaty title	Agreement between the People's Republic of Bulgaria and the Republic of Turkey concerning cooperation ...
Basins involved	Maritsa/Marica, Tundzha, Veleka, Rezovska
Main basin	Marica
Date signed	10/23/1968
Signatories	Bilateral
Parties	Turkey, Bulgaria
Principal focus	Water supply
Non-water linkages	None
Comments on above	
Monitoring	Yes
Allocations	
Enforcement	None
Unequal power relationship	Not available
Information sharing	Yes
Conflict resolution	Council
Method for water division	None
Negotiations	
Additional comments	Created a commission and a means to share data and cooperate on developing the common rivers

MEKONG BASIN

Treaty title	Convention between Laos and Thailand for the supply of power
Basins involved	Mekong, Nam Pong, Nam Ngum
Main basin	Mekong
Date signed	08/12/1965
Signatories	Bilateral
Parties	Laos, Thailand
Principal focus	Hydropower
Non-water linkages	Money
Comments on above	Payment for any power transferred
Monitoring	Yes
Allocations	
Enforcement	Not available
Unequal power relationship	Not available
Information sharing	Yes
Conflict resolution	Not available
Method for water division	Not available
Negotiations	
Additional comments	The two states agreed to interconnect their electric grids between two hydropower plants

Treaty title	Joint declaration of principles for utilization of the waters of the lower Mekong basin, signed by [Cambodia], [Laos], [Thailand], and [Vietnam]
Basins involved	Mekong
Main basin	Mekong
Date signed	01/31/1975
Signatories	Multilateral
Parties	Cambodia, Laos, Thailand, Vietnam
Principal focus	Industrial uses
Non-water linkages	None
Comments on above	
Monitoring	Not available
Allocations	
Enforcement	Not available
Unequal power relationship	Not available
Information sharing	Not available
Conflict resolution	Not available
Method for water division	Not available
Negotiations	
Additional comments	Agreement is only a statement of principles and reinforcement of the 1957 creation of the Mekong Committee

Treaty title	Agreement on the cooperation for the sustainable development of the Mekong river basin
Basins involved	Mekong
Main basin	Mekong
Date signed	04/05/1995
Signatories	Multilateral
Parties	Cambodia, Lao PDR, Thailand, Vietnam
Principal focus	Water supply
Non-water linkages	None
Comments on above	
Monitoring	Yes
Allocations	The treaty provides for the Joint Committee to "prepare and propose ... Rules for Water Utilization and Inter-Basin Diversions ..."
Enforcement	None
Unequal power relationship	Not available
Information sharing	Yes
Conflict resolution	Council
Method for water division	Not available
Negotiations	
Additional comments	

MEMEL BASIN

Treaty title	Convention between the German Reich and the Lithuanian Republic regarding the maintenance and administration of the frontier waterways ...
Basins involved	Memel, Kurische Haff
Main basin	Memel
Date signed	01/29/1928
Signatories	Bilateral
Parties	Germany, Lithuania
Principal focus	Flood control
Non-water linkages	None
Comments on above	The parties agreed to divide the costs of ice-breaking: 80% to Germany, 20% to Lithuania
Monitoring	Yes
Allocations	Germany is entitled to the use of Wystit lake for hydropower
Enforcement	Not available
Unequal power relationship	Not available
Information sharing	Yes
Conflict resolution	UN/third party
Method for water division	Not available
Negotiations	
Additional comments	

MEUSE BASIN

Treaty title	Exchange of notes constituting an agreement concerning the treaty of 12 May 1863 to regulate the diversion of water from the River Meuse and the ...
Basins involved	Meuse
Main basin	Meuse
Date signed	02/24/1961
Signatories	Bilateral
Parties	Netherlands, Belgium
Principal focus	Water supply
Non-water linkages	Money
Comments on above	Costs associated with dismantling lock 19 fall to the Netherlands
Monitoring	No
Allocations	
Enforcement	None
Unequal power relationship	No
Information sharing	Not available
Conflict resolution	None
Method for water division	Not available
Negotiations	
Additional comments	The treaty is more about rebuilding diversion works but apparently there is no problem with doing so

MIRIM LAGOON BASIN

Treaty title	Exchange of notes constituting an agreement ... for the development of the Mirim lagoon
Basins involved	Mirim lagoon
Main basin	Mirim lagoon
Date signed	04/26/1963
Signatories	Bilateral
Parties	Brazil, Uruguay
Principal focus	Navigation
Non-water linkages	None
Comments on above	
Monitoring	No
Allocations	
Enforcement	None
Unequal power relationship	Not available
Information sharing	None
Conflict resolution	Other government agency

Method for water division	None
Negotiations	
Additional comments	Development and navigation concerns were paramount

NAATAMO BASIN

Treaty title	Agreement between [Finland] and [Norway] on the transfer from the course of the Näätämo (Neiden) river to the course of the Gandvik river ...
Basins involved	Näätämo, Gandvik
Main basin	Näätämo
Date signed	04/25/1951
Signatories	Bilateral
Parties	Norway, Finland
Principal focus	Hydropower
Non-water linkages	Money
Comments on above	N Kr 15,000 paid to Finland as compensation for lost power generation
Monitoring	Not available
Allocations	Water diverted between basins for power generation in Norway, which agrees to compensate Finland for lost water power
Enforcement	Not available
Unequal power relationship	Not available
Information sharing	Not available
Conflict resolution	Not available
Method for water division	Complex/clear
Negotiations	
Additional comments	

NIGER BASIN

Treaty title	Act of Berlin
Basins involved	Niger
Main basin	Niger
Date signed	02/26/1885
Signatories	Bilateral
Parties	Great Britain, France
Principal focus	Navigation
Non-water linkages	None
Comments on above	
Monitoring	Not available
Allocations	No allocations

Enforcement	Not available
Unequal power relationship	No
Information sharing	Not available
Conflict resolution	Not available
Method for water division	None
Negotiations	Channel modifications and canals were provided for (with no funding, apparently) and "should be considered in their quality of means of communication" as part of the river and subject to the treaty's governing principles
Additional comments	This treaty laid the foundation for many future treaties, including the landmark treaty signed at Niamey, 1963/4

Treaty title	Convention of Barcelona
Basins involved	Niger
Main basin	Niger
Date signed	04/20/1921
Signatories	Multilateral
Parties	Great Britain, France, "among others" (riparians?)
Principal focus	Navigation
Non-water linkages	None
Comments on above	
Monitoring	Not available
Allocations	
Enforcement	Not available
Unequal power relationship	Not available
Information sharing	Not available
Conflict resolution	Not available
Method for water division	None
Negotiations	
Additional comments	Also provided for works undertaken for irrigation or hydropower, unless such works infringed on "vital interests"

Treaty title	Act ... states of the Niger basin
Basins involved	Niger
Main basin	Niger
Date signed	10/26/1963
Signatories	Multilateral
Parties	Cameroon, Chad, Dahomey, Guinea, Cote D'Ivoire, Mali, Niger, Nigeria, Upper Volta
Principal focus	Industrial uses
Non-water linkages	None
Comments on above	
Monitoring	Yes
Allocations	

Enforcement	Council
Unequal power relationship	No
Information sharing	Yes
Conflict resolution	Council
Method for water division	None
Negotiations	
Additional comments	

Treaty title	Agreement concerning the River Niger Commission and the navigation and transport on the River Niger
Basins involved	Niger
Main basin	Niger
Date signed	11/25/1964
Signatories	Multilateral
Parties	Benin, Cameroon, Chad, Cote D'Ivoire, Guinea, Mali, Niger, Nigeria, Upper Volta
Principal focus	Industrial uses
Non-water linkages	None
Comments on above	
Monitoring	Not available
Allocations	
Enforcement	Council
Unequal power relationship	No
Information sharing	Not available
Conflict resolution	Council
Method for water division	None
Negotiations	
Additional comments	

Treaty title	Convention creating the Niger Basin Authority
Basins involved	Niger
Main basin	Niger
Date signed	11/21/1980
Signatories	Multilateral
Parties	Benin, Cameroon, Chad, Cote D'Ivoire, Guinea, Mali, Niger, Nigeria, Upper Volta
Principal focus	Industrial uses
Non-water linkages	None
Comments on above	
Monitoring	Yes
Allocations	
Enforcement	Not available
Unequal power relationship	Not available
Information sharing	Yes
Conflict resolution	Council
Method for water division	None

Negotiations
Additional comments   Changed the River Niger Commission to the Niger Basin Authority. Created a Council of Ministers, Technical Committee of Experts, and the Executive Secretariat. Treaty to last for 10 years

NILE BASIN

Treaty title   Protocol between Great Britain and Italy for the demarcation of their respective spheres ...
Basins involved   Nile
Main basin   Nile
Date signed   04/15/1891
Signatories   Bilateral
Parties   Great Britain, Italy
Principal focus   Water supply
Non-water linkages   None
Comments on above
Monitoring   Not available
Allocations   Obviously, nearly all of the water goes to Britain (Egypt) since Italy agreed to not construct any significant diversions
Enforcement   Not available
Unequal power relationship   Not available
Information sharing   Not available
Conflict resolution   Not available
Method for water division   None
Negotiations   Italy, at this point in the century, was "given" Libya and Ethiopia for colonization as some concession by the major powers. This treaty reflects the strength of Britain's position, even as downstream riparian
Additional comments   Italy agreed not to construct any works on the Atbara that would affect its flow into the Nile (mainly pertained to irrigation works)

Treaty title   Exchange of notes between Great Britain and Ethiopia
Basins involved   Nile
Main basin   Nile
Date signed   03/18/1902
Signatories   Bilateral
Parties   Great Britain, Ethiopia
Principal focus   Water supply
Non-water linkages   None

Comments on above
Monitoring   Not available
Allocations   Great Britain receives all of the waters of the Blue
                                    Nile, unless they agree with Ethiopia on a given
                                    project
Enforcement   Not available
Unequal power relationship   Yes
Information sharing   Not available
Conflict resolution   Not available
Method for water division   None
Negotiations   Closely resembles UK-Italy treaty of 1891
Additional comments   Ethiopia agreed not to interfere with the flow of
                                    the Blue Nile (or lake Tsana) "except in consulta-
                                    tion with His Britannic Majesty's government and
                                    the government of the Sudan ..."

Treaty title   Agreement ... modifying the agreement signed at
                                    Brussels, 12 May 1894
Basins involved   Nile
Main basin   Nile
Date signed   05/09/1906
Signatories   Bilateral
Parties   Great Britain, Independent Congo
Principal focus   Water supply
Non-water linkages   None
Comments on above
Monitoring   Not available
Allocations   Britain gets 100% until it says otherwise
Enforcement   Not available
Unequal power relationship   Yes
Information sharing   Not available
Conflict resolution   UN/third party
Method for water division   None
Negotiations
Additional comments   Congo agreed to not construct any work (or allow
                                    any work to be constructed) which would diminish
                                    the flow into Lake Albert

Treaty title   Exchange of notes between Great Britain and Italy
Basins involved   Nile
Main basin   Nile
Date signed   12/20/1925
Signatories   Bilateral
Parties   Great Britain, Italy
Principal focus   Water supply
Non-water linkages   Political concessions
Comments on above   Italy given economic exclusivity in Ethiopia in ex-

	change for limiting river development so that Britain could build a barrage at Lake Tsana
Monitoring	Not available
Allocations	Italy relinquished all water rights except reasonable use, extending to small hydropower projects and reservoirs
Enforcement	Not available
Unequal power relationship	Not available
Information sharing	Not available
Conflict resolution	Not available
Method for water division	None
Negotiations	
Additional comments	Italy recognized prior hydraulic rights of Egypt and the Sudan

Treaty title	Exchange of notes ... in regard to the use of ... the river Nile for irrigation purposes
Basins involved	Nile
Main basin	Nile
Date signed	05/07/1929
Signatories	Bilateral
Parties	Great Britain, Egypt
Principal focus	Water supply
Non-water linkages	Other linkages
Comments on above	British agreed to lend technical support
Monitoring	Yes
Allocations	Egypt accepted the findings of the 1925 Nile commission restricting the amount of water impounded by Sudan except during the flood period
Enforcement	Not available
Unequal power relationship	Yes
Information sharing	Not available
Conflict resolution	UN/third party
Method for water division	Complex but clear
Negotiations	Egypt and Sudan would agree before any new construction took place to increase local water supply
Additional comments	

Treaty title	Agreement ... regarding water rights on the boundary between Tanganyika and Ruanda-Urundi
Basins involved	Nile
Main basin	Nile
Date signed	11/22/1934
Signatories	Bilateral
Parties	Great Britain, Belgium

Principal focus	Water supply
Non-water linkages	None
Comments on above	
Monitoring	Yes
Allocations	
Enforcement	Not available
Unequal power relationship	Not available
Information sharing	Not available
Conflict resolution	UN/third party
Method for water division	Equal parts
Negotiations	
Additional comments	Industrial/mine pollution also addressed strongly. Inhabitants of either territory "should be permitted to navigate any river or stream forming the common boundary and take therefrom fish and aquatic plants and water ... for any purposes conforming with their customary rights"

Treaty title	Exchanges of notes ... regarding the construction of the Owen Falls dam, Uganda
Basins involved	Nile
Main basin	Nile
Date signed	05/31/1949
Signatories	Bilateral
Parties	Great Britain, Egypt
Principal focus	Hydropower
Non-water linkages	None
Comments on above	
Monitoring	Not available
Allocations	
Enforcement	Not available
Unequal power relationship	Yes
Information sharing	Not available
Conflict resolution	UN/third party
Method for water division	None
Negotiations	
Additional comments	Provided that Uganda (although not a signatory) could build (or contract to build) a hydroelectric dam so long as the dam "did not adversely affect the discharges of water to be passed through the dam ..."

Treaty title	Exchange of notes constituting an agreement between [Great Britain] ... and [Egypt] regarding the construction of the Owen Falls dam, Uganda
Basins involved	Nile

Main basin	Nile
Date signed	12/05/1949
Signatories	Bilateral
Parties	Egypt, Great Britain (Uganda)
Principal focus	Hydropower
Non-water linkages	Money
Comments on above	Contract for building the dam amounted to £3,639,540 5s. Sluices contract is £124,866
Monitoring	Yes
Allocations	See 1929 Nile Waters Agreement
Enforcement	None
Unequal power relationship	Yes
Information sharing	Yes
Conflict resolution	Council
Method for water division	None
Negotiations	
Additional comments	

Treaty title	Exchange of notes constituting an agreement between [Great Britain] (on behalf of ... Uganda) and [Egypt] regarding cooperation in meteorological ...
Basins involved	Nile
Main basin	Nile
Date signed	01/19/1950
Signatories	Bilateral
Parties	Egypt, Great Britain (Uganda)
Principal focus	Hydropower
Non-water linkages	Money
Comments on above	£E4,200 to pay for meteorological and hydrologic data, maximum of £E4,500
Monitoring	Yes
Allocations	
Enforcement	Not available
Unequal power relationship	Not available
Information sharing	Yes
Conflict resolution	Not available
Method for water division	None
Negotiations	
Additional comments	

Treaty title	Exchange of notes constituting an agreement between the [UK/Uganda] and [Egypt] regarding the construction of the Owen Falls dam in Uganda
Basins involved	Nile
Main basin	Nile

Date signed	07/16/1952
Signatories	Bilateral
Parties	Egypt, Great Britain (Uganda)
Principal focus	Hydropower
Non-water linkages	Money
Comments on above	Egypt pays Uganda £980,000 (loss of hydroelectric power) and also flood compensation (later)
Monitoring	Not available
Allocations	
Enforcement	Not available
Unequal power relationship	Yes
Information sharing	Not available
Conflict resolution	None
Method for water division	Not available
Negotiations	
Additional comments	Lake Victoria was to be used for the storage of additional water but would reduce flow to the Owen Falls dam

Treaty title	Agreement between the government of the United Arab Republic and the government of Sudan
Basins involved	Nile
Main basin	Nile
Date signed	11/08/1959
Signatories	Bilateral
Parties	Sudan, Egypt
Principal focus	Water supply
Non-water linkages	Money
Comments on above	£E15 million would be paid by Egypt for inundated lands; the Aswan, Roseires dams to be built
Monitoring	Yes
Allocations	Egypt: 48 BCM. Sudan: 4 BCM. Agreement on the evaporative losses in Sudanese swamps would be reflected in equal shared costs. Net benefit of Sudd el Aali reservoir shared at a 14.5/7.5 ratio
Enforcement	Not available
Unequal power relationship	Not available
Information sharing	Yes
Conflict resolution	Council
Method for water division	Complex/clear
Negotiations	
Additional comments	Technical committees set up. An agreed-upon view would be shown to other riparian states. Flow reductions to other riparians would be shared equally. Projects to reduce evaporative losses in the swamps will be begun

ODER BASIN

Treaty title	Agreement between [Czechoslovakia] and [Poland] concerning the use of water resources in frontier waters
Basins involved	Oder
Main basin	Oder
Date signed	03/21/1958
Signatories	Bilateral
Parties	Czechoslovakia and Poland
Principal focus	Water supply
Non-water linkages	None
Comments on above	
Monitoring	Yes
Allocations	
Enforcement	Council
Unequal power relationship	No
Information sharing	Yes
Conflict resolution	None
Method for water division	None
Negotiations	"The Contracting parties shall come to an agreement on the amount of water to be taken from frontier waters for domestic, industrial, power generation, and agricultural requirements and on the discharge of waste water"
Additional comments	

PAATSJOKI BASIN

Treaty title	Convention between [Norway] and [Finland] concerning the international legal regime of the waters of the Pasvik (Paatsjoki) and the Jakobselv ...
Basins involved	Paatsjoki, Vuoremajoki
Main basin	Paatsjoki
Date signed	02/14/1925
Signatories	Bilateral
Parties	Finland, Norway
Principal focus	Water supply
Non-water linkages	None
Comments on above	
Monitoring	No
Allocations	Each party received half of the flows of the river, and all of the river flows where the party owned both banks of the river

Enforcement	None
Unequal power relationship	Not available
Information sharing	Yes
Conflict resolution	UN/third party
Method for water division	Equal parts
Negotiations	
Additional comments	

Treaty title	Treaty between the [USSR] and [Finland] on the transfer to the territory of the Soviet Union of part of the state territory of Finland in the region of …
Basins involved	Paatsjoki
Main basin	Paatsjoki
Date signed	02/03/1947
Signatories	Bilateral
Parties	USSR, Finland
Principal focus	Hydropower
Non-water linkages	Land
Comments on above	Finland cedes 176 km^2 near the Jäniskoski hydropower station and the Niskakoski control dam
Monitoring	Not available
Allocations	
Enforcement	Not available
Unequal power relationship	Yes
Information sharing	Not available
Conflict resolution	Not available
Method for water division	Not available
Negotiations	
Additional comments	

Treaty title	Convention between [Finland] and the [USSR] concerning the maintenance of river channels and the regulation of fishing on water courses …
Basins involved	Multiple rivers
Main basin	None
Date signed	10/28/1922
Signatories	Bilateral
Parties	USSR, Finland
Principal focus	Fishing
Non-water linkages	None
Comments on above	
Monitoring	Not available
Allocations	
Enforcement	Not available
Unequal power relationship	Not available

Information sharing	Yes
Conflict resolution	Not available
Method for water division	None
Negotiations	
Additional comments	

Treaty title	Agreement between the [USSR], [Norway], and [Finland] concerning the regulation of Lake Inari by means of the Kaiakoski ... dam
Basins involved	Paatsjoki
Main basin	Paatsjoki
Date signed	04/29/1959
Signatories	Multilateral
Parties	USSR, Finland, Norway
Principal focus	Hydropower
Non-water linkages	Money
Comments on above	USSR paid Finland 75,000,000 Finnish markaa for damages associated with Lake Inari
Monitoring	Yes
Allocations	Daily discharge of the reservoir can range from 80 to 240 cm. If floods threaten to overtop the dam, discharge may rise to 500 cm. If the reservoir falls below 115.83 msl, discharge may fall to 45 cm
Enforcement	None
Unequal power relationship	Yes
Information sharing	Yes
Conflict resolution	Council
Method for water division	None
Negotiations	Finland agreed not to undertake (or authorize another to undertake) "any measures likely to affect the regime of Lake Inari or the river Paatsjoki"
Additional comments	

PARANA BASIN

Treaty title	Agreement concerning cooperation between [Brazil] and [Paraguay] in a study on the utilization of the water power of the Acaray and Monday ...
Basins involved	Acaray, Monday
Main basin	Paraná
Date signed	01/20/1956
Signatories	Bilateral
Parties	Brazil, Paraguay

Principal focus	Hydropower
Non-water linkages	None
Comments on above	
Monitoring	Yes
Allocations	Brazil will have the right to purchase 20% of the power from the generating stations
Enforcement	Not available
Unequal power relationship	Yes
Information sharing	Yes
Conflict resolution	Not available
Method for water division	Not available
Negotiations	
Additional comments	
Treaty title	Agreement between [Argentina] and [Paraguay] concerning a study of the utilization of the water power of the Apipe Falls
Basins involved	Paraná
Main basin	Paraná
Date signed	01/23/1958
Signatories	Bilateral
Parties	Argentina, Paraguay
Principal focus	Hydropower
Non-water linkages	None
Comments on above	
Monitoring	Yes
Allocations	
Enforcement	Not available
Unequal power relationship	Yes
Information sharing	Yes
Conflict resolution	None
Method for water division	Not available
Negotiations	
Additional comments	Established a Joint Argentine-Paraguayan Technical Commission to make a survey of hydroelectric potential. Later the two parties were to pay equal shares of the cost of construction
Treaty title	Treaty between [Brazil] and [Paraguay] concerning the hydroelectric utilization of the water resources of the Paraná river ...
Basins involved	Paraná, Iguassu
Main basin	Paraná
Date signed	04/26/1973
Signatories	Bilateral
Parties	Brazil, Paraguay
Principal focus	Hydropower

Non-water linkages	Money
Comments on above	Payments for use of hydroelectric potential; also infrastructure-building
Monitoring	Not available
Allocations	
Enforcement	Council
Unequal power relationship	Not available
Information sharing	Yes
Conflict resolution	None
Method for water division	Unclear
Negotiations	
Additional comments	

Treaty title	Agreement on Paraná river projects
Basins involved	Paraná
Main basin	Paraná
Date signed	10/19/1979
Signatories	Multilateral
Parties	Argentina, Brazil, Paraguay
Principal focus	Hydropower
Non-water linkages	None
Comments on above	
Monitoring	Yes
Allocations	
Enforcement	None
Unequal power relationship	Yes
Information sharing	Yes
Conflict resolution	None
Method for water division	None
Negotiations	
Additional comments	Itaipu dam project agreed to and technical cooperation established

PASVIK BASIN

Treaty title	Agreement between Norway and the Union of Soviet Socialist Republics on the utilization of water power on the Pasvik (Paatso) river
Basins involved	Pasvik
Main basin	Pasvik
Date signed	12/18/1957
Signatories	Bilateral
Parties	USSR, Norway
Principal focus	Hydropower

Non-water linkages	Money
Comments on above	USSR pays Norway NKr1 million for "unavoidable damage caused ... in connexion with construction ...
Monitoring	Yes
Allocations	Apportions water between the river mouth to the 70.32 m contour for use in hydropower. The USSR may use waters from 0 to 21 m and from 51.87 m (Fjaer lake) to 70.32 m (where the river intersects the Soviet-Norwegian border).
Enforcement	Not available
Unequal power relationship	Yes
Information sharing	Yes
Conflict resolution	None
Method for water division	Complex but clear
Negotiations	
Additional comments	Russia also made available (but did not cede) territory for operation of the hydropower plant, totalling 6.7 ha

PILCOMAYO BASIN

Treaty title	Supplementary boundary treaty between [Argentina] and [Paraguay] on the river Pilcomayo
Basins involved	Pilcomayo
Main basin	Pilcomayo
Date signed	06/01/1945
Signatories	Bilateral
Parties	Argentina, Paraguay
Principal focus	Water supply
Non-water linkages	None
Comments on above	
Monitoring	Yes
Allocations	
Enforcement	Not available
Unequal power relationship	Yes
Information sharing	Yes
Conflict resolution	Not available
Method for water division	Not available
Negotiations	
Additional comments	Created a joint technical commission that proposed entrainment and storage works for the Pilcomayo river. Also, reservoirs and canals could firmly establish the border between the two countries

RHINE BASIN

Treaty title	State Treaty concerning the construction of a hydroelectric power-plant on the Sauer at Rosport/Ralingen
Basins involved	Rhine
Main basin	Rhine
Date signed	04/25/1950
Signatories	Bilateral
Parties	Luxembourg, Germany (FRG)
Principal focus	Hydropower
Non-water linkages	None
Comments on above	
Monitoring	Not available
Allocations	Luxembourg owns 100% of power produced at the dam, and "the removal of water on the German side of the river above the dam shall be permitted only if an equivalent quantity of water is introduced above the said dam"
Enforcement	Not available
Unequal power relationship	Yes
Information sharing	Yes
Conflict resolution	Council
Method for water division	Complex but clear
Negotiations	
Additional comments	

Treaty title	State treaty between [Luxembourg] and [West Germany] concerning the construction of hydroelectric power-installations on the Our
Basins involved	Our
Main basin	Rhine
Date signed	07/10/1958
Signatories	Bilateral
Parties	Luxembourg, Germany (FRG)
Principal focus	Hydropower
Non-water linkages	None
Comments on above	
Monitoring	Yes
Allocations	
Enforcement	Not available
Unequal power relationship	Yes
Information sharing	Yes
Conflict resolution	Not available
Method for water division	Not available
Negotiations	

Additional comments	The power plant was to generate (net) 960,000 kW when complete
Treaty title	Agreement between [West Germany], [Austria], and [Switzerland] relating to the withdrawal of water from Lake Constance
Basins involved	Rhine
Main basin	Rhine
Date signed	04/30/1966
Signatories	Multilateral
Parties	Federal Republic of Germany, Austria, Switzerland
Principal focus	Water supply
Non-water linkages	None
Comments on above	
Monitoring	Yes
Allocations	Up to 750 l/sec used by any party outside the catchment area without notification. Up to 1500 l/sec used by any party inside the catchment area without notification
Enforcement	None
Unequal power relationship	No
Information sharing	Yes
Conflict resolution	Council
Method for water division	Unclear
Negotiations	
Additional comments	Withdrawals of water over 750 l/sec for use outside the catchment must be reported and authorized. Withdrawals of water over 1,500 l/sec for use inside the catchment must be reported and authorized. Withdrawals of water do not justify any claim to a specific quantity in the future
Treaty title	Convention concerning development of the Rhine between Strasbourg and Lauterbourg
Basins involved	Rhine
Main basin	Rhine
Date signed	07/04/1969
Signatories	Bilateral
Parties	Germany (FRG) and France
Principal focus	Hydropower
Non-water linkages	Money
Comments on above	Each party agreed to pay half the costs of construction, amounting to DM90 million up to DM100 million
Monitoring	Yes

Allocations	Each party receives half of the estimated 1,280 GWh annually from the two hydropower stations
Enforcement	None
Unequal power relationship	No
Information sharing	Yes
Conflict resolution	Council
Method for water division	Equal parts
Negotiations	
Additional comments	

RIO GRANDE BASIN

Treaty title	Agreement relating to the construction of Amistad Dam on the Rio Grande to form part of the system of international storage dams provided for by the ...
Basins involved	Rio Grande
Main basin	Rio Grande
Date signed	10/24/1960
Signatories	Bilateral
Parties	USA, Mexico
Principal focus	Hydropower
Non-water linkages	None
Comments on above	
Monitoring	Not available
Allocations	
Enforcement	Not available
Unequal power relationship	Yes
Information sharing	None
Conflict resolution	None
Method for water division	None
Negotiations	
Additional comments	

ROYA BASIN

Treaty title	Franco-Italian convention concerning the supply of water to the Commune of Menton
Basins involved	Roya
Main basin	Roya
Date signed	09/28/1967
Signatories	Bilateral
Parties	France, Italy

Principal focus	Water supply
Non-water linkages	Money
Comments on above	A deposit of 10 million lire for obligations deriving from the use of the water
Monitoring	Yes
Allocations	France gets 400 l/sec from the Roya, of which 100 l/sec continues on to Ventimiglia (back to Italy). When the Roya's flow falls below 5,600 l/sec, flows are reduced proportionally
Enforcement	Council
Unequal power relationship	No
Information sharing	Yes
Conflict resolution	UN/third party
Method for water division	Complex but clear
Negotiations	Materials for construction of the water diversion and pumping are not subject to tariffs
Additional comments	The treaty signed for a 70 year-duration. Pumping stations also included in the treaty, to be built at communal expense for both towns receiving water, one on each side of the border

RUYUMA BASIN

Treaty title	Exchange of notes ... regarding the boundary between Tanganyika territory and Mozambique
Basins involved	Ruvuma
Main basin	Ruvuma
Date signed	05/11/1936
Signatories	Bilateral
Parties	Great Britain, Portugal
Principal focus	Water supply
Non-water linkages	None
Comments on above	
Monitoring	Not available
Allocations	
Enforcement	Not available
Unequal power relationship	Not available
Information sharing	Not available
Conflict resolution	Not available
Method for water division	None
Negotiations	
Additional comments	River bank inhabitants were given unrestricted rights to draw water, fish, and remove saliferous sand for salt extraction

SENEGAL BASIN

Treaty title	Convention of Bamako
Basins involved	Senegal
Main basin	Senegal
Date signed	07/26/1963
Signatories	Multilateral
Parties	Senegal, Mali, Mauritania, Guinea
Principal focus	Industrial uses
Non-water linkages	None
Comments on above	
Monitoring	Yes
Allocations	
Enforcement	Not available
Unequal power relationship	Not available
Information sharing	Yes
Conflict resolution	Council
Method for water division	None
Negotiations	
Additional comments	

Treaty title	Convention of Dakar
Basins involved	Senegal
Main basin	Senegal
Date signed	01/30/1970
Signatories	Multilateral
Parties	Senegal, Mali, Mauritania, Guinea
Principal focus	Hydropower
Non-water linkages	None
Comments on above	
Monitoring	Not available
Allocations	
Enforcement	Council
Unequal power relationship	Not available
Information sharing	Not available
Conflict resolution	Not available
Method for water division	None
Negotiations	
Additional comments	Dam agreed to be built. Also, ports and channels improved and the channel discharge established at 300 cm/sec

SENQU BASIN

Treaty title	Treaty on the Lesotho Highlands water project between [Lesotho] and [South Africa]

Basins involved	Senqu/Orange
Main basin	Senqu/Orange
Date signed	10/01/1986
Signatories	Bilateral
Parties	South Africa, Lesotho
Principal focus	Hydropower
Non-water linkages	Money
Comments on above	Loans for construction. Payment by each party reflects their percent of the benefits, although RSA's benefit is water supply; Lesotho's is electricity
Monitoring	Yes
Allocations	RSA receives an increasing amount of water as the project moves forward: from 57 MCM in 1995 to 2208 MCM after 2020
Enforcement	Council
Unequal power relationship	Yes
Information sharing	Yes
Conflict resolution	Council
Method for water division	Complex/clear
Negotiations	
Additional comments	South Africa wants this treaty for the water, and Lesotho will get the hydropower from the reservoir – almost as an afterthought

SEPIK BASIN

Treaty title	Agreement between [Australia/Papua New Guinea] and [Indonesia] concerning administrative border arrangements
Basins involved	Sepik, Fly
Main basin	Sepik, Fly
Date signed	11/13/1973
Signatories	Bilateral
Parties	Papua New Guinea, Indonesia
Principal focus	Pollution
Non-water linkages	None
Comments on above	Mostly cross-border rights, not water supply
Monitoring	No
Allocations	Natives given traditional rights for withdrawals, fishing, and social customs/ceremonies
Enforcement	None
Unequal power relationship	Not available
Information sharing	None
Conflict resolution	None
Method for water division	Unclear

Negotiations

Additional comments   Agreement not to pollute waters that will flow into the other country, among other agreements; mostly concerning native peoples and traditional rights, especially social and fishing

SYR DARYA

Treaty title   Agreement on joint activities in addressing the Aral sea ...

Basins involved   Aral sea, Syr Darya, Amu Darya

Main basin   Syr Darya

Date signed   03/26/1993

Signatories   Multilateral

Parties   Kazakhstan, Kyrgyzstan, Tajikstan, Turkmenistan, Uzbekistan

Principal focus   Pollution

Non-water linkages   Other linkages

Comments on above   The Russians promised financial support and technical support, although they are not signatories

Monitoring   Yes

Allocations   No allocative amounts are available. In fact, this treaty seems to have non-allocation or non-natural increases of water supply in mind

Enforcement   None

Unequal power relationship   Not available

Information sharing   Yes

Conflict resolution   None

Method for water division   None

Negotiations   Treaty lasts for 10 years with optional 10-year extension. It creates the Interstate Council for the Aral Sea Basin Crisis with three committees under it: Executive Committee, Coordinating Commission on Water Resources, and the Commission of [Development and Cooperation]

Additional comments   There is an earlier agreement (18 February 1992, signed in Almaty) referred to in this document. The "Russian Federation" agrees to lend technical and financial support (no figures given) for water treatment/supply, measures to fight desertification, "environment monitoring system," and training

Treaty title   Resolution of the Heads of States of the Central Asia [sic] on work of the EC of ICAS on implementation ...

Basins involved	Aral sea, Syr Darya, Amu Darya
Main basin	Syr Darya
Date signed	03/03/1995
Signatories	Multilateral
Parties	Kazakhstan, Kyrgyzstan, Tajikistan, Turkmenistan, Uzbekistan
Principal focus	Pollution
Non-water linkages	Money
Comments on above	Agreed to deposit funds for the IFAS
Monitoring	Not available
Allocations	None
Enforcement	None
Unequal power relationship	Not available
Information sharing	Yes
Conflict resolution	None
Method for water division	None
Additional comments	This treaty exists to set up the members of the ICAS [International Council on the Aral Sea] more than anything else

URUGUAY BASIN

Treaty title	Agreement concerning the utilization of the rapids of the Uruguay river in the Salto Grande area
Basins involved	Uruguay
Main basin	Uruguay
Date signed	12/30/1946
Signatories	Bilateral
Parties	Argentina, Uruguay
Principal focus	Hydropower
Non-water linkages	Money
Comments on above	Costs of hydropower systems will be equally shared
Monitoring	Yes
Allocations	
Enforcement	Council
Unequal power relationship	Yes
Information sharing	Yes
Conflict resolution	Council
Method for water division	None
Negotiations	
Additional comments	

VISTULA BASIN

Treaty title	Agreement between [Poland] and the [USSR] concerning the use of water resources in frontier waters
Basins involved	Vistula
Main basin	Vistula
Date signed	07/17/1964
Signatories	Bilateral
Parties	USSR, Poland
Principal focus	Flood control
Non-water linkages	None
Comments on above	
Monitoring	Yes
Allocations	
Enforcement	Not available
Unequal power relationship	Yes
Information sharing	Yes
Conflict resolution	None
Method for water division	Not available
Negotiations	Each party agrees to not undertake works which may affect the use of resources by the other
Additional comments	The treaty discusses cooperation on many subjects, not just flood control. They will work out standards relating to water purity and establish procedures for controlling pollution

VUOKSA BASIN

Treaty title	Agreement between [Finland] and the [USSR] concerning the production of electric power in the part of the Vuoksi river bounded by the Imatra ...
Basins involved	Vuoksa
Main basin	Vuoksa
Date signed	07/12/1972
Signatories	Bilateral
Parties	Russia, Finland
Principal focus	Hydropower
Non-water linkages	Other linkages
Comments on above	The loss of 19,900 MWH will be compensated to Finland in perpetuity
Monitoring	Yes
Allocations	
Enforcement	Council

Unequal power relationship	Yes
Information sharing	Yes
Conflict resolution	Other government agency
Method for water division	Complex but clear
Negotiations	
Additional comments	A five-year treaty, considered to be extended in five-year increments unless otherwise dissolved by one or the other party

ZAMBESI BASIN

Treaty title	Exchange of notes constituting an agreement between [Great Britain] and [Portugal] providing for the Portuguese participation in the Shir valley ...
Basins involved	Zambesi
Main basin	Zambesi
Date signed	01/21/1953
Signatories	Bilateral
Parties	Portugal, Great Britain
Principal focus	Hydropower
Non-water linkages	Money
Comments on above	One-third costs of dam construction borne by Portugal
Monitoring	Not available
Allocations	
Enforcement	Not available
Unequal power relationship	Not available
Information sharing	Not available
Conflict resolution	Not available
Method for water division	None
Negotiations	
Additional comments	Irrigation also considered, as well as land reclamation

Treaty title	Agreement between [Great Britain/Rhodesia-Nyasaland] with regard to certain ... natives living on the Kwando river
Basins involved	Kwando
Main basin	Zambesi
Date signed	11/18/1954
Signatories	Bilateral
Parties	Great Britain (Rhodesia, Nyasaland), Portugal
Principal focus	Water supply
Non-water linkages	None
Comments on above	

Monitoring	Not available
Allocations	
Enforcement	Not available
Unequal power relationship	Not available
Information sharing	Not available
Conflict resolution	Not available
Method for water division	None
Negotiations	
Additional comments	Natives were allowed use of the Kwando for water supply, irrigation, and fishing during the dry season
Treaty title	Agreement … Relating to the Central African Power Corporation
Basins involved	Zambesi
Main basin	Zambesi
Date signed	11/25/1963
Signatories	Bilateral
Parties	Southern Rhodesia, Northern Rhodesia
Principal focus	Hydropower
Non-water linkages	None
Comments on above	
Monitoring	Not available
Allocations	A cooperative corporation of the two nations regulates the water level in the reservoir "in the interests of the operation of the installations and of the safety of the dam …"
Enforcement	Not available
Unequal power relationship	Not available
Information sharing	Yes
Conflict resolution	Council
Method for water division	Complex but clear
Negotiations	Unknown
Additional comments	25-year duration of the treaty. It is one of only three dams in Africa, the totality of which uses 5% of estimated hydroelectric power
Treaty title	Untitled: Agreement between South Africa and Portugal
Basins involved	Zambesi
Main basin	Zambesi
Date signed	04/01/1967
Signatories	Bilateral
Parties	South Africa, Portugal
Principal focus	Hydropower
Non-water linkages	Money
Comments on above	Malawi agreed to purchase electricity from the dam

Monitoring	Not available
Allocations	
Enforcement	Not available
Unequal power relationship	Not available
Information sharing	Not available
Conflict resolution	Not available
Method for water division	None
Negotiations	
Additional comments	

9

Annotated literature

To aid those in their research on the topics found in this document, we have included some abstracts and annotations for materials found in the reference list. There was no systematic judgement made as to what would be included or excluded; it was based on availability at time of completion of this document.

af Ornas, A.H. and S. Lodgaard. 1992. *The Environment and International Security*. Oslo, Norway: Uppsala University, Department of Human Geography/International Peace Research Institute. The book consists of proceedings of a workshop held in Uppsala in 1991. It aimed to explore the many links between environment and security, looking at two types of environmental conflict: conflicts between man and nature, centring on sustainability, and conflicts between man and man, centring on development. Nine short papers examine environmental destruction as a method of warfare, cases of social conflict arising from ecological destruction, the internationalization of the Finnish forest conflict, and water resource conflicts. (M. Amos)

Alemu, Senai. 1995. The Nile Basin: Data Review and Riparian Issues. Final Draft Report. Washington, DC: AGRPW, The World Bank. This report highlights the background information essential to evaluate the prevailing characteristics of the Nile basin as a unified system. Population trends, flow regimes of the Nile and its main tributaries, water use patterns, future and potential water requirements and water availability in the 10 riparian countries are discussed. The report makes a review and assessment of Nile water treaties and agreements entered between different parties in the last 100 years, including the recent endeavours for regional cooperation. Efforts were made to make an overview of the dynamics of riparian positions on certain key issues of the Nile. The report also narrates previous suggestions forwarded with respect to Nile water allocation criteria and specific proposals. Special emphasis is made of the key riparians of the Nile basin – Egypt, Ethiopia, Sudan, and Uganda. It is believed that the report may be useful to Nile basin policy-makers, the donor agencies, professionals engaged in water projects in the basin, and above all to those initiators of change who are committed to resolve conflicts and bring about change – change for a better quality of life for the present, and a better world for tomorrow's generations of the Nile basin countries.

Alheritiere, Dominique. 1985. Settlement of Public International Disputes on Shared Resources. *Natural Resources Journal* 25(3): 701–11. A literary and ideological overview of the negotiation of existing resource conflicts. The author begins with a discussion of some treaties and history of UN-era resource conflicts. The author continues with a catalogue of the various means of peaceful dispute resolution as encouraged by the United Nations texts.

Ali, Mohammed. 1965. *The River Jordan and the Zionist Conspiracy*. Cairo: Information Department. Short history of the Middle East and water politics since 1916, legal aspects, and some technical data.

Allan, J.A. 1992. Substitutes for Water Are Being Found in the Middle East and North Africa. *GeoJournal* 28(3): 375–85. Evident mismatches exist between the demand and supply of water in many countries particularly in the semi-arid and arid worlds. The resulting food gaps which concern both the national governments of these countries as well as the international agencies which extend assistance to them, appear at first sight to pose challenges beyond the economic and political capacity of peoples and institutions to make the necessary adjustments. It will be argued that everywhere there are examples of conflict over water being avoided, and while the avoidance of conflict can in many cases be calculated to have been at a cost to the environment, nevertheless, to date serious conflict has been avoided. Case studies from a region seen to have the worst water resource future outside the industrialized world, the Middle East and North Africa, will be discussed, which exemplify the numerous strategies adopted by countries in their various ecological, economic, and political circumstances. A fortunate few governments have substituted oil capital for water while others have filled the food gap, which is generally an expression of the water gap, by ceding economic and political autonomy. (abstract)

Allee, D.J. 1993. Subnational Governance and the International Joint Commission: Local Management of United States and Canadian Boundary Waters. *Natural Resources Journal* 33 (Winter): 133–51. This article explores the subnational governance and the International Joint Commission on the local management of the shared waters of the Great Lakes. The focus, by both United States and Canada, for water management issues is on the states and the provinces involved, not just country to country. Allee suggests generalizations about some roles subnational governments have played, how they have been involved in Commission decisions, blocked or facilitated them, enjoyed access, been frustrated in meeting local objectives, and were able to use the process to complement other relations with their federal governments.

Amy, Douglas J. 1987. *The Politics of Environmental Mediation*. New York: Columbia University Press. An in-depth view of alternate dispute resolution, environmental conflicts, and mediation in the United States. Includes a grid of conflict types and the best situations in which to seek an ADR.

Andah, K. and F. Siccardi. 1991. Prediction of Hydrometeorlogical Extremes in the Sudanese Nile Region: A Need for International Co-operation. In *Hydrology for the Water Management of Large River Basins*, ed. F.H.M. Van de Ven et al., 3–12. International Association of Hydrological Sciences. Analysis of the 1988 floods in Sudan have shown that a lead time of at least one month is required for mitigation interventions. The present five days' lead time for prediction of Nile floods in Khartoum are not adequate for an efficient advance warning. The rainfall over Sudan exhibits frequent positive and negative anomalies. Two approaches are suggested in the present work. The first involves further analysis on the possible spatial coherence and teleconnections of rainfall anomalies over

Africa to enhance the lead time based on prediction of extreme events in other regions. The second requires an integrated real time flood forecasting and an effective weather monitoring system covering the countries sharing the Nile basin. These demand technical cooperation between the countries within the basin. (abstract)

Anderson, E.W. 1991. Hydropolitics. *Geographical Magazine* (February): 10–14. Discusses the likelihood of conflict as it governs some riparian politics. He examines all aspects of a water conflict: geographical location, national interest, military and/or political power to bear on the issue, and money.

—— 1991. Making Waves on the Nile. *Geographical Magazine* (April): 31–4. The Nile's course and geography creates heavy difficulties when anyone tries to divide it fairly. Sudan and Egypt use most of the water but contribute almost none in rainfall, whereas Ethiopia contributes a great deal of water but cannot use much at all due to technical considerations and war.

—— 1991. The Source of Power. *Geographical Magazine* (March): 12–15. Discusses Turkey and her relations with Iraq and Syria. Turkey controls both oil and water conduits to several nations, causing concerns on all sides. However, "friction is not the same thing as war."

Anderson, Kirsten Ewers. 1995. Institutional Flaws of Collective Forest Management. *Ambio* 24 (6, September): 349–53. This article examines the institutional development within the afforestation of village revenue lands in India as well as institutions set up for management of already existing forests. These institutions are "collective or joint forest-management committees" indicating a joint government-village arrangement for forest protection. However, it is often unclear or skewed who has which rights and to what. The present article argues for a careful analysis of the kinds of rights, of the categories of rightholders as well as the biophysical character of the resource itself. The degree of sociopolitical overlap between new induced institutions and the existing ones of local government such as the village panchayat is counterproductive. Another critical issue is the match between the institutional set-up and the biophysical characteristics of the resource itself. Both issues are examined in the article. (abstract)

Anon. 1994. Protection and Use of Transboundary Watercourses and International Lakes in Europe. *Natural Resources Forum* 18(3): 171–80. This paper analyses recent developments in national strategies and policies in Europe for the protection and use of transboundary waters, and describes progress made by European states in cooperating on these issues, in particular under the Convention on the Protection and Use of Transboundary Watercourses and International Lakes (Helsinki, 1992) pending its entry into force. (abstract)

Anonymous. 1987. Where Dams Can Cause Wars. *The Economist* 18 (July): 37–38. Discusses Turkey's place in the Middle East water issue. Turkey's dams will increase salt content of the Euphrates waters and reduce hydroelectric output of Syrian and Iraqi dams. Quotes Turgut Ozal (Prime Minister of Turkey), who says that competition over water will help foster peace.

Ashworth, John and Ivy Papps. 1991. Equity in European Community Pollution Control. *Journal of Environmental Economics and Management* 20 (1, January): 46–54. This paper examines the impact of an EEC pollution control directive on the profits of the firms in the chlor-alkali industry. Various definitions of equal treatment are developed and it is argued that knowledge of the relationship between discharges and various methods of abatement (the discharge function) is crucial to an investigation of the extent to which the directive meets the equity criteria. Estimation of total, average, and marginal costs implies that these criteria are not met by this directive. (abstract)

Associated Press. 1995. Water Crisis Looms, World Bank Says. *The Washington Post*, Monday, August 7. This newspaper article is a warning by Ismail Serageldin, a World

Bank vice president, on the water crisis that is looming. Serageldin discusses in general the areas that the Bank will be looking to fund in the future, indicating that the Bank will not fund projects that look only at irrigation or only at municipal use, but the plans impacting on the resource.

Azar, Christian and John Holmberg. 1995. Defining the Generational Environmental Debt. *Ecological Economics* 14 (1, July): 7–20. Assume that we have borrowed the earth from our children, and that we one day shall give it back to them and account for what we have done to it. Then we would have to try to restore the damage we have caused. Further, we would have to offer compensation for the damage we have done that we cannot repair at a lower cost. The generational environmental debt (GED) is a measure of these costs. In this paper we define and discuss the concept of GED and calculate GED for emissions of the greenhouse gas CO_2. The global GED for CO_2 emissions is estimated at 10,000 billion US dollars and the Swedish GED for CO_2 emissions is estimated at 60 billion US dollars. (abstract)

Azar, Edward E. 1990. *The Management of Protracted Social Conflicts*. London: Aldershot. Discusses framework and backgrounds on the social implications of protracted conflicts, as well as Track Two diplomacy. Also extensive discussion of global problem solving: Lebanon, Sri Lanka, and the Falklands/Malvinas issues.

Bakour, Y. 1991. *Planning and Management of Water Resources in Syria*. Damascus, Syria: Arab Organization for Agricultural Development. A detailed account of Syrian water resources and policy bodies.

Barnard, William S. 1994. From Obscurity to Resurrection: The Lower Oranger River as International Boundary. In *Political Boundaries and Coexistence: Proceedings of the IGU-Symposium, Basle/Switzerland, 24–27 May*, ed. Werner A. Galluser, 125–34. New York, NY: Peter Lang. Situated in a remote part of Southern Africa, the lower course of the Orange river forms the international boundary between South Africa and Namibia for some 670 km. Boundary evolution commenced in 1847 when the river was proclaimed the northern limit of the Cape Colony; in 1890 the lower course became an international boundary between the British and German Empires, to be inherited and resurrected by their respective successor states in 1990. The spatial differentiation of the boundary area is controlled by the incised character of the river valley rather than its human occupancy; a border region never developed. For the 75 years when Namibia was a South African dependency, the separating functions of the boundary were suppressed although not entirely obliterated. Integrating functions were strengthened by a growing demand for the river's water in the hyper-arid boundary environment. Boundary resurrection comes at a time when South Africa is planning to tap up to 90% of the mean annual run-off of 11,480 million m^3 in the upper basin. Future trends suggest an intensification of irrigation and tourism on both banks of the lower course, but riparian users will have to accept reduced, stringently budgeted allocations of water. (abstract)

Barret, S. 1994. *Conflict and Cooperation in Managing International Water Resources*. Centre for Social and Economic Research on the Global Environment, Vol. Wm 94094. This article is written from the premise that conflicts can arise about shared water resources and the question whether negotiated treaties can ensure that nations which share bodies of water share gains from cooperation. Three case studies are examined, the Columbia River Treaty, the Indus Waters Treaty, and the Convention on the Protection of the Rhine Against Pollution Chlorides. Analysis of game theory is done exploring primarily the Coase Theorem.

Bau, Jao. 1995. Cooperation Among Water Research and Development Institutions of Europe. *Water International* 20: 129–35. This article discusses the problem of environmental degradation in central and eastern Europe. The obstacles facing East-West cooperation are analysed, and Bau presents a programme of cooperation offered by the European Communities Commission.

Baumann, Duane D., John J. Boland, and John H. Sims. 1984. Water Conservation: The Struggle Over Definition. *Water Resources Research* 20 (4, April): 428–34. During the 1970s much attention was focused on the role of water conservation in the planning and management of urban water supplies, and actions to implement water conservation were taken at all levels of government. Yet many policies and programmes appear to rely on conceptions of water conservation which confuse supply and demand oriented strategies or which inefficiently conserve water at the expense of other scarce resources. Through an analysis of the underlying concepts and values, a definition of water conservation is reached which is both precise and practical and which provides a sound basis for the development of water conservation policies. Examples drawn from recent field studies illustrate the application of the definition.

Beaumont, Peter. 1994. The Myth of Water Wars and the Future of Irrigated Agriculture in the Middle East. *Water Resources Development* 10(1): 9–22. Beaumont argues that water tensions will not result in war because irrigation water will be diverted to higher value urban uses before people will fight over crop-watering.

Beckerman, Wilfred. 1992. Economic Growth and the Environment: Whose Growth? Whose Environment? *World Development* 20(4): 481–96. The widespread clamour for immediate draconian action to reduce the danger of global warming is an unjustifiable diversion of attention from the far more serious environmental problems facing developing countries. Resource constraints do not consitute limits to growth, and the likely economic damage done by climate change would be a negligible proportion of world output. The loss of welfare for the population in developing countries today as a result of inadequate access to safe drinking water and sanitation, or of urban air pollution, is far greater, and should be given priority over the interests of future generations. The "sustainable" growth concept is either morally indefensible or totally non-operational. (abstract)

Bell, F.C. 1988. The Sharing of Scarce Water Resources. *Geoforum* 19(3): 353–66. Australians have been sensitive about water issues from the beginning, and like the United States, many engineering projects were undertaken in the 1950s and 1960s when a true cost-benefit analysis would not have permitted the projects to go on. The author seeks to prove that the idea of using such a high number of engineering "fixes" came from the perception that Australia's water resources were scarce, but that in a physical sense water is not scarce. Economically, water is scarce because of the difficulty of containing and managing it.

Benvenisti, E. and H. Gvirtzman. 1993. Harnessing International Law to Determine Israeli-Palestinian Water Rights: The Mountain Aquifer. *Natural Resources Journal* 33 (Summer): 543–67. Focusing on the question of joint management of the water of the Mountain Aquifer, this article applies the generally accepted principles of international law to a peaceful arrangement regarding the management of this resource. This study is limited to examining the implications of the management of the Mountain Aquifer for only those options which would establish a separate legal entity for the Palestinians of the area, be it an autonomous territory, an independent state, or a Jordanian-Palestinian confederation.

Bercovitch, Jacob. 1992. Mediators and Mediation Strategies in International Relations. *Negotiation Journal* 8 (2, April): 99–112. Discusses kinds of mediation; discusses strategies and behaviour among nations resolving differences and among those trying to help them resolve those differences. The author provides frameworks for possible negotiation strategies as well as some possible actions for the mediator. Ways to evaluate the mediator and success of the negotiations are also offered.

Berthelot, R. 1989. The Multidonor Approach in Large River and Lake Basin Development in Africa. *Natural Resources Forum*, August, 209–15. From the outset of its efforts to promote the socio-economic development of river and lake basins in developing countries, UNDP took the position that in most cases the river basin was the appropriate goegraphical entity for development. However, the amount of technical assistance required to evaluate the potentialities of a large river or lake basin is beyond the capabilities of any single donor of assistance. Similarly, the capital investment required to implement a basin development programme, ranging from US$ several hundred million to possible US$ several thousand million, is usually beyond the capabilities of any single financial institution. On the basis of the experience gained in the development of the Mekong river and the Senegal river, UNDP advocates concerted and coordinated cooperation among donors interested in different aspects of one given large river or lake basin development, for which it coined the phrase "multidonor approach". (abstract)

Bhatti, Neeloo, David G. Streets, and Wesley K. Foell. 1992. Acid Rain in Asia. *Environmental Management* 16(4): 541–62. Acid rain has been an issue of great concern in North America and Europe during the past several decades. However, due to the passage of a number of recent regulations, most notably the Clean Air Act in the United States in 1990, there is an emerging perception that the problem in these western nations is nearing solution. The situation in the developing world, particularly in Asia, is much bleaker. Given the policies of many Asian nations to achieve levels of development comparable with the industrialized world – which necessitate a significant expansion of energy consumption (most derived from indigenous coal reserves) – the potential for the formation of, and damage from, acid deposition in these developing countries is very high. This article delineates and assesses the emissions patterns, meteorology, physical geology, and biological and cultural resources present in various Asian nations. Based on this analysis and the risk factors to acidification, it is concluded that a number of areas in Asia are currently vulnerable to acid rain. These regions include Japan, North and South Korea, southern China, and the mountainous portions of southeast Asia and southwestern India. Furthermore, with accelerated devlopment (and its attendant increase in energy use and production of emissions of acid deposition precursors) in many nations of Asia, it is likely that other regions will also be affected by acidification in the near future. Based on the results of this overview, it is clear that acid deposition has significant potential to impact the Asian region. However, empirical evidence is urgently needed to confirm this and to provide early warning of increases in the magnitude and spread of acid deposition and its effects throughout this part of the world. (abstract)

Bingham, Gail and Suzanne Goulet Orenstein. 1991. The Role of Negotiation in Managing Water Conflicts. Submitted to American Society of Civil Engineers for publication. This article seeks to address the role of negotiation in solving water resource problems which inevitably in some cases have led to conflict. Starting first with a section entitled Defining Negotiation and Mediation, the authors then turn to the History of Mediation in Environmental Conflicts, Creating Efficient and Effective Problem-solving Processes for Water Conflicts, Requirements for Successful Use of Conflict Resolution Processes, and then concludes with the recommendation that in negotiation one must remember that effective negotiation requires facing the differences that divide groups and developing strategies that allow constructive solutions in spite of those differences.

Biot, Yvan, Piers M. Blaikie, Cecile Jackson, and Richard Palmer-Jones. 1995. *Rethinking Research on Land Degradation in Developing Countries*. World Bank Discussion Papers, Vol. 289. Washington, DC: World Bank. This paper critically reviews the three main approaches to land degradation and conservation – the classic, populist, and neo-liberal. The implications of these paradigm shifts are examined in terms of research needs. Next, the paper discusses the role of science and technology, and the origins and substance of differences in the perception, evaluation, and diagnosis of degradation. Focus is then shifted to analysing how farmers and pastoralists make decisions about resource use and management, and a research approach is suggested for analysing decision-making. Two case studies illustrate the approach. (abstract)

Biswas, Asit K. 1992. Indus Water Treaty: The Negotiating Process. *Water International* 17:

201–09. This paper is an analysis of the negotiating process that resulted in the Indus River Treaty between India and Pakistan. This analysis stems from the issues surrounding arid and semiarid countries international water bodies. Biswas contends that not enough attention has been paid to review the negotiating processes that have led to successful water treaties.

—— 1993. Management of International Waters: Problems and Perspective. *Water Resources Development* 9(2): 167–88. The purpose of this paper is to review objectively some of the major developments on the increasingly critical issue of management of international water resources during the past two decades. It is the author's opinion that the management of international water resources has not received adequate attention in the recent past. He is calling for a change to the current "softly, softly" approach by international organizations in dealing with these complex issues.

—— 1995. Institutional Arrangements for International Cooperation in Water Resources. *Water Resources Development* 11(2): 139–45. Three papers were specially commissioned for a Special Session on "Institutional Arrangements for International Cooperation in Water Resources" during the 8th World Congress on Water Resources in Cairo, Egypt, in November 1994. The Session also included a panel discussion in which presidents or senior officials from six major water-related professional associations made brief presentations on the desirability of establishing a World Water Council. This paper is a summary of the entire Special Session. (abstract)

Blake, Gerald. 1994. International Transboundary Collaborative Ventures. In *Political Boundaries and Coexistence: Proceedings of the IGU-Symposium, Basle/Switzerland, 24–27 May*, ed. Werner A. Galluser, 359–71. New York, NY: Peter Lang. International boundary studies generally focus upon the delineation and function of boundaries, and their impact upon borderlands and borderlanders. Alternatives to full state territorial sovereignty such as international zones, neutral zones etc. deserve more attention, not least because of their potentially useful role in a world of geopolitical change. Similarly, more systematic evaluation of a whole range of transboundary collaborative ventures should be attempted. As a start, attention is drawn in this paper to the range of such activities, associated with resource management, economic development, transport management, and environmental protection. The focus is upon cases which involve genuine cooperative effort, and which to some degree mean a diminution of national sovereignty in a specific region. An attempt is made to devise a classification of the types of collaboration currently being undertaken in various parts of the world, on land and sea. Finally, consideration is given to the significance of trends towards transboundary cooperation for international boundary studies. (abstract)

Bochniarz, Zbigniew. 1992. Water Management Problems in Economies in Transition. *Natural Resources Forum*, February, 55–63. Ambitious programmes of reform in Central and Eastern Europe (CEE) are threatened by a serious deterioration of the environment. Large-scale damage of the natural environment in many forms, including water pollution has created development barriers which adversely affect the living conditions of current and future generations. Despite similar patterns of environmental policy in the CEE countries compared with their western counterparts, neither environmental legislation nor economic incentives have produced any significant improvement in water quality over the last 10–20 years. For that very reason in the transition period, it is necessary to identify existing deficiencies in the system and to build up a mixed system of new institutions of water management, more realistic legislation with a strong enforcement system, and market-based incentives for water conservation and protection against pollution. (abstract)

Bolin, I. 1990. Upsetting the Power Balance: Cooperation, Competition, and Conflict Along an Andean Irrigation System. *Human Organization* 49(2): 140–48. The purpose of this study, centred on an Andean irrigation system in Peru, is to show that in-depth research on patterns of social interaction prior to the initiation of a project will achieve greater satisfaction and sustained cooperation among all groups affected by development. This paper examines the effects of a development project which provides a plentiful water supply along an entire canal system done through canal construction and improvement.

Brooks, David B. 1993. Adjusting the Flow: Two Comments on the Middle East Water Crisis. *Water International* 18 (1, March): 35–39. Whereas almost all nations in the Middle East face a chronic problem of water shortage, the riparians of the Jordan River (mainly Israel, Jordan, and Palestine) are close to crisis. To deal with this situation, it is suggested that the emphasis in water management be shifted from supply to demand and from quantity to quality. This approach emulates alternative energy analysis, dubbed the "soft path," which has demonstrated that it is typically economically cheaper and ecologically less damaging to approach problems from the demand than from the supply side. Major opportunities exist to increase efficiency of use in the Jordan river basin, particularly for irrigation, which is by far the main consumer. Other opportunities lie in avoiding the degradation arising from excessive use of pesticides and fertilizers, inadequate treatment of sewage, and industrial dumping. Continuing the analogy with energy, policy analysis should be recast in terms of normative scenarios so as to determine the feasibility and impacts of alternative policies and reactions. All sides see close linkage between water and security. Therefore, only through exploration of alternative futures, not simply protection of the present into the future, can we find ways to minimize the potential for conflict. (abstract)

Bruhacs, J. 1992. Evaluation of the Legal Aspects of Projects in International Rivers. *European Water Pollution Control* 2(3): 10–13. International law involves legal norms relating to the management of international rivers. Therefore there are procedural obligations for the evaluation of possible impacts resulting from projects. This paper attempts to clarify some implied difficulties of this legal regime. (abstract)

Buck, S.J., G.W. Gleason, and M.S. Jofuku. 1993. "The Institutional Imperative": Resolving Transboundary Water Conflict in Arid Agricultural Regions of the United States and the Commonwealth of Independent States. *Natural Resources Journal* 33: 595–628. The hypothesis of this study is that conflict created by disjunction in intranational contexts is resolved through multi-level institutional interaction. The results of the study indicate there is a phenomenon which suggests an "institutional imperative" of maintaining the vitality of subnational and supranational institutions to resolve international transboundary water conflict. The case studies for this report are the arid agricultural regions of the United States and the Commonwealth of Independent States.

Butts, Kent Hughes. 1993. *Environmental Security: What is DOD's Role?* Carlisle, PA: Strategic Studies Institute, U.S. Army War College. When newly appointed Secretary of Defense Les Aspin reorganized his principal staff, he created the position of Deputy Under Secretary of Defense for Environmental Security. The creation of this position draws attention to an issue that has a powerful following in Congress and the current administration – the use of the military for environmental security missions. This study examines important environmental roles and missions currently being executed by the Department of Defense (DOD), provides an assessment of their contributions to national security, and makes recommendations concerning DOD's future environmental peacetime role. (abstract)

Calleros, J. Roman. 1991. The Impact on Mexico of the Lining of the All-American Canal. *Natural Resources Journal* 31 (Fall): 829–38. This article, written from Mexico's perspective, urges Mexico to take a strong position in opposing the proposed lining of the All-American canal. If this project is developed, the groundwater recharge will be significantly reduced affecting 121 wells in Mexico and more than 33,000 acres of farmland.

Cano, Guillermo. 1989. The Development of the Law in International Water Resources and the Work of the International Law Commission. *Water International* 14: 167–71. Provides an overview of international law, including NGOs.

Capistrano, Ana Doris and Clyde F. Kiker. 1995. Macro-scale Economic Influences on Tropical Forest Depletion. *Ecological Economics* 14 (1, July): 21–29. The paper examines the influence of global and domestic factors on forest depletion in 45 tropical developing countries from 1967 to 1985. It links forest depletion with conditions in the international markets and with domestic macroeconomic and demographic factors. Calculated elasticities suggest that real exchange rate devaluation, debt service, food self-sufficiency, income, and export prices of forestry and agricultural output had significant influence on the depletion of tropical forests during the study period. (abstract)

Caponera, Dante A. 1987. International Water Resources Law in the Indus Basin. In *Water Resources Policy for Asia*, ed. Mohammed Ali et al. Boston, MA: A.A. Balkema. Discusses Indo-Pakistani/Indo-Bangladeshi water disputes, and the international agreements between states. Caponera emphasizes the need for multilateral negotiations.

—— 1993. Legal Aspects of Transboundary River Basins in the Middle East: The Al Asi (Orontes), the Jordan, and the Nile. *Natural Resources Journal* 33: 629–63. This article presents an analysis of the legal situation of three international river basins in the Middle East: the Al Asi (Orontes), the Jordan, and the Nile. This analysis is used to determine the rights and obligations of the co-basin states to the waters available in each basin. The rivers were chosen for this study because of their similarity in being located in an arid area. Caponera tentatively presents regional or local solutions for cooperative arrangements from the legal standpoint.

Chaube, U.C. 1990. Water Conflict Resolution in the Ganga-Brahmaputra Basin. *Water Resources Development* 6(2): 79–85. This article analyses interstate water disputes within India and the conflict over sharing of water between India and Bangladesh over the Ganga-Brahmaputra basin. This was done using a simplified two-level decomposition – coordination study of the Indo-Nepal region of the Ganga basin.

—— 1992. Multilevel Hierarchical Modelling of an International Basin. In *Proceedings of the International Conference on Protection and Development of the Nile and Other Major Rivers,* Vol. 2/2. International Conference on Protection and Development of the Nile and Other Major Rivers. Cairo, Egypt, 3–5, February. This article discusses the results of a simplified two-level decomposition-coordination study of the Indo-Nepal region of the Ganga basin which was conducted using deterministic linear programming models to provide a scientific basis for resolution of conflicts and explanation of interlinkages. The issues for study of the Ganga-Brahmaputra water resources (GBWR) system is viewed in terms of the geopolitical, temporal and goal/functional requirements and multiplicity of independent decision authorities.

Chitale, M.A. 1995. Institutional Characteristics for International Cooperation in Water Resources. *Water Resources Development* 11(2): 113–23. There has been a steady evolution in the arrangements for cooperation in the management of international river basins over the last two centuries. Rather than aiming at a standardized set-up for all the international river basins, basin organizations can best be allowed to grow in phases according to the emerging needs of the respective basins. The river basin organizations and regional water management bodies will need a global common platform for exchanging their experiences and for developing common global strategies. A World Water Council can provide such an umbrella set-up with its General Assembly comprising international basin entities, regional bodies, international professional associations, and the UN agencies dealing with water. (abstract)

Clarke, R. 1991. *Water: The International Crisis.* London: Earthscan. Describes the world's freshwater shortage and examines both the economic and political factors which have led to it, then discusses how climatic conditions and poverty, leading to poor land management and overpopulation, have contributed to water scarcity. Many of the world's major rivers, including the Rhine, Danube, Nile, and Niger meet the water requirements of several countries, however, agreements for managing these supplies are, in many cases, fragile. The author considers the possibilities of international conflict over water control and outlines the effect of water availability on development in poor countries. The book concludes with a number of solutions, both traditional and technological, which could lead the way towards ensuring world water security. (M.Z. Barber)

Cohen, Saul B. 1992. Middle East Geopolitical Transformation: The Disappearance of a Shatterbelt. *Journal of Geography* 91 (1, January/February): 2–10. The Cold War's end has brought about major global geopolitical restructuring. It is also cause for regional geopolitical reordering. The Gulf War and its aftermath are but one expression of Middle Eastern disequilibrium. This shatterbelt region has been caught up in both intraregional tensions and the post-World War II history of competition between the Maritime and Eurasian continental realms. Now the Middle East is becoming strategically reoriented to the West. While powerful centrifugal forces still prevail, the reduction of external competitive pressures permits centripetal forces to become more salient. In addition to Arabism and Islan, these include migration and capital flows and water and oil transportation lines. A new balance among the Middle East's six regional powers can be fostered but not dictated by the outside world. Equilibrium can best be promoted, not by a Pax Americana, but by the United States and the European Community acting as two competitive but allied stabilizers. (abstract)

Conant, M.A. 1990. The Middle East Agenda. *Geopolitics of Energy* 12(6): 3–7. Change has come to key relationships in the Gulf as a consequence of the Iraqi-Iranian stalemate, the suspected ambitions of Hussein, and the return of Egypt to prominence among Arab states. An Iran-Iraq war is still a concern for regional relationships as there are worries about Iraqi intentions. Other factors discussed included the position of Saudi Arabia and Kuwait, the oil price policy, nuclear weapons availability, regional stability, and political change, as well as internal disputes over water, and trade and relations with the "Superpowers." (abstract)

Crane, M. 1991. Diminishing Water Resources and International Law: U.S.-Mexico, A Case Study. *Cornell International Law Journal* 24(2): 299–323. The objective of this study is to illustrate the need to improve the international legal regime governing transboundary groundwater. This is achieved by analysing the recent dispute between the United States and Mexico over shared groundwater resources. Included is a presentation of the applicable international law, an explanation of why the customary international law of groundwater is in such a rudimentary stage, an application of the definitive international environmental law to the US-Mexican dispute, concluding by borrowing from US domestic law to resolve the dispute between Mexico and the US.

Crow, B. and A. Lindquist. 1990. Development of the Rivers Ganges and Brahmaputra: the Difficulty of Negotiating a New Line. *DPP Working Paper – Open University* 19: 1–46. Periodic floods making international news are only one symptom of the untamed state of the major rivers of South Asia. With the declaration of the 1989 Group of Seven Summit, and subsequent agreement on an Action Plan for Floods in Bangladesh, some taming is now planned. The slow development of irrigation and of hydroelectric power in the region nevertheless remains a major factor keeping living standards low in an area with perhaps the largest concentration of poverty in the world. Disagreement between the governments of India and Bangladesh constitutes a serious constraint contributing to that slow pace of development. This paper examines the last major round of negotiations. New ideas on both sides and a thorough re-evaluation by a group in the Bangladesh government brought the negotiators close to agreement. The paper indicates the need for interdisciplinary perspectives on international development negotiations. (abstract)

Cummings, R.G. and V. Nercissiantz. 1992. The Use of Water Pricing as a Means for Enhancing Water Use Efficiency in Irrigation: Case Studies in Mexico and the United States.
Natural Resources Journal 32: 731–55. The purpose of this article is to examine the role of
water markets in enhancing irrigation water use efficiency. Particular concern in this paper
was focused on water pricing as a means for improving water use efficiency in irrigation.
The authors explore a number of alternative ways in which water prices might be used to
provide farmers with incentives to collaborate in programmes to enhance efficiency. These
issues are examined in the context of the case studies of Mexico and the United States.

Curtuis, M. 1995. More Contentious Than West Bank is Resource Below: Water. *Los Angeles Times*, Saturday, 15 July. This newspaper article is an exploration of the dispute over
water between Israel and Palestine.

Day, R. and J.V. Day. 1977. A Review of the Current State of Negotiation Order Theory; an
Appreciation and a Critique. *The Sociological Quarterly* 18 (Winter): 126–42. It is argued
that the theory of negotiated order, which has remained largely submerged within a series
of fairly specific "grounded" case studies of occupations, professions, and complex organizations in the health field, has a number of attractive qualitites to it. As such, it represents one of the more exciting recent developments in the study of organizations. The
present paper examines the more recent historical origins of the theory, summarizes its
major points, and critically assesses its strengths and weaknesses. Since most of the substance of the theory is found in studies of health professionals working in hospital settings,
the examples used here are also derived from this particular body of literature. Some future possible directions which we believe will improve the overall explanatory power of
the perspective are presented in the concluding section. (abstract)

Delli Priscoli, Jerome. 1988. Conflict Resolution in Water Resources: Two 404 General
Permits. *Journal of Water Resources Planning and Management* 114 (1, January): 66–77.
The use of alternative dispute resolution techniques in water resources is demonstrated
and experience evaluated against current theory of bargaining and negotiating. Conflicts
among environmentalists, developers, and government agencies are well known; they involve planning, constructing, operating, and regulating water-resources projects. Two
Section 404 permit cases are compared. One in 1980, involves issuing a general permit
(GP) for wetland fill on Sanibel Island, Florida. The other, in 1987, involves issuing a GP
for hydrocarbon exploration drilling throughout Louisiana and Mississippi. Generally,
permits are granted on a case-by-case basis, but Corps district engineers may also issue
GPs for activities that produce no negative cumulative impacts. In these cases the Corps
adopted a revolutionary approach to GPs. Rather than writing the permit in-house, the
Corps suggested that the parties who conflict over permit applications get together and
write the technical specification for a GP. The Corps told environmentalists, citizens,
contractors, industrialists, developers, and representatives of government agencies if they
agree to the specifications of a permit within the broad legal constraints of the 404 law, the
Corps would confirm the agreement and call it a GP. The price of such an agreement is
consensus among the parties normally in conflict over permit applications. In this way the
Corps becomes the facilitator of consensus among interested parties by using its authority.
The Sanibel permit operated unchallenged for five years, the legal life of such a permit.
The Mississippi/Louisiana permit was just issued. These cases both confirm and question
some propositions emanating from the fields of negotiating and bargaining. (abstract)

—— 1989. Public Involvement, Conflict Management: Means to EQ and Social Objectives.
Journal of Water Resources Planning and Management 115 (1, January): 31–42. Engineers,
scientists, and even some social scientists prefer to look at water resources planning and
management as primarily analytical. However, more and more of the water professionals'
analytical work depends on people-oriented techniques either to relate their activities to
outside interests or to build better internal team relationships. Frequently, the major
problems that engineers and scientists face are not technical. They are problems of

reaching agreement on facts, alternatives, or solutions. Public involvement and conflict management techniques are key to servicing such needs. After briefly describing public involvement and conflict management techniques, seven observations on why incorporating social and environmental objectives into water resources planning and management require these process techniques are presented. (abstract)

Derman, Bill and Anne Ferguson. 1995. Human Rights, Environment, and Development: The Dispossession of Fishing Communities on Lake Malawi. *Human Ecology* 23 (2, June): 125–42. In a growing number of cases throughout Africa, communities' resource bases are being undermined or appropriated by outsiders, a process which seriously threatens the continuation of local cultures and livelihoods. In this article, we use a political ecology framework to examine how the linked processes of economic development, political power, and environmental change are transgressing the rights of fishing communities on the shores of Lake Malawi. In the cases described, these communities, or community members within them, find themselves powerless to prevent the expropriation of the resources over which they previously had either legal or customary control. Thus, it is not the economic processes of dispossession alone which lead to human rights violations but rather dispossession combined with an authoritarian political context. (abstract)

Deshan, T. 1995. Optimal Allocation of Water Resources in Large River Basins: I, Theory. *Water Resources Management* 9: 39–51. The purpose of this paper is to present the techniques of optimal allocation of water resources (OAWR) and to demonstrate how these methods can be employed in practice for studying both simple and complex water resources problems. Used as an application, the techniques and methods are applied to the OAWR of the Yellow river basin.

—— 1995. Optimal Allocation of Water Resources in Large River Basins: II, Application to Yellow River Basin. *Water Resources Management* 9: 53–66. This paper is an application of the theoretical approach of the optimal allocation of water resources in large river basins to the Yellow river basin. The objective of using a case study is to attain the optimization of water allocation to achieve the maximum national economic benefits and the optimal reservoir storage required to maintain the long-term balance of water resources.

Devlin, John F. 1992. Effects of Leadership Style on Oil Policy. *Energy Policy*, November, 1048–54. Neighbours, with similar systems of government and sharing a common ideological origin, Syria and Iraq have none the less had markedly different oil histories since 1980. Syria's oil production has tripled since 1982; Iraq's has yet to return to the 3.5 million barrels per day it reached in 1979 and 1980. The reason lies in their markedly different leadership. Asad in Syria has, in his patient, cautious style, virtually eliminated the country's reliance on oil or oil product imports; oil exports are earning substantial sums annually. Saddam Hussain in Iraq, ambitious to make himself and his country leaders in the Arab world, has made a series of decisions having disastrous consequences for Iraq's oil industry – invasion of and war with Iran, seizure of Kuwait, and rejection of UN terms for resumption of oil exports. Contrary to conventional wisdom, he has not authorized exports despite his country's real need for the earnings they would bring. (abstract)

Dewitt, David. 1994. Common, Comprehensive, and Cooperative Security. *The Pacific Review* 7(1): 1–15. An overview of the changing idea of security to one not dependent on military hardware or nuclear deterrence.

De Silva, K.M. 1994. Conflict Resolution in South Asia. *International Journal on Group Rights* 1(4): 247–67. South Asia has some of the most intractable political conflicts in the world today, and at three levels: international, national, and subnational. Conflict resolution in South Asia has three unusual features, beginning with the Sino-Indian dispute over their common border along the Himalayas. Second, the principal asymmetrical feature of the South Asian political system, the overwhelming dominance of India, makes multilateral negotiations over issues that involve India's vital interests. Third, separatist agitation, politicized religion and ethnic conflicts disturb the peace in many parts of South Asia. Potential conflicts of the future include disputes over the sharing of scarce resources, especially water and irrigation works; and the problem of refugees and displaced persons arising from the region's many disputes, as well as its problem of severe overpopulation. (abstract)

Dinar, Ariel, P. Seidl, H. Olem, V. Jorden, A. Duda, and Johnson R. 1995. *Restoring and Protecting the World's Lakes and Reservoirs*. (Technical Paper Number 289). Washington, DC: World Bank. The purpose of this report is to call on specialized UN agencies and other international support organizations and governments to take the necessary actions to make development activities more sustainable with regard to lakes and reservoirs. Aspects of threats to beneficial uses, ecosystem degradation as well as economics, regulations, institutions, and a comprehensive approach to managing, protecting, and rehabilitating lakes and reservoirs are examined. A series of 14 lake pollution and restoration case studies are included as appendixes to the paper.

Dorman, S. 1991. Who Will Save the Aral Sea? *Environmental Policy Review* 5(2): 45–54. The purpose of this article is to examine the problems facing Central Asia and Kazakhstan, economic, social, and environmental. The author's position is that many of these problems have been caused, at least in part, by the water management policies implemented in the region. Dorman suggests that a review of the literature indicates that the Central Asians and the Kazakhs believe that it is Moscow's responsibility to solve the water-resources problem in the region.

Druckman, D. 1993. The Situational Levers of Negotiating Flexibility. *Journal of Conflict Resolution* 37(2): 236–76. This study was conducted by examining the effects of a number of situational variables on decisions to be flexible or inflexible in a simulation of an international negotiation on the regulation of gases contributing to the depletion of the ozone layer. Druckman created four negotiating-stage scenarios using a combination of variables. The experiment was conducted using two international samples, one group of scientists, the other of diplomats from different countries.

Drysdale, A. 1992. Syria and Iraq – the Geopathology of a Relationship. *GeoJournal* 28(3): 347–55. The decision by Syria to support the anti-Iraq coalition during the Gulf War has to be understood within the wider political geographic and historical regional context. Mutual feelings of territorial injury in the demarcation of the states' territory and centrifugal forces threatening the continued integrity of the state, intense ideological and territorial disputes have resulted in bitter interstate enmity. Disputes have arisen over Syrian closure of Iraqi oil pipelines, the allocation of water flowing from the Euphrates and Syrian support of Iran during the Iraq-Iran war. The collapse of Syria's superpower ally, the Soviet Union, resulted in Syria supporting the anti-Iraq international coalition during the 1991 Gulf War, in an attempt to regain wider international legitimacy. (abstract)

Dryzek, John S. and Susan Hunter. 1987. Environmental Mediation for International Problems. *International Studies Quarterly* 31: 87–102. Environmental mediation has found some success as a mechanism for dispute settlement and problem solving in domestic settings. The prospects for its application in the international polity are explored here in the context of both localized transboundary issues and global environmental problems. Though facing a formidable set of necessary conditions, international environmental mediation could prove efficacious. Indeed, mediation offers one of the few methods for coping with environmental problems that retains the essentially decentralized character of the contemporary international political system. (abstract)

Dudley, N.J. 1992. Water Allocation by Markets, Common Property and Capacity Sharing: Companions or Competitors? *Natural Resources Journal* 32 (Fall): 766–78. This article explores the concept of capacity sharing, a new way of defining and allocating rights to flowing and stored surface water in a river valley. Also explored, as a point of comparison, is the literature on common property approaches to resource management.

Dudley, R.L. 1990. A Framework for Natural Resource Management. *Natural Resources Journal* 30 (Winter): 107–22. The actual practice of setting up multiple-use natural resource management on federal lands has been a case of "muddling through." No overall theoretical rationale exists on a nationwide basis. However, it is possible to envision a theoretical framework for managing natural resources based on a political economy paradigm of market failure/government failture. This paper proposes such a framework involving three parts. The first identifies resources and areas best served by private ownership or by public ownership. The second identifies principles and standards which would guide management. The third develops the bureaucracy needed to set up and manage the identified areas. The paper also looks at theoretical requirements for effective implementation of the proposed framework. (abstract)

Dufournaud, Christian. 1982. On the Mutually Beneficial Cooperative Scheme: Dynamic Change in the Payoff Matrix of International River Basin Schemes. *Water Resources Research* 18(4): 764–72. Utilizing metagame theory (game theory not dependent on rational behaviour of those involved), the author addresses the point at which one party will leave the agreement because it is advantageous at that point.

Dworsky, L.B. and A.E. Utton. 1993. Assessing North America's Management of its Transboundary Waters. *Natural Resources Journal* 33 (Spring): 413–59. This report is a summarization of the results and recommendations of a project entitled "The North American Experience Managing International Transboundary Water Resource: International Joint Commission and the International Boundary and Water Commission." This article presents nine issues the authors feel are of overriding importance to prepare the International Joint Commission (US-Canada) and the International Boundary and Water Commission (US-Mexico) for future challenges of change due to population growth, industrialization, greater demands on jointly owned resources, and shifting trading patterns.

Eagleson, Peter S. 1986. The Emergence of Global-Scale Hydrology. *Water Resources Research* 22 (9, August): 6S–14S. Emerging problems of environmental change and of long-range hydrologic forecasting demand knowledge of the hydrologic cycle at global rather than catchment scale. Changes in atmosphere and/or landscape characteristics modify the earth's metabolism through changes in its biogeochemical cycles. The most basic of these is the water cycle which directly affects the global circulation of both atmosphere and ocean and hence is instrumental in shaping weather and climate. Defining the spatial extent of the environmental impact of a local land surface change, or identifying, for forecasting purposes, the location and nature of climatic anomalies that may be casually linked to local hydrologic persistence requires global scale dynamic modelling of the coupled ocean-atmosphere-land surface. Development, evaluation, verification, and use of these models requires the active participation of hydrologists along with a wide range of other earth scientists. The current state of these models with respect to hydrology, their weaknesses, data needs, and potential utility are discussed. (abstract)

Eden, S. 1988. Negotiation and the Resolution of Water Allocation Disputes. M.Sc. thesis. Tucson: University of Arizona. Negotiation as a process for resolving water allocation disputes has advantages and disadvantages with respect to other dispute resolution methods. The principal advantages are derived from direct participation of interested parties. The chief disadvantages are that it cannot produce agreement in all conflicts and that such agreements as are reached may not adequately consider the public interest. No satisfactory method was found to evaluate the public interest content of negotiated settlements, although several paradigms are examined. Instead, the public interest was assumed to receive adequate protection in negotiations when all the parties with a stake in the outcome participate or are represented. (abstract)

El-Yussif, Faruk. 1983. Condensed History of Water Resources Development in Mesopotamia. *Water International* 8: 19–22. El-Yussif begins with water projects in Mesopotamia from 1900 BCE and continues forward.

Falkenmark, Malin, J. Lundqvist, and C. Widstrand. 1989. Macro-scale Water Scarcity Requires Micro-scale Approaches: Aspects of Vulnerability in Semi-Arid Development.
Natural Resources Forum (November): 258–67. This paper shows that water scarcity is a
complex problem when it affects countries with a semi-arid climate, ie countries for which
there are fluctuations between a dry season and a season when rain occurs. The paper
discusses the general vulnerability of the semi-arid zone in terms of four different types of
water scarcity, the effects of which are being superimposed on each other: two are natural
(type A, arid climate, type B, intermittent drought years) and two are man-induced (type
C, desiccation of the landscape driven by land degradation, and type D, population-driven
water stress). When fuelled by a rapid population increase, a risk spiral develops, mani-
festing itself in social and eoncomic collapse during intermittent drought years. The paper
concludes that many countries in Africa are heading for severe water scarcity – in fact
two-thirds of the African population will live in severely water-stressed countries within a
few decades. This severe water stress will largely be the result of unfettered population
growth. (abstract).

Fashchevsky, B. 1992. Ecological Approach to Management of International River Basins.
European Water Pollution Control 2(3): 28–31. The problems of an ecological approach to
water-resources utilization of transboundary rivers are discussed in this report. The need
for consideration of ecological factors, particularly qualitative ones, is illustrated by a
number of river basins. The importance of water regime and floodplain for a river's life
and the need to consider these factors in water resources management are pointed out.
Methods are suggested for ecological flow assessment in different flow probability years.
(abstract)

Fearnside, Philip M. 1993. Deforestation in Brazilian Amazonia: The Effect of Population
and Land Tenure. *Ambio* 22 (8, December): 537–45. LANDSAT data for 1978, 1988,
1989, 1990, and 1991 indicate that by 1991 the area of forest cleared had reached 426,000
km^2 (10.5% of the 4 million km^2 originally forested portion of Brazil's 5 million km^2 Legal Amazon Region). Over the 1978–1988 period, forest was lost at a rate of 22,000 km^2
yr-1 (including hydroelectic flooding), while the rate was 19,000 km^2 yr-1 for 1988–1989,
14,000 km^2 yr-1 for 1989–1990 and 11,000 km^2 yr-1 for 1990–1991. The reduction in the
rate since 1987 has mostly been due to Brazil's economic recession rather than to any
policy changes. The number of properties censused in each size class explains 74 per cent
of the variation in deforestation rate among the nine Amazonian states. Multiple re-
gressions indicate that 30 per cent of the clearing in 1991 can be attributed to small
farmers (properties <100 ha in area), and the remaining 70 per cent to either medium or
large ranchers. The social cost of reducing deforestation rates would therefore be much
less than is implied by frequent pronouncements that blame "poverty" for environmental
problems in the region. (abstract)

Fishelson, Gideon. 1992. Solutions for the Scarcity of Water in the Middle East in Times of
Peace. In *Kfar Blum Conference*. Kfar Blum Conference. Tel Aviv: Armand Hammer
Fund for Economic Cooperation in the Middle East, December. Tradeable water rights/
permits only occur at the local level, whereas nations use a quota system for water allocations. The author addresses water sharing, cooperation, conservation and augmentation.

Fisher, Franklin M. 1993. An Economic Framework for Water Negotiation and Management. Massachusetts Institute of Technology, 7 November. The author has created a
proposal, broken into small sections and based on economic analysis, to mitigate water
conflicts. He discusses Resource Allocation, Negative Externalities, Joint Management,
National Water Policies, and Property Rights.

Flack, J.E. and D.A. Summers. 1971. Computer-Aided Conflict Resolution in Water Resource Planning: An Illustration. *Water Resources Research* 7(6): 1410–14. The computer system illustrated in this article is called Cognograph. It is used to make explicit to water planners their sources of agreement and disagreement and to aid them in resolving their difference. The authors contend that conflict analysis that uses an interactive computer graphics system holds promise of helping decision-makers resolve the judgemental differences that arise in the planning process.

Flatters, Frank and Theodore Horbulyk, M. 1995. Water and Resource Conflicts in Thailand: An Economic Perspective. Prepared for Natural Resource and Environment Program, Thailand Development Research Institute, April. This paper, from the economist's perspective, is a study on recent conflicts over the use of water in Thailand. The examples used were drawn from case studies commissioned by the Natural Resources and Environment Program of the Thailand Development Research Institute (TDRI) for its study of water and resource conflict. A number of the important water-resource issues such as the allocation of surface and groundwater within and between river basins; and water quality and the discharge of effluent are surveyed in this paper. The paper concludes with suggesting some data and research needs that must be met to speed the adoption of economic criteria and instruments to resolve water use conflicts.

Flint, C.G. 1995. Recent Developments of the International Law Commission Regarding International Watercourses and Their Implications for the Nile River. *Water International* 20(4): 197–204. The International Law Commission of the United Nations has recently presented a set of Rules on the Non-Navigational Uses of International Watercourses to the General Assembly for approval. This article evaluates the potential implications of this new legal regime for the unusual but important case of the Nile river, which is dominated by the strength and control of its furthest downstream state, Egypt. The relationship between upstream and downstream riparian states is critical in determining the applicability of any new legal regime as is shown by the Nile river example. This article begins with a presentation and evaluation of the International Law Commission's new legal framework for international watercourses with primary focus on the General Principles. Following an evaluation of these principles, the international situation of the Nile river is described, illuminating the need for new legal solutions to historically troublesome water conflicts. The article concludes with an evaluation of the potential implications of the International Law Commission's Rules on the Non-Navigational Uses of International Watercourses for the Nile River and more broadly for international water law in general. (abstract)

Folk-Williams, J.A. 1982. Negotiation Becomes More Important in Settling Indian Water Rights Disputes in the West. *Resolve* (Summer): 1–5. The US Department of the Interior announced it would negotiate all pending disputes over Native American water rights and forego litigation. The author examines the role of the federal government and discusses several concepts of water rights such as the Winters Doctrine and reserved water rights.

—— 1988. The Use of Negotiated Agreements to Resolve Water Disputes Involving Indian Rights. *Natural Resources Journal.* 28 (Winter): 63–103. The purpose of this article is to summarize key problems and offer suggestions about negotiation processes that might be helpful to parties considering or engaged in negotiating Indian water cases. Also emphasized is that each case must be examined independently before a judgement is reached on whether or not negotiation can be used successfully.

Forster, Bruce. 1989. The Acid Rain Games: Incentives to Exaggerate Control Costs and Economic Disruption. *Journal of Environmental Management* 28 (4, June): 349–60. This paper contends that the lack of developments in acid rain control in the United States may be approached by considering Hartle's Intersection of Games: the special interest group game; the political game; and the bureaucratic game. The special interests groups – electric utilities, coal companies, coal miners and electricity consumers in the mid-west and east – have an incentive to emphasize scientific uncertainty of benefits of control while exaggerating the costs of control and economic disruption that would result in order to avoid or at least delay any legislation that would reduce their economic welfare. It is shown in this paper, using estimates from various sources, that the cost and disruption is not as firm as the opponents claim. The views of the politicians are consistent with the special interests for the regions they represent, but couch their opposition in the more altruistic sounding argument that benefits are uncertain while costs are high and known. The scientists and bureaucrats may also be caught in a bureaucratic game in which research activity is forced to justify policy positions rather than improve knowledge. This paper concludes using an insurance analogy that the various uncertainties provide a reason for acid rain control action rather than inaction in order to avoid potential irreversibilities associated with acid rain impacts. (abstract)

Fox, Irving K. and David LeMarquand. 1979. International River Basin Co-operation: The Lessons from Experience. *Water Supply and Management* 3: 9–27. This article is based on a report prepared for the United Nations Secretariat in which 10 case studies of successful arrangements for managing and using international water resources were summarized and assessed to provide a basis for arriving at some conclusions about how to create constructive action, to manage and utilize international water resources efficiently and fairly. In addition, drawing upon the experiences reported in the papers found in the Appendix, the authors attempt to illustrate how cooperative action has been achieved and the kinds of institutions which have facilitated cooperative action. The concluding section is a summarization of the kinds of measures and institutional arrangements that experience suggests will foster the best use of international water resources.

Fox, Jefferson, John Krummel, Sanay Yarnasarn, Methi Ekasingh, and Nancy Podger. 1995. Land Use and Landscape Dynamics in Northern Thailand: Assessing Change in Three Upland Watersheds. *Ambio* 24 (6, September): 328–34. This study analysed human-induced loss and fragmentation of tropical forests in three upland watersheds in northern Thailand between 1954 and 1992. During this 38-year period, forest cover declined, agricultural cover increased, population and population density grew, and agriculture changed from subsistence to cash crops. These changes resulted in a spatially diverse landcape with implications for biological and cultural diversity, sustainable resource use, and the economic conditions of the region. By qunatifying the spatial and temporal patterns of tropical forest change, we have attempted to show how the landscape in these upland tropical forests is controlled by physical and biological, as well as social and economic, parameters. The study illustrates the hierarchy of temporal and spatial events that result in global biome changes. (abstract)

Frankel, N. 1991. Water and Turkish Foreign Policy. *Political Communication and Persuasion* 8: 257–311. This study examines Turkish perspectives on water management and the influence of water on Turkish domestic and foreign affairs. The purpose of Turkey's water management projects and their influence on Turkish relations with her neighbours are specific issues addressed in the article.

Fraser, N.M. and K.W. Hipel. 1984. *Conflict Analysis: Models and Resolutions, Series Volume 11.* New York, NY: North-Holland. Chapter 2, Garrison Diversion Unit is a large water-resources project in North Dakota that may eventually cause environmental damage in Canada and also in the United States. The purposes of this chapter are to provide a comprehensive example of a complex conflict study and thereby introduce some important considerations and features in conflict analysis. The authors contend to show how risk can be realistically considered in the conflict analysis of environmental controversies. Chapter 7, Computer Assistance in Conflict Analysis, presents algorithms for outcome removal which were implemented in FORTRAN and APL. An analysis of the Popular River conflict using the Conflict Analysis Program (CAP) is performed.

—— K.W. Hipel, J. Jaworsky, and R. Zuljan. 1990. A Conflict Analysis of the Armenian-

Azerbaijani Dispute. *Journal of Conflict Resolution* 34 (3, December): 652–77. The authors use a computer program to research the conflict models the authors created. The authors give a history of the conflict 1905–1990, their analysis of the conflict, and the computer-generated equilibria for their model.

Frederick, K.D. 1993. *Balancing Water Demands with Supplies, The Role of Management in a World of Increasing Scarcity.* (Technical Paper Number 189). Washington, DC: World Bank. This paper deals with how to balance water demands with supplies. The experience of OECD countries are examined. While one cannot apply the exact same management practices to all water demand problems, the author suggests that the principles described should be considered a value to those who wish to introduce demand management techniques into their nation's water equation.

Frederiksen, H.D. 1992. *Water Resources Institutions: Some Principles and Practices.* (Technical Paper Number 191). Washington, DC: World Bank. This article examines the history and state of resources development of variations in application of particular situations by water resources institutions. The nation's institutions are defined by the author as laws, customs, organizations, and all that is associated. This article summarizes the principles followed by many countries in dealing with the situations they are confronting.

—— 1996. Water Crisis in the Developing World: Misconceptions About Solutions. *Journal of Water Resources Planning and Management* (March/April): 79–87. The premise of this article is that there are severe consequences of the international community's confined perspectives on solutions of the water crisis in the developing world. Frederiksen suggests that there are four constraints that are inadequately considered in the debate dealing with the water crisis: (1) scarce time to act to meet the pending needs; (2) the limited measures available for securing essential water supplies; (3) the competing demands for funds to provide the means; and (4) minimal ability to manage unpredicted droughts. The case study used to show these consequences is India's Sardar Sarovar project.

Frey, Fred W. 1993. The Political Context of Conflict and Cooperation Over International River Basins. *Water International* 18(1): 54–68. The purpose of this article is to present a power-analytic framework and some initial steps toward such a theory. Examined in this paper are the basic concepts of conflict and cooperation, contributions of international law, and the best known typologies of conflict. Analysis is also included of the actors involved, relevant motivational factors, perceptual (cognitive) processes, and features of power structures. The setting for the application are examples from the Middle East.

Frisvold, George B. and Margriet F. Caswell. 1994. Transboundary Water Agreements and Development Assistance. International Conference on Coordination and Decentralization in Water Resources Management. Rehovot, Israel, 3–6 October. This paper considers the impact of international development assistance on transboundary water transfer and pollution abatement agreements. Section two of this paper presents an example of two countries applying for an assistance grant to develop jointly surface water resources and reduce water pollution. A game-theoretic model was developed to represent the negotiations of countries over the particulars of the development plan. In section three, the model results are used to examine how the bargaining outcome is affected. Section four examines the impact of how allocation of water rights or environmental regulations affect the distribution of gains from bargaining. Section five considers how alternative opportunities and pre-existing development plans affect optimal grant programme design.

Galtung, Johan. 1994. Coexistence in Spite of Borders: On the Borders in the Mind. In *Political Boundaries and Coexistence: Proceedings of the IGU-Symposium, Basle/Switzerland, 24–27 May,* ed. Werner A. Galluser, 5–14. New York, NY: Peter Lang. The basic thesis of the paper is that as geographical, political borders are gradually losing their significance, for many places in the world the borders in the mind, based on the nation (which in turn is based on culture and particularly the macro-cultures known as civilizations) become more salient. Concretely this means that goods and services, labour and capital, people and ideas will flow across old borders, but mainly within the same culture. At that point new goegraphical borders will rapidly increase in salience, defining super-states harbouring super-nations. An example is, of course, the European Union based on Catholic-Protestant Christianity and Latin-Germanic languages. Similar developments will probably take place in Orthodox-Slavic and Muslim-Turkish regions. The paper then discusses the problems that arise when such nations see themselves as chosen by God (and others as chosen by the Devil), and indicates some possible solutions. (abstract)

Gleick, Peter. 1992. Effects of Climate Change on Shared Fresh Water Resources. In *Confronting Climate Change*, ed. I.M. Mintzer, 127–40. Cambridge University Press, for Stockholm Environment Institute. Climate change will not only affect the level and location of the seas. It will also alter the timing, and distribution of precipitation and runoff – the renewable sources of freshwater on which human societies and natural ecosystems depend. The paper analyses the implications of general circulation modelling experiments for rainfall, soil moisture, and streamflows, and notes that the potential impacts in some regions may be severe. Growing populations may add more demand for this water. These effects may be particularly important where two or more nations depend heavily on shared rivers or lakes. Where water resources are already tightly stretched, freshwater availability could become a military security development for some struggling nations. Even if climate change does not take place, concerns over shared water resources will probably become an increasingly important part of international relations in the future. Minimizing potential damage and conflict will require changes in water management strategies, improved water efficiencies (in supply and use), and a systematic evolution of international strictures on military/environmental aggression. Ultimately, sharing water resources will require unprecedented levels of regional cooperation. (Editor)

—— 1992. *Water and Conflict*. Environmental Change and Acute Conflict, Vol. 1. Toronto, Canada: University of Toronto and the American Academy of Arts and Sciences. Gleick's article of 1993 (below) is a modification and updated version of this paper.

—— 1993. Water and Conflict: Fresh Water Resources and International Security. *International Security* 18(1): 79–112. This article explores the relationship between water resources and international security. The author provides ways in which water resources have historically been the objectives of interstate conflict and how they have been used as instruments of war. Several quantitative indices for measuring the vulnerability of states to water-related conflict are presented. The sections of this paper include: Environment, Resources, and International Security; The Geopolitics of Shared Water Resources; Resource Inequities and the Impacts of Water Developments; Future Conflicts over Water; Indices of Water-resources Vulnerability; Reducing the Risks of Water-Related Conflicts; and Conclusions.

—— 1993. Water and War in the Middle East. Briefing for U.S. Congress. Washington, DC: Energy and Environmental Study Institute, 5 November. A relatively general briefing for the US Congress. Discusses water as a flashpoint (along with other related tensions) in the Middle East.

—— 1993. *Water in Crisis: A Guide to the World's Fresh Water Resources*. New York, NY: Oxford University Press. This book is a guide to the world's freshwater resources providing background information on critical water issues and water data on both global and regional scales. This volume consists of nine essays on freshwater issues by various authors: An introduction to global freshwater issues; World freshwater resources; Water quality and health; Water and ecosystems; Water and agriculture; Water and energy; Water and economic development; Water, politics, and international law; and Water in the 21st century. Data are also included on a variety of subjects: global and regional fresh water resources; rivers, lakes, and waterfalls; sanitation and water-related disease; water quality and contamination; water and agriculture; water and ecosystems; water and energy; water and human use; water policy and politics; and units, data conversions, and constants.

—— 1994. Water, War and Peace in the Middle East. *Environment* 36 (3, April): 6–42. The purpose of this article is to explore the water resource problems in the Middle East. This article stresses the need to manage jointly the shared water resources of the region which Gleick suggests as an unprecedented opportunity to move toward an era of cooperation and peace. Some historic information as well as general data about the various water resources in the Middle East are discussed.

Goldie, L.F. 1985. Equity and the International Management of Transboundary Resources. *Natural Resources Journal* 25 (3, July). Discusses how equitable principles of international law can provide guidelines and benchmarks for treaty makers and policy-makers. Addresses cooperative management in place of competitive management to create economic efficiency, and criteria for such measures.

Gradus, Y. 1994. The Israel-Jordan Rift Valley: A Border of Cooperation and Productive Coexistence. In *Political Boundaries and Coexistence: Proceedings of the IGU-Symposium, Basle/Switzerland, 24–27 May*, ed. Werner A. Galluser. New York, NY: Peter Lang. This article is a proposal for binational cooperation between Israel and Jordan in the areas of transportation, natural resource exploitation, water, agriculture, and tourism. The proposed water issue cooperation is joint management of the water table and the Dead Sea.

Grover, B. and M. Jefferson. 1995. A World Water Council: One Possible Model. *Water Resources Development* 11(2): 125–38. Discusses the World Energy Council as a possible model for a new Water Council. An annual budge of US$2 million is likely to be adequate. It is somewhat unlikely that any existing water-related international organization can be transformed successfully into a World Water Council. (abstract)

Guariso, Giorgio, Dale Whittington, Baligh Shindi Zikri, and Khalil Hosny Mancy. 1981. Nile Water for Sinai: Framework for Analysis. *Water Resources Research* 17 (6, December): 1585–93. The purpose of this article is to explore the concept of transporting Nile water to the Sinai. To complete this analysis, the authors formulate a multi-objective programming model in order to examine the trade-offs between the economic and political objectives and to study their interrelationships with such variables as water quantity, water quality, water transport costs, crop rotations, and irrigation technology. The analysis results provides evidence that shows that official plans largely ignore the sensitivity of the value of water in the rest of the country with regards to the reclamation efforts in the Sinai.

Handley, Paul. 1993. River of Promise. *Far Eastern Economic Review*, 16 September, 68–72. View of the economic efforts of Mekong basin countries to benefit trade and development along the river.

Hayes, Douglas L. 1991. The All-American Canal Lining Project: A Catalyst for Rational and Comprehensive Groundwater Management on the United States-Mexico Border. *Natural Resources Journal* 31 (Fall): 803–27. This article is a call for cooperation between the United States and Mexico to look at the current dispute over the All-American Canal Lining Project as an opportunity to mutually develop these groundwater resources instead of producing short-term results to the detriment of the sharing neighbour. Both the United States and Mexico claim rights to the groundwater currently leaking from the All-American Canal under the 1944 Colorado River Treaty.

Haynes, Kingsley and Dale Whittington. 1981. International Management of the Nile – Stage Three? Water Supply. *Geographical Review* 71 (1, January): 17–32. The purpose of this article is to show that a new management stage may soon be required to recognize the shift from concerns about water quantity to water quality consideration and from single large-scale project development to small-scale project planning. The authors suggest that the new third stage of basinwide management will be an expansion of international riparian interests and increased operations complexity of the Nile.

Hayton, R.D. 1993. The Matter of Public Participation. *Natural Resources Journal* 33 (2, Spring): 275–81. Examines the current status of and recent progress in cooperative arrangements for the development of water resources shared by two or more countries. Such arrangements may range from the simple exchange of data to the design and implementation of major projects and formal resolutions of disputes. Topics which are of growing concern to river-basin organizations include integrated development and management of shared water resources and dealing with the freshwater-maritime interface. The article indicates that optimum coordinated development, use and protection of shared water resources is still a distant goal, and that increased institutionalization of cooperation is required. (abstract)

—— and A.E. Utton. 1989. Transboundary Groundwaters: The Bellagio Draft Treaty. *Natural Resources Journal* 29 (Summer): 663–722. This article presents the Bellagio Draft Treaty, a draft international groundwater treaty developed by a group of multidisciplinary specialists over an eight-year period. Added commentary by the authors provides greater understanding to the 20 articles found in the treaty. The purpose of the Bellagio Draft Treaty was to provide a system for the mutually agreed management of international aquifers in critical areas.

Hennessy, J. and N. Widgery. 1995. River Basin Development – the Holistic Approach. *International Water Power and Dam Construction* 47(5): 24–26. The holistic approach to water management advocated by the International Committee on Irrigation and Drainage (ICID) is discussed. Appropriate water management is defined as the use of the right solution to meet development needs in a particular environment sustainably. Underpinning principles are outlined. Institutional changes and a radical reappraisal of management practices will be required. Some initiatives and examples are presented. These include the Lesotho Highlands water project in South Africa; multi-purpose resource development of the Komati river basin in Swaziland; and the project to heighten the Roseires Dam in Sudan. (J.M. McLaughlin)

Heraclides, Alexis. 1989. Conflict Resolution, Ethnonationalism and the Middle East Impasse. *Journal of Peace Research* 26(2): 197–212. From the perspective of "ethnic" conflict resolution, the author categorizes (14 separate categories) state-ethnonationalist conflicts. Identifies 10 problems whose resolution is necessary for forward movement of the peace process.

Hill, Barbara J. 1982. An Analysis of Conflict Resolution Techniques: From Problem-Solving Workshops to Theory. *Journal of Conflict Resolution* 26 (1, March): 109–38. Discusses problem-solving workshops as developed by Burton, Doob, and Kelman, and suggests how such theories might be implemented.

Hipel, K.W. and Niall M. Fraser. 1980. Metagame Analysis of Garrison Conflict. *Water Resources Research* 16(4): 629–37. Metagame analysis is a type of game theory which can be employed for assessing the political feasibility of large-scale water-resources projects. The Garrison Diversion Unit, a planned irrigation scheme in North Dakota has caused a complex international controversy, which is analysed by using metagame analysis. Based upon the options available to the participants in the conflict and also the preferences of the participants, metagame analysis is used to predict possible feasible political solutions to the Garrison conflict. In addition, metagame analysis provides a framework for systematically studying the Garrison dispute so that the political complexities of the problem can be understood and put into proper perspective. (abstract)

—— R.K. Ragade, and T.E. Unny. 1976. Metagame Theory and Its Applications to Water Resources. *Water Resources Research* 12 (3, June): 331–39. The authors use metagame theory to examine water resource conflicts because it is a positive descriptive approach for political resolution. This article illustrates that resource problems that are formulated using classical game theory, based on the theory of rationality, do not work. The results show that the conflicts examined in this article portray the potential of metagame theory for analysing water-resource management problems.

Hof, Frederick. 1995. The Yarmouk and Jordan Rivers in the Israel-Jordan Peace Treaty. *Middle East Policy* 3 (4, April): 47–56. Following a short history of Israel-Jordan water politics and conflict, Hof discusses key elements of the 1994 peace treaty. Syria and the Palestinian Entity are given a brief treatment as well.

Hofius, K. 1991. Co-operation in Hydrology of the Danube Basin Countries. In *Hydrology for the Water Management of Large River Basins*, ed. F.H.M. Van de Ven et al., 37–43. International Association of Hydrological Sciences. Since 1965 the Danube basin countries, ie the FRG, Austria, Czechoslovakia, Hungary, Yugoslavia, Rumania, Bulgaria, and the Soviet Union, have been cooperating under the International Hydrological Decade and the International Hydrological Programme. The type and the results of that cooperation are discussed. The first part of the paper deals with the organization of the cooperation, which at the beginning was difficult because of the different social systems of the Danube Basin countries. The cooperation under the International Hydrological Programme includes joint conferences at two-year intervals on hydrological forecasts and the preparation of a monograph on the hydrology of the Danube basin. After the completion and publication of the monograph, the Danube countries now continue work on four specific projects. In the second part of the paper the hydrological characteristics of the flow regime of the Danube are presented. The author of the paper is at present the co-presenter, and is at present coordinator of the Danube countries for the follow-up programme to the monograph. The paper is submitted by order of the Danube basin countries. (abstract)

—— 1991. Cooperation in Hydrology of the Rhine Basin Countries. In *Hydrology for the Water Management of Large River Basins*, ed. F.H.M. Van de Ven et al., 25–35. International Association of Hydrological Sciences. Since 1970 the Rhine basin countries, ie Austria, Switzerland, FRG, France, Luxembourg and the Netherlands, are cooperating under the International Hydrological Decade and the International Hydrological Programme. The type and the results of that cooperation are discussed. The first part of the paper deals with the administrative possibilities of implementing the cooperation of several states bordering a large river basin. It is important that the programmes to be set up are not too comprehensive so that they lead to results within a reasonable time. Particular problems arise in integrating the results at the borders. Isolines, for instance, do not agree with one another etc. Through the cooperation of the Rhine basin countries experience has been gained, which may be of use for other states in their cooperation for large river basins. The results of the work achieved so far by the International Commission for the Hydrology of the Rhine Basin established in 1970 are presented. In the second part of the paper the most striking hydrological characteristics of the flow regime of the Rhine are presented. (abstract)

Hori, Hiroshi. 1993. Development of the Mekong river basin, Its Problems and Future Prospects. *Water International* 18: 110–15. This paper describes the necessity, history, and problems of the development and the status of the entire Mekong river basin, and describes prerequisites for sustainable development.

Hosh, Leonardo and Jad Isaac. 1992. Roots of the Water Conflict in the Middle East. In *The Middle East Water Crisis*. The Middle East Water Crisis. University of Waterloo, 7–9 May. Discusses political boundaries and water sources, paths, and springs. "Today's boundaries in the Middle East are, primarily, artificial frontiers imposed within the past 75 years by distant foreign powers." The authors then examine previous proposals for Middle East peace, identifying mistakes and defining portions of a new plan.

Howell, P.P. and J.A. Allan. 1994. *The Nile: Sharing a Scarce Resource. An Historical and Technical Review of Water Management and of Economic and Legal Issues.* Cambridge, MA: Cambridge University Press. Following recent climatic fluctuations, and consequent fluctuations in river flow in the Nile basin, forecasts of water availability will have to be greatly revised and new management strategies formulated to adapt to the new conditions. This book contains 19 papers presented at a conference held to address these issues. Papers are divided by theme into four sections. Section I contains five papers outlining the environmental history of the Nile and its past management. Section II contains two papers addressing the environmental data needs for predicting environmental change and devising future management strategies. The third section contains six papers proposing future strategies for managing water supply in the Nile basin. Individual papers discuss strategies for Egypt, Sudan, and Uganda. The final section contains six papers addressing international agreements, legal issues, and economic considerations for the effective management of the Nile Basin. (N. Davey)

Huddle, Franklin P. 1972. *The Mekong Project: Opportunities and Problems of Regionalism.* Report to House Committee on Foreign Affairs. Washington, DC: US Government Printing Office. Explores the concept of regionalism as a technique for systematically applying science and technology to bring Southeast Asia out of its series of conflicts. Gives detailed history of Mekong development.

Hulme, Mike and Mick Kelly. 1993. Exploring the Links Between Desertification and Climate Change. *Environment* 35 (6, July/August): 4–11. The African Sahel and many of the people it has tenuously supported are retreating southward, giving way to the inhospitable Sahara Desert. Why is the Sahara spreading? Is unsustainable land use or climate change to blame? And, if climate change is a factor, how much of that change is natural and how much results from all the greenhouse gases people have pumped into the atmosphere? Before an effective attempt can be made to stop desertification, its causes – and particularly any self-reinforcing cycles – must be understood. (abstract)

Ingram, H., and D.R. White. 1993. International Boundary and Water Commission: An Institutional Mismatch for Resolving Transboundary Water Problems. *Natural Resources Journal* 33 (Winter): 153–200. The purpose of this article is to analyse the performance of the International Boundary and Water Commission (IBWC) on the basis of its established record in resolving water problems. The authors recognize the usefulness of the IBWC model for other nations that share boundaries. However, criticisms of the IBWC are also explored through a review of scholarly and professional evaluations of the IBWC, particularly related to the Commission's United States section's ability to respond to state and local problems.

Islam, N. 1992. Indo-Bangladesh Common Rivers: The Impact on Bangladesh. *Contemporary South Asia* 1(2): 203–25. This article deals with the environmental and legal issues arising from Indo-Bangladesh common rivers, the diplomatic problems over water usage and the impact on Bangladesh's economy and society and indeed its security in the long term. It finds that the problem of sharing the water of the common rivers has soured the relationship between India and Bangladesh. The competing claims for land where meandering rivers have altered land structure have already taken the form of armed skirmishes between the security forces of the two countries. Failure to tackle the problem could have a disastrous impact on the environment. The multinational character of the river system necessitates multinational cooperation, aiming at securing bilateral and regional cooperation and security. (abstract)

Jacobs, Jeffrey W. 1994. Toward Sustainability in the Lower Mekong River Basin Development. *Water International* 19: 43–51. Discusses deforestation, water development projects, and the Mekong Committee.

Jamail, M.H. and Stephen Mumme. 1982. The International Boundary and Water Commission as a Conflict Management Agency in the US-Mexico Borderlands. *The Social Science Journal* 19(1): 45–60. The analysis completed in this article is a concentration on the development of the International Boundary and Water Commission (IBWC) since the signing of the US-Mexican Water Treaty in 1944. New problems facing the IBWC and the growing political dominance in US-Mexican relations are examined.

Jordan, Jeffrey L. 1992. *Resolving Intergovernmental Water Disputes Through Negotiation.* Athens, GA: University of Georgia. Examines the conflict between a city and county government in the southern United States. An 11-step process was used to solve the dispute under consideration.

Jovanovic, D. 1985. Ethiopian Interests in the Division of the Nile River Waters. *Water International* 10(2): 82–85. Jovanovic points out the flaws in the current Nile water allocations. Then he offers some possible solutions to future division of the waters. One must consider rain-fed agriculture, population, and maximum return per hectare of irrigated land. He also addresses non-water natural resources within a country that could support the economy.

Judge, Shana. 1994. The Nile: River of Hope or Conflict? *Transboundary Resources Report* 8 (2, Summer): 1–3. This report is a brief look into the current legal regime governing the Nile river. It is also a call for cooperation among all riparian states to form a comprehensive legal regime.

Just, Richard E., John K. Horowitz, and Sinaia Netanyahu. 1994. Problems and Prospects in the Political Economy of Trans-Boundary Water Issues. University of Maryland, College Park, MD, September. The main focus of this paper is the development and applied framework used to evaluate the potential for transboundary cooperation for water-resource sharing. The authors identify two distinct stages in water allocation decisions, the design and payment for large-scale projects such as reservoirs and dams, and the problem of who is allocated the use of the water must be solved. The case study used in this article is the relationship between Israel and its neighbours.

Kahhaleh, Subhi. 1981. *The Water Problem in Israel and its Repercussion on the Arab-Israeli Conflict.* Beirut: Institute for Palestine Studies. History and political analysis of Jordan river issues. Contains good figures and numerical references to dams and rivers in the area.

Kally, Elisha. 1989. The Potential for Cooperation in Water Projects in the Middle East at Peace. In *Economic Cooperation in the Middle East,* ed. Gideon Fishelson, 303–25. Boulder, CO: Westview Press. This article suggests that every project is independently feasible, but technical, economic, and political conditions as well as other projects being done in this framework will determine whether a project is implemented. The projects used in this study are the conveyance of Nile water eastward to the Gaza Strip, and a joint Jordanian-Israeli project for utilizing the Yarmouk river. Kally suggests that it is possible to envisage different combinations of various projects, and only the concrete shape of the peace will determine the "basket" of projects that could be undertaken.

Karsh, Efraim. 1991. Neutralization: The Key to an Arab-Israeli Peace. *Bulletin of Peace Proposals* 22(1): 11–24. Discusses a strategy for neutralizing the entire Arab-Israeli "sector." Provides US and Russian views of neutrality. The author realizes that the scheme is almost unbelievable.

Kassem, Atef M. 1992. The Water Use Analysis Model (WUAM): A River Basin Planning Model. In *Proceedings of the International Conference on Protection and Development of the Nile and Other Major Rivers,* Vol. 2/2. International Conference on Protection and Development of the Nile and Other Major Rivers. Cairo, Egypt, 3–5, February. This article presents the Water Use Analysis Model (WUAM) which unlike most river basin planning models built around the concept of "supply management," was developed by placing a special emphasis on water demand modelling and incorporating the key variables affecting water use, which are missing in the other planning models. The paper explores all aspects of WUAM and its potential for other water applications.

Kattelmann, Richard. 1990. Conflicts and Cooperation over Floods in the Himalaya-Ganges Region. *Water International* 15(4): 189–94. Flood damage near the Ganges has increased due to population pressures. Unsupported scientific theories are fuelling a conflict unnecessarily. Some engineering approaches have merit, but the more effective strategies are flood forecasting and warning.

Kaye, Lincoln. 1989. Resources and Rights: Rivalries Hamper Indo-Bangladesh Water Sharing. *Far Eastern Economic Review* 2 (February): 19–22. Discusses floodwaters in Bangladesh and the 1970 talks between India and Bangladesh, including some discussion of the water-sharing agreements.

—— 1989. The Wasted Waters. *Far Eastern Economic Review* 2 (February): 16–18. Kaye discusses India and the water needs of that country. He offers advice based on India's behaviour in the past when negotiating with riparian neighbours, including issues of economics of scale, information sharing, and multilateral negotiations.

Kelman, Herbert C. 1982. Creating the Conditions for Israeli-Palestinian Negotiations. *Journal of Conflict Resolution* 26 (1, March): 39–75. Kelman addresses necessary conditions for a win-win settlement of Israeli-Palestinian issues, working to overcome zero-sum ideas.

Kershner, Isabel. 1990. Talking Water: Secret U.S.-mediated Negotiations Could Herald Regional Cooperation in the Middle East. *The Jerusalem Report*, 25 October, 44–45. Kershner downplays the idea that the next Middle East war will arise out of a water dispute. She reports on behind-the-scenes mediation by the United States and water issues on the Yarmouk river.

Khan, M. Yunus. 1990. Boundary Water Conflict Between India and Pakistan. *Water International* 15 (4, December): 195–99. Details the agreements between India and Pakistan over the Indus system. Water was first apportioned with "preferential right to existing uses." Also includes information on the Permanent Indus Commission.

Khan, Tauhidul Anwar. 1994. Challenges Facing the Management and Sharing of the Ganges. *Transboundary Resources Report* 8 (1, Spring): 1–4. Problems of managing and sharing the Ganges. Includes negotiation history 1975–1992 and discussion of several proposed projects.

Khan, Z.A. 1976. *Basic Documents on Farakka Conspiracy From 1951 to 1976*. Dacca: Khoshroz Kitab Mahal. This book primarily uses newspaper references. It highlights the economic and transportation stresses that the Farakka diversion would place upon Bangladesh.

Kirmani, S. 1990. Water, Peace and Conflict Management: The Experience of the Indus and Mekong River Basins. *Water International* 15 (December): 200–05. Includes some history of the Indus and Mekong river problems. Proposes three points for negotiation strategy: understanding that without negotiations, one or more nations will be harmed; third-party negotiation; and financial support for management of the problem.

—— and R. Rangeley. 1994. *International Inland Waters: Concepts for a More Active World Bank Role*. Technical Paper, Vol. 239. Washington, DC: World Bank. This article is a review of the World Bank's role in international water affairs. The purpose of this report is to address concerns expressed by the delegates of an international workshop on comprehensive water resources management organized by the Bank in June 1991. The paper recommends that the Bank should play a more proactive role in international water affairs, its policy should be flexible, and its immediate focus should be to assist riparian countries in their efforts to establish cooperative arrangements.

Kishel, J. 1993. Lining the All-American Canal: Legal Problems and Physical Solutions. *Natural Resources Journal* 33 (Summer): 697–726. The purpose of this article is to explore the potential solutions to a developing transboundary conflict between the groundwater laws of the United States and Mexico. Kishel suggests that the conflict can be solved with physical and institutional solutions, which are developed outside of conventional transboundary dispute resolution mechanisms.

Kliot, Nurit. 1995. Building a Legal Regime for the Jordan-Yarmouk River System – Lessons from Other International Rivers. *Transboundary Resources Report* 9(1): 1–3. The purpose of this paper is to present possible regimes for the Jordan-Yarmouk river based on international regimes. These regimes are established by exploring treaties and agreements in other international rivers with particular attention to those rivers where there is conflict and in which the regime or institutional arrangement deal with consumptive water.

—— and Yoel Mansfeld. 1994. The Dual Landscape of a Partitioned City: Nicosia. In *Political Boundaries and Coexistence: Proceedings of the IGU-Symposium, Basle/Switzerland, 24–27 May*, ed. Werner A. Galluser, 151–61. New York, NY: Peter Lang. The 1974 Turkish invasion and occupation of North Cyprus imposed a new border on Cyprus and divided the small island into two entitites separated by a demilitarized zone and by the UN peacekeeping forces. The "Green Line" is a defacto border with very restricted cross-border transactions as the government of Cyprus acknowledges the northern territory as an occupied area and not as the (self-proclaimed) Turkish Republic of North Cyprus. Cooperation between the two partitioned parts is limited to sharing water. (abstract)

Kolb, Deborah M. and Susan S. Silbey. 1990. Enhancing the Capacity of Organizations to Deal with Disputes. *Negotiation Journal* 6 (4, October): 297–304. Deals with conflict within existing businesses or other such organizations. They address prevention, the capacity to handle disputes and barriers to a larger such capacity. The authors affirm that conflicts are inevitable and one key to resolving them amicably is to make sure that all involved feel their needs will be addressed.

Kowalok, Michael E. 1993. Common Threads: Research Lessons from Acid Rain, Ozone Depletion, and Global Warming. *Environment* 35 (6, July/August): 12–20. Research on environmental hazards is often haphazard: studies from different disciplines suddenly fit together in an unexpected way; scientists geographically far apart come to similar conclusions; and what initially appears to be a minor effect turns out to be critical. This brief account of the research leading to the discovery of three major environmental threats demonstrates that successful environmental research, despite its unpredictability, has several important characteristics. (abstract)

Kremenyuk, Victor, ed. 1991. *International Negotiation: Analysis, Approaches, Issues*. San Francisco: Jossey-Bass. This book intentionally concentrates on aspects of negotiation with least relation to political considerations. It traces the role of applied systems analysis, history of conflict negotiation at the International Institute for Applied Systems Analysis, and several approaches to international negotiation.

Kriesberg, Louis. 1988. Strategies of Negotiating Agreements: Arab-Israeli and American-Soviet Cases. *Negotiation Journal* (January): 19. The author compares situations of international conflict that do and that do not result in de-escalation of the problem, from the 1947 Austrian State US-Soviet talks to the 1986 PLO-Hussein-Peres talks.

Kuffner, Ulrich. 1993. Water Transfer and Distribution Schemes. *Water International* 18 (1, March): 30–34. Wherever water is needed, solutions have been sought to bring it to cities, industries, and dry fertile lands. Large water transfer schemes have a long history dating back thousands of years. In modern times, water transfer projects have been built in several countries, including Israel. They have found wide interest and have drawn attention to similar solutions for the Middle East. The gradual transformation of transfer schemes into large-scale distribution networks has found less attention. In Israel and, to a lesser extent in Jordan, extensive distribution systems bring water over large distances from various sources to the main consumption centres. The advantages of these networks include the possibility of balancing supply and demand over large distances, of pooling the financial resources of communities and regions, of sharing important structures such as reservoirs and treatment facilities, and providing greater security against local supply failures. It is suggested that consideration should be given to the expansion of large-scale distribution networks across national borders. The gradual acceptance of water as an economic good in the international community may facilitate the acceptance of such a solution. Reference is made to World Bank assistance to water projects in the Middle East and to World Bank support for international river basin development in the Indus basin and, more recently, in Southern Africa. (abstract)

Landau, George D. 1980. The Treaty for Amazonian Cooperation: A Bold New Instrument for Development. *Georgia Journal of International and Comparative Law* 10 (3, Fall): 463–489. All eight countries of the Amazon basin signed the TAC (Treaty for Amazonian Cooperation) on 3 July 1978. This article details the aspects of the treaty, a history of the negotiations, issues raised by the treaty, and the strategy for development that the treaty proposes.

Lee, Kai N. 1982. Defining Success in Environmental Dispute Resolution. *Resolve* (Spring): 1–3. The author seeks to find a measure or benchmark of success when using non-adjudicatory methods for environmental disputes.

Lee, T. 1992. Water Management Since the Adoption of the Mar del Plata Action Plan: Lessons for the 1990s. *Natural Resources Forum* (August): 202–11. In the 13 years since the United Nations Water Conference, the policies applied to the administration of water resources have undergone considerable modification in most countries. For most of this period no overall trend in the direction of change can be easily seen. Recently, however, with the general adoption of policies decentralizing water management responsibilities away from central governments, an opportunity has been presented for the general application of some of the basic management principles enunciated in the Mar del Plata Action Plan. This paper presents a review of current water administration policies and of water management problems in Africa, Asia, Latin America, and the Caribbean. (abstract)

LeMarquand, David. 1989. Developing River and Lake Basins for Sustained Economic Growth and Social Progress. *Natural Resources Forum* (May): 127–38. The purpose of this article is to provide a context for evaluating how best to advance well-conceived water-resources development in Africa. The author looks at: river basin development planning concepts and experiences as they have evolved; their application and aptness to developing countries, particularly in Africa; the role of foreign assistance with emphasis on multidonor financing; and the complications of developing water resources shared by two or more states. The conclusion suggests that concepts of multipurpose projects, river basin planning, river basin institutions and regional development provides a useful framework for planning the various water-related issues facing Africa.

Lesser, J.A. 1990. Resale of the Columbia River Treaty Downstream Power Benefits: One Road from Here to There. *Natural Resources Journal* 30 (Summer): 609–28. Joint Canada-United States development of the Columbia river system was made possible by the Columbia River Treaty. To finance the construction of the storage projects it would be required to build, Canada sold its share of the additional hydroelectric power benefits made possible by the treaty to the United States. That power will be completely repatriated to Canada by 2003. Unless a new sale is arranged, the Pacific Northwest may have to replace as much as 600 average megawatts of energy, and 1,400 megawatts of capacity. The issues associated with any potential resale, however, will be complex. This paper presents estimates of the value of the Canadian share of energy from the perspective of the Pacific Northwest, and discusses policy issues that eventually will have to be addressed by both countries. (abstract)

Libiszewski, Stephan. 1994. Sources of Life, Sources of Strife. *Swiss Review of World Affairs*

6: 8–10. Water is being wasted and polluted everywhere, while population growth strains the supply. International tensions over this vital resource are increasing so much that in a number of places where important rivers cross borders, war could break out. This article focuses on the Tigris-Euphrates basin, where ethnic hostilities and power politics among nations only amplify the conflicts over water. The author is a scholar working on an international research project called "The Environment and Conflict" at the Research Institute for Security Policy and Conflict Analysis at the Swiss Institute of Technology (ETH) in Zurich. (abstract)

Linnerooth, J. 1990. The Danube River Basin: Negotiating Settlements to Transboundary Environmental Issues. *Natural Resources Journal* 30: 629–58. This article explores the environmental degradation factors as major issues surrounding the Danube river. Detailed are both the scientific and institutional complexities involved in negotiating agreements among the Danube riparian nations. Forms of cooperative action are suggested as well as the potential role for an independent analyst in the negotiation process.

Linnerooth-Bayer, Joanne. 1993. Current Danube River Events and Issues. *Transboundary Resources Report* (Winter): 7. A brief discussion of the Gabcikovo-Nagymaros hydroelectric project and the Rhine-Main-Danube canal.

Lonergan, Stephen C. and David B. Brooks. 1994. *Watershed: The Role of Fresh Water in the Israeli-Arab Conflict.* Ottawa: International Development and Research Centre. A detailed and comprehensive view of the Arab-Israeli water conflict. Of particular note is Chapter 8, "Water and Security in Israel."

Lord, W. 1980. Water Resource Planning: Conflict Management. *Water Spectrum* 12: 2–10. The purpose of this article is to explore whether there are institutional obstacles to effective conflict management within the planning process. There is a focus on the institutional aspects of water-resources decision-making. The author's conclusion is that there are aspects of existing institutions which militate against effective conflict management.

Lynne, Gary D., J. Walter Milon, and Michael E. Wilson. 1990. Identifying and Measuring Potential Conflict in Water Institutions. *Water Resources Bulletin* 26 (4, August): 669–76. Scarcity combined with differences in values, beliefs, and attitudes can lead to behaviour differences and conflicts over water. This paper develops an index for measuring potential conflict using survey information about water attitudes and beliefs of individuals in three groups in a Florida case study. The index helps in assessing the current capability of the institution to reduce conflict. The results suggest that the current institution is effective, but changes may help to streamline the consumptive-use permitting process, to improve educational programmes, and to seek improved institutional arrangements to reduce future conflict over economic uses of water. (abstract)

MacAvoy, Peter V. 1986. The Great Lakes Charter: Toward a Basinwide Strategy for Managing the Great Lakes. *Case Western Reserve Journal of International Law* 18(49): 49–65. A history of the Great Lakes Charter – an eight-state, two-province pact signed in February 1985. It outlines tenets of the charter and offers some next steps in basinwide strategy.

MacDonnell, L.J. 1988. Natural Resources Dispute Resolution: An Overview. *Natural Resources Journal* 28 (Winter): 5–19. This article is the introductory piece in an issue of *Natural Resources Journal* devoted to emerging alternative approaches to addressing natural resources-based disputes. Under consideration are the sources and types of conflict in natural resources, general approaches to dispute resolution, and choices in determining which approach to use. Included is a table of the distribution of environmental disputes.

Mageed, Y.A. and G.F. White. 1995. Critical Analysis of Existing Institutional Arrangements. *Water Resources Development* 11(2): 103–11. The need for improved institutional arrangements in water resources management has been recognized for many years; prior to Mar del Plata, and most recently at Dublin and Rio. Consensus seems to be emerging that a new global organization should be created from representatives of local, national., regional, and international organizations embracing environmental, economic, and political concerns. It would promote exchange of information and experience to define issues and methods deserving of attention, and would critically appraise previous actions in selected sectors and areas of water management. It should not duplicate existing organizations. (abstract)

Mahendrarajah, S. and P.G. Warr. 1991. Water Management and Technological Change: Village Dams in Sri Lanka. *Journal of Agricultural Economics* 42(3): 309–24. This paper studies the intertemporal allocation of monsoonal water storage in village dam-based irrigation systems in the dry zone of Sri Lanka. The tradiational water management practices observed in these villages are based on common property access and serve to minimize social conflict over water rights. They are also acceptably efficient in economic terms, given the water demands of the traditional rice production technology. Adoption of high-yielding variety (HYV) rice technology produces a dramatic increase in rice output, but the traditional water-management practices then become less efficient. The paper demonstrates a method for determining the nature of an efficient water-management system and for estimating the economic magnitude of the inefficiency arising from the traditional practices. In the case study, efficient water management increases the gains available from HYV adoption by a further one-fourth. (abstract)

Maluwa, T. 1988. Legal Aspects of the Niger River Under the Niamey Treaties. *Natural Resources Journal* 28 (Fall): 671–97. The primary conclusion of this study is that the West African conventions, the Niamey Treaties, should be regarded as constituting an undeniable contribution to the development of international fluvial law in general. As such, concepts embodied in these treaties have been found to be expresses in preliminary drafts being considered by the International Law Commission in a codification programme. This article contains primarily legal terminology.

Marr, P. and W. Lewis, eds. 1993. *Riding the Tiger: The Middle East Challenge After the Cold War*. Boulder, CO: Westview Press. A collection of essays dealing with negotiation and policy issues in the aftermath of the USSR's collapse. Includes analysis of military, resource, ethnic, and religious issues, by various authors.

Mateo, R.M. 1992. Administration of Water Resources: Institutional Aspects and Management Modalities. *Natural Resources Forum* (May): 117–25. While water has always been important to people, recent pressures of population, ecology, geography, and economic development have created new demands for water and call into question the old institutional arrangements for the administration of water resources. Drawing upon experience, largely from Spain and Latin America but including some examples from Europe, this paper examines the role that private enterprise and water cooperatives could play in promoting more efficient water use. The paper concludes that private management techniques under public control may prove useful but are no panaceas. (abstract)

Mather, T. 1989. The Planning and Management of African River and Lake Basin Development and Conservation. *Natural Resources Forum* (February): 59–70. The purpose of this article is to identify the constraints which impede development of African river basin resources. The author suggests these identifications should be included to aid the predicted outcomes of planned proposals for the ultimate successes of these plans. The identification of these constraints include physical and climatic reasons, socio-cultural characteristics and priorities of national economies.

Matsuura, Shigenori. 1995. China's Air Pollution and Japan's Response to It. *International Environmental Affairs* 7 (3, Summer): 235–48. Long-range transport of air pollutants has been a serious global environmental problem. Japan's Central Research Institute of Electric Power Industry (CRIEPI) researchers are investigating the possibility of acid rain from China. So far, the acid rain that has appeared in Japan has not had a significant effect. However, because of the high economic growth projected for the coming years in China, acidic fallout from that country is likely to cause serious damage there and in Japan as well. China is one of the largest coal users in the world. One of the biggest factors making air quality worse is the lack of appropriate technologies. Although Japanese desulphurization technologies are well advanced, they cannot be directly applied to stationary pollution sources in China, because of their costs. Japan has sophisticated technologies, but lacks experience in and knowledge of alternative, intermediate, and low-technology solutions. Given this, Japan may attempt to buy pollution control technology equipment from the United States, which provides a wider range of desulphurization technologies, as a part of the Official Development Assistance programme. At the same time, international support that includes the establishment of additional global environmental monitoring states in China are also necessary. (abstract)

McKinney, Matthew. 1992. Designing a Dispute Resolution System for Water Policy and Management. *Negotiation Journal* 8 (2, April): 153–64. Examines the case of Montana's Department of Natural Resources and Conservation and its attempt to implement a state water-planning process. Includes history 1988–1992, an evaluation of the process, and more general guidelines for dispute resolution systems.

Megahan, Walter F. and Peter N. King. 1985. Identification of Critical Areas on Forest Lands for Control of Nonpoint Sources of Pollution. *Environmental Management* 9 (1, January): 7–18. Most non-point source pollution problems on forest lands can be controlled by careful planning and management of specific critical areas. Critical areas include sites with high mass and surface erosion hazards, overland flow areas, and the riparian zone. Some guides for identifying critical areas are presented along with examples of land-use constraints that might be applied. (abstract)

Meier, Richard L. 1991. *A Global Role for the Palestine Arabs: The Integrated Agro-Industrial Complex*. University of California, Berkeley, April. Discusses the Oak Ridge Nuclear labs attempt to provide a solution to the Palestinian problem with desalination plants and fertilizer. Then the author moves on to discuss some of his work that followed the Oak Ridge proposal.

Miller, Morris. 1995. Transformation of a River Basin – Case of the Mekong Committee. Asian Water Forum. Bangkok, Thailand: United Nations University, 30 January–1 February. Answers the questions "Why examine the functions of the Mekong Committee?" and "What should and could be done to enable the MC to more effectively promote the development of the Mekong region?"

Moore, J. 1992. *Water-Sharing Regimes in Israel and the Occupied Territories*. Ottawa: Department of National Defence, Canada. This study outlines an approach for equitably apportioning the transboundary groundwaters shared between Israel and the Palestinians of the occupied territories.

Mumme, Stephen. 1993. Innovation and Reform in Transboundary Resource Management: A Critical Look at the International Boundary and Water Commission, United States and Mexico. *Natural Resources Journal* 33 (Winter): 93–132. The purpose of this study is to examine the reform needed of the International Boundary and Water Commission (US-Mexico) in this time of numerous changes along the border, demographic, political, and attitudinal. These changes are impacting the Commission's ability to manage transboundary resource problems of the region. The author indicates that there are limitations on the Commission detailed in the treaty mandate, however, there are several areas where development and improvements can be made, including, sanitation and water quality, instream flow, and creative approaches to project financing.

Murakami, M. and A.T. Wolf. 1995. Techno-political Water and Energy Development Alternatives in the Dead Sea and Aqaba Regions. *Water Resources Development* 11(2): 163–

83. Water and energy will be key elements in any regional development schemes in arid regions, being the limiting factors for planned tourism/resorts, industry, and commerce. Two regions may be particularly attractive for regional economic development planning: the Dead Sea region, including the territory of Israel, Palestine, and Jordan; and the Aqaba/Eilat area, which includes Egypt, Israel, Jordan, and Saudi Arabia. These two regions could act as showcases of cooperation between the countries of the Middle East. This study describes both technical and political priorities for water and energy development projects, including non-conventional alternatives, particularly proposed hydro-solar and seawater pumped-storage schemes with hydropowered reverse osmosis (RO). Technical and political implications of these projects are examined in a framework of interstate regional economic cooperation. (abstract)

Murphy, I.L. and J.E. Sabadell. 1986. International River Basins: A Policy Model for Conflict Resolution. *Resources Policy* 12 (June): 133–44. The need to facilitate agreements between countries in dispute over the management and use of a shared river basin increases as water needs and new pollution issues emerge. Develops a policy model to track the decisions made within countries to resolve international differences. The decision process is better understood and its outcome more predictable when data about the joint goals and decisions of governments are matched with the components of the policy model. Three case studies are summarized and conclusions with respect to the impact of political processes on the resolution of international river basin conflicts are presented. (abstract)

Murray, John S. 1990. Dispute Systems: Design, Power, and Prevention. *Negotiation Journal* 6 (2, April): 105–09. Report on a conference of the Society of Professionals in Dispute Resolution. Accents the dispute resolution system designer's responsibility and the role of prevention.

Naff, Thomas and Ruth C. Matson. 1984. *Water in the Middle East: Conflict or Cooperation?* Boulder, CO: Westview Press. A comprehensive work that addresses water issues along the Jordan, Litani, Euphrates, Orontes, and Nile rivers. Legal aspects and the potential for cooperation are also discussed. Each river chapter details hydrology, technical aspects, history, conclusions, and additional references.

Nazem, Nurul Islam and Mohammad Humayun Kabir. 1986. *Indo-Bangladesh Common Rivers and Water Diplomacy*. Dacca: The Bangladesh Institute of International and Strategic Studies. Discusses such topics as Farakka, Indo-Bangladesh relations since 1971 (when Bangladesh became independent), augmentation of dry-season flows, smaller rivers, and perceived policy options for both states.

Newson, M. 1992. Water and Sustainable Development: the "Turn Around Decade?". *Journal of Environmental Planning and Management* 35(2): 175–83. The paper comments on the development of an international movement towards both sustainability and subsidiarity in the management of large river basins, and also stresses the significance of river basins for their sense of place and symbolism as environmental territory. Two contrasting efforts are compared, in the developed world (European Community) through the Freshwater Europe Campaign, and in the developing world through the lead-in to the Earth Summit. Finally, suggestions are made that the best manifesto for the problem of water and development is an ecosystems approach to the river basin, modified by a greater use of approaches from social science and objectives of social justice. (abstract)

Nickum, J.E. and K.W. Easter. 1990. Institutional Arrangements for Managing Water Conflicts in Lake Basins. *Natural Resources Forum* 14 (3, August): 210–21. Water conflicts may arise from market failures caused by (i) poor specification or transferability of water rights; (ii) incentive problems such as rent seeking, open access or free riding; or (iii) transaction costs. They may also occur because of failures in non-market alternatives such as government management. Recent recognition of non-market failures has led to greater consideration of market-based approaches to conflict resolution such as tradeable permits or bargaining. It has also enriched our understanding of government-based approaches. All approaches would appear to benefit from greater attention to promoting collective action by the users of water. For illustration we refer to the problems of lake basins, using several specific examples from Asia. (abstract)

Nishat, Aminun. 1995. Impact of Ganges Water Dispute on Bangladesh. Asian Water Forum. United Nations University, 30 January-1 February. Offers a brief history of Ganges negotiations and adverse effects of the Farakka's reduction of the Ganges flow, including economic detriment due to Ganges diversions.

Nomas, H.B. 1988. The Water Resources of Iraq: An Assessment. diss, 486 pp. University of Durham. The Euphrates and Tigris are international rivers with renewable but finite resources, shared by four riparian states, Turkey, Syria, Iraq, and Iran. At present, only Iraq is a major water user and, due to the dominant arid to semi-arid climate, relies almost entirely upon this drainage system. In recent decades, the upstream riparian states have planned for large-scale irrigation and hydroelectric power developments. These will certainly affect water availability and quality in the lower riparian state, Iraq, if they are fully implemented as planned. Thus, the objective of this study is to emphasize the Euphrates and Tigris as an integrated crucial shared resource and to determine the whole aspect of the present and prospective water development situations. The main solution to the problem is achieving a mutual international agreement to secure the appropriate share for each state in terms of water quality and quantity. It is hoped that the riparian states will promote these cooperative efforts to achieve lasting agreements in order to avoid potential conflict. (abstract)

North, Ronald M. 1993. Application of Multiple Objective Models to Water Resources Planning and Management. *Natural Resources Forum*, August, 216–27. This paper provides a self admitted brief and incomplete example of a macroeconomic-based, multiple objective, water-resource planning and management model. The authors approach is to use generally accepted principles of strategic planning and management. In addition, there is a recommended goal programming algorithm which is capable of integrating the combinations of structural and management solutions. This is achieved by a comparison of results in terms of incommensurate values for economic, environmental, and social indicators.

Okada, Norio, Keith W. Hipel, and Yoshiharu Oka. 1985. Hypergame Analysis of the Lake Biwa Conflict. *Water Resources Research* 21 (7, July): 917–26. Hypergame modelling of conflicts and predicting the compromise solutions. The authors expect that their conflict analysis can be used in any conflict, and have had success with predicting the actual outcomes of historical conflicts.

Okidi, C.O. 1988. The State and the Management of International Drainage Basins in Africa. *Natural Resources Journal* 28 (Fall): 645–69. The purpose of this paper is to examine some of the macro-policy questions of state involvement in actual management and utilization of the waters of any basin in Africa. The author suggests that whether the management and utilization of the water is done within a national or regional framework, the role of the state will evince common conditions which are similar when looked at from the perspective of promotion of the socio-economic well-being of the human population.

Ozawa, Connie P. and Lawrence Susskind. 1985. Mediating Science-Intensive Policy Disputes. *Journal of Policy Analysis and Management* 5(1): 23–39. Proposes means to not let scientific issues mask underlying distributional issues.

O'Connor, David. 1992. The Design of Self-Supporting Dispute Resolution Programs. *Negotiation Journal* 8 (2, April): 85–91. A report on the experience of the State of Massachusetts' Office of Dispute Resolution. It relates history of the following topics: generating business, programme goals, design considerations, the referral of disputes (from other agencies), dispute acceptance screening procedures, selection and performance of the dispute resolvers, and the quality with which these duties were performed.

Paul, T.V. 1994. *Asymmetric Conflicts: War Initiation by Weaker Powers.* Cambridge, MA: Cambridge University Press. This book examines the question of why militarily and economically weaker states initiate war on stronger ones? The author questions the idea that a stronger power (with a suitable retaliatory capability) can maintain peace by remaining stronger. Case studies include Japan-Russia (1904), Japan-US (1941), Chinese-UN (Korea, 1950), Pakistan-India (1965), Egypt-Israel (1973), and Argentina-UK (1982).

Paulsen, C.M. 1993. Policies for Water Quality Management in Central and Eastern Europe. In *Two Essays on Water Quality in Central and Eastern Europe. ENR93-20*, 15–25. Washington, DC: Resources for the Future. Charles M. Paulsen, of Resources for the Future, examines the design of source-control policies for point sources of water pollution in Central and Eastern Europe (CEE) with an emphasis on policies that will meet ambient water quality targets cost-effectively. Paulsen argues that cost-effective policies are difficult to implement, and that the direction of water pollution control policy is made unclear by economic restructuring in CEE. Nevertheless, the cost savings from more efficient policies can be large, so much policy warrants serious consideration. (abstract)

Perritt, R. 1989. African River Basin Development: Achievements, the Role of Institutions, and Strategies for the Future. *Natural Resources Forum*, August, 204–08. This article provides a summary of the presentations and themes of the international conference on the African Experience with River Basin Development. Included is a list produced by the conference of general priorities for improving river basin development in Africa. The purpose of the conference was to learn from the past, suggest appropriate corrective measures, and broaden the horizons for river basin development in the form of proposals envisioning a more comprehensive framework for institutional involvement.

Platter, Adele G. and Thomas F. Mayer. 1989. A Unified Analysis of International Conflict and Cooperation. *Journal of Peace Research* 26(4): 367–83. Mathematical modelling of structural components international relations. Addresses proximity, history, tendency to initiate conflict or negotiations.

Precoda, Norman. 1991. Requiem for the Aral Sea. *Ambio* 20 (3–4, May): 109–14. Heavy withdrawals of irrigation water from the Syr and Amu, the Aral sea's two main tributaries, have for all practical purposes eliminated their spills and led to a sharp decrease in the level of the sea. This and the disruption of ecological equilibrium in this immense region have had catastrophic consequences for both the inhabitants of the region and for the environment. The circumstances leading up to and important features of some of the principal consequences are described. (abstract)

Priest, J.E. 1992. International Competition for Water and Motivations for Dispute Resolution. *Agricultural Water Management* 21(1–2): 3–11. Disputes regarding water allocation among nations have affected and distorted water use and development across the South Asian subcontinent, Middle East, and Africa ever since decolonization following World War II. Issues, posturing, developments, and attempts at dispute resolution are reviewed for six great rivers. Demographics, physical developments, the political environment, and motivations that influence the process of dispute resolution are identified. (abstract)

Quigg, Phillip W. 1977. A Water Agenda to the Year 2000. *Common Ground* 3(4): 11–16. The author, arguing that water should be regarded as a vulnerable and finite resource, along the same lines as food and energy, presents a comprehensive summary of current water problems and issues. He considers the development of water resources, the goal of pure drinking water for all, more efficient irrigation, recharge and water mining, industrial recycling, the protection of watersheds and wetlands and the problems of arid lands. Then, under the heading of waste water and treatment he looks at discharge standards, urban and agricultural runoff, toxic wastes, groundwater, disposal of sewage sludge and sanitation for the Third World. He concludes by reviewing the techniques and institutions in water law, water economics, water as a source of energy, water sharing and water disputes, water and the environment and water management. (M. Higgins)

Quinn, J.T. and J.J. Harrington. 1992. Generating Alternative Designs for Interjurisdictional Natural Resource Development Schemes in the Greater Ganges River Basin. *Papers in Regional Science* 71(4): 373–91. Planning for the development of regional water resources is often complicated by severe disputes. For example, in the Greater Ganges river basin, there are disagreements between India and Bangladesh over sharing the low river flows during the dry season and over controlling the potentially destructive large river flows during the monsoon. This paper illustrates an approach for providing the two riparian nations with distinct water resources plans to help solve their regional water conflicts. More specifically, a linear programming model representing a multipurpose river basin system is presented. The concept of near optimality is employed to generate a variety of solutions, in contrast to searching only for a global optimum. These solutions are grouped into similar project designs by applying a cluster analysis, which is a multivariate technique. Several project designs are graphically displayed, and their implications for national and international agreements are discussed. The range of regional alternatives available to India and Bangladesh could aid in their negotiations. (abstract)

Radosevich, G.E. 1995. The Mekong – A New Framework for Development and Management Under a Renewed Spirit of Cooperation. In *Asian Water Forum*. Asian Water Forum. Bangkok, Thailand: United Nations University, 30 January–1 February. Discusses legal and institutional aspects of the Mekong river development efforts. Focuses more on what has been accomplished rather than what needs to be done, but it does have a discussion of a new framework for cooperation.

Rangeley, Robert, Bocar M. Thiam, Randolph A. Andersen, and Colin A. Lyle. 1994. *International River Basin Organizations in Sub-Saharan Africa*. World Bank Technical Paper, Africa Technical Department Series, Vol. 250. Washington, DC: World Bank. This report has been prepared as a contribution to a concerted effort within the World Bank to define more explicitly its policies towards water-resources management. In particular, it aims to assist the World Bank and other international orgnaizations in their pursuit of a more active role in helping sub-Saharan African countries improve the management of shared water resources. At the same time, the report contains some generic findings that should be of direct value to decision-makers in riparian states involved in such issues. (abstract)

Raskin, P., E. Hansen, Z. Zhu, and D. Stavisky. 1992. Simulation of Water Supply and Demand in the Aral Sea Region. *Water International* 17(2): 55–67. The Aral sea, a huge saline lake located in the arid south-central region of the former USSR, is vanishing because the inflows from its two feed rivers, the Amudar'ya and Syrdar'ya, have diminished radically over the past three decades. The loss of river flow is the result of massive increases in river withdrawals, primarily for cotton irrigation in the basins. A microcomputer model, the Water Evaluation and Planning System (WEAP), has been developed for simularing current water balances and evaluating water management strategies in the Aral sea region. WEAP treats water demand and supply issues in a comprehensive and integrated fashion. The scenario approach allows flexible representation of the consequences of alternative development patterns and supply dynamics. For the Aral region's complex water systems, a detailed water demand and supply simulation was performed for the 1987–2020 period, assuming that the current practices continue. The analysis provides a picture of an unfolding and deepening crisis. Policy scenarios incorporating remedial actions will be reported in a separate paper. (abstract)

Redclift, M. 1991. The Multiple Dimensions of Sustainable Development. *Geography* 76, part 1(330): 36–42. The problem with referring to "sustainable development" is that its very appeal is its vagueness. Sustainable development means different things to different people: ecologists, environmental planners, economists, and activists. Part of the interest in the discussion of sustainable development lies in the way the concept has been borrowed from both the natural and social sciences. This paper examines the contribution that a broadly-based concept of sustainable development can make: focusing attention on poor peoples' use of sustainability in seeking livelihoods from resources-poor areas of the South. (abstract)

Reguer, S. 1993. Controversial Waters: Exploitation of the Jordan River, 1950–80. *Middle Eastern Studies* 29(1): 53–90. This article examines the growing importance of Jordan water to both Israel and Jordan as the two countries planned for expanding agriculture, industry and population in the three decades following the end of the British mandate over Palestine. It will describe what happened when politics infringed on the realization of the development plans and how these two countries had to contend with political barriers in order to exploit the Jordan river. The primary sources for this study were provided mainly by the Jordan Valley Authority in Amman and Tahal in Tel Aviv. This study points out the great strides that can be made without international cooperation; however, there are limits to national river development. These limits have been reached in the Jordan-Yarmuk system. Competition for water usage has resumed, and if no solution is found, water scarcity is likely to be one of the reasons for sparking the next regional conflict. Water scarcity should now become an imperative for political cooperation. (abstract)

Rhodes, Thomas C. and Paul N. Wilson. 1995. Sky Islands, Squirrels, and Scopes: The Political Economy of an Environmental Conflict. *Land Economics* 71 (1, February): 106–21. Siting conflicts involving endangered species create economic and political pressures on existing property rules. A welfare model of institutional change with endogenous transaction costs is used to analytically describe the environmental conflict surrounding the Mount Graham International Observatory project. Emphasis is placed on the evolution of property rights under conditions of imperfect information. This case study illustrates (a) the nature of scientific and economic incentives for institutional change due to technological innovation and (b) the process of institutional choice in a policy environment characterized by biological uncertainty and economic power. (abstract)

Richards, Alan. 1993. Strengthening Markets to Build Peace: The General Case, Illustrated by the Example of Agriculture and Water. Conference on the Middle East Multilateral Conference. Los Angeles: University of California, Los Angeles, June. Proposes a statement of principles and policy suggestions for agricultural/water use in Israel-Jordan-Palestine. Strengthening the markets in the region (and working around the contradictory politics and economics already in place) will better the standard of living, but require more cooperation and more contact among peoples.

Rizk, Edward. 1964. *The River Jordan.* Information Paper, Vol. 23. New York: Arab Information Center. Like other books on the dispute, the short history is punctuated by the plans for division of the Jordan, and then Rizk follows with brief chapters entitled "The Dangers of Israel's Jordan Diversion Project" and "Analysis of Some Israeli and Zionist Arguments."

Rogers, Peter. 1969. A Game Theory Approach to the Problems of International River Basins. *Water Resources Research* 5 (4, August): 749–60. The lower Ganges and the Brahmaputra rivers flow out of India and join together in the province of East Pakistan to form one of the major river systems in the world. Each year the rivers flood during the monsoon causing loss of life and great damage to crops and property. The problem is further exacerbated by the fact that the best flood control points in the basin are not under the political control of Pakistan, which suffers most from the floods. The number of possible combinations of structural and nonstructural variables for this river system is so large that conventional methods of analysis are inadequate. We demonstrate that by use of systems analysis techniques, including linear programming and game theory, we are able to consider a rational plan to control the floods while taking advantage at the same time of the complementalities which exist between flood control and other possible uses for the river, such as power production, irrigation, navigation, and salinity control. More-over using the concepts of game theory we are able to investigate a range of strategies for cooperation between the two riparian nations which will result in significant benefits to each. (abstract)

——— 1991. International River Basins: Pervasive Unidirectional Externalities. The Eco-nomics of Transnational Commons. Universita di Siena, Italy, 25–27 April. The paper reviews the phenomenon of international river basins and concludes that sharing of river basins between and among countries is the rule rather than the exception for the major river systems of the world. More than 200 river basins, accounting for more than 50% of the land area of the earth, are shared by two or more nation-states, powerful and often jealous social units that dominate what is still an age of nationalism. When population densities were low there was plenty of water for all and major conflicts were avoided. With the rapid population and economic growth experienced in the past few decades conflicts over use of water are becoming more important. It is expected that in the near future these water conflicts will become much more severe. The paper reviews the litera-ture on attempts to analyse the conflicts and negotiate solutions. One interesting finding is that the upstream-downstream externalities are not always negative; there are many cases where upstream development of water resources leads to increased benefits to down-stream users. Some rudimentary game models are examined and some tentative con-clusions, based upon various game theory concepts of stability are presented. The paper ends with suggestions on how to plan Pareto-admissible outcomes for international basins. (abstract)

——— 1993. Integrated Urban Water Resources Management. *Natural Resources Forum* (February): 34–42. The focus of this article is the organizations and institutions of water-resources management strategies. Economic aspects of planning for urban water use are concentrated on heavily. Ensuring that each utility covers operating costs as well as capital costs by the economic pricing of water use is stressed as a needed policy improvement.

——— 1993. The Value of Cooperation in Resolving International River Basin Disputes. *Natural Resources Forum* 17(2): 117–31. The purpose of this paper is to review the phe-nomenon of international river basins, and concludes that the sharing of river basins be-tween and among countries is the rule rather than the exception for the major river sys-tems of the world. Included is a review of the literature which attempts to analyse river basin conflicts and negotiated solutions. Some game models and tentative conclusions are made on the basis of various game theory concepts of stability in an application to the Ganges-Brahmaputra basin.

——— R. Burden, and C. Lotti. 1978. Systems Analysis and Modeling Techniques Applied to Water Management. *Natural Resources Forum* 2: 349–58. Discusses modelling of multi-ple-use water management, including benefit-costs analysis and overlapping institutional-political boundaries.

Ross, Lee, and Constance Stillinger. 1991. Barriers to Conflict Resolution. *Negotiation Journal* 7 (4, October): 389–404. Discussion of barriers to conflict resolution: Secrecy and deception, intransigence, psychological barriers, and additional complexities. Also a dis-cussion of improving negotiation techniques ("pump-priming").

Rowley, Gwyn. 1993. Multinational and National Competition for Water in the Middle East: Towards the Deepening Crisis. *Journal of Environmental Management* 39: 187–97. This brief paper introduces the highly sensitive and contentious subject of international com-petition over water resources within the Middle East. Attention is directed to the general aridity of the Middle East, the multinational nature of the region's three major river basins – the Jordan, the Tigris-Euphrates and the Nile, the lack of regulatory authority to mediate upon water-resource allocations, and the mounting internecine animosities between the various states within the region. The potential for major larger-scale conflict over water in the Middle East is seen to be increasing. (abstract)

Sadler, B. 1990. Sustainable Development and Water Resource Management. *Alternatives* 17(3): 14–24. Focuses on evaluating the contributions that sustainability concepts can make to the theory and practice of water-resource management. Water-resource management should be used as a basis for the unification of environmental, economic, and social concerns, however, it is hampered by incomplete understanding of resource systems and their productivity, resilience, and vulnerability to cumulative impacts. The paper discusses the limitations of the Canadian Federal Water Policy and the planning framework being pursued in the Great Lakes basin by the International Joint Commission which is based on an ecosystem perspective of the Lakes as a functional entity. The author follows the decision-making process that led to the construction of the Dixon dam on the Red Deer river, Canada, and concludes with some lessons for the future. (M.Z. Barber)

Salewicz, K.A. 1991. Management of Large International Rivers – Practical Experience from a Research Perspective. In *Hydrology for the Water Management of Large River Basins*, ed. F.H. Van de Ven et al., 57–69. International Association of Hydrological Sciences. Experience gained from research conducted at the International Institute for Applied Systems Analysis is presented. Water management issues for two case study rivers – the Danube and the Zambezi – are briefly presented, and then experiences in dealing with institutional and organizational aspects of the research are examined. Conclusions summarize both positive and negative experiences and recommendations are made concerning further activities in this field. (abstract)

Saliba, Samir N. 1968. *The Jordan River Dispute*. The Hague: Martinus Nijhoff. Discusses history, including details of pre- and post-partition plans for the Jordan by Ionides, Lowdermilk, Hays, McDonald, Bunger, Israel, and Arabs. Discusses Syria, Lebanon, Jordan, and Israel. Covers international law, attempts at resolutions such as Johnston, Baker-Harza, Unified plan, and the Cotton plan. Looks at other regional disputes as well, including the Rio Grande, Columbia, Indus, Danube, and legal solutions.

Schaake, J.C., R.M. Ragan, and E.J. Vanblargan. 1993. GIS Structure for the Nile River Forecast Project. In *Application of Geographic Information Systems in Hydrology and Water Resources Management*, ed. K. Kovar and H.P. Nachtnebel, 427–31. International Association of Hydrological Sciences. The Nile Forecast System is being developed to predict inflows into the Aswan dam in Egypt. The system contains a set of hydrological models that are structured as a grid point, distributed modelling system. The system also contains a geographic information system (GIS) component in order to create map displays and to define the parametric hydrological input for each grid cell. The terrain analysis software developed uses elevation and stream data to define important hydrological parameters such as the gridded flow connectivity, drainage areas, flow lengths, and slopes. (abstract)

Schmida, Leslie. 1983. *Keys to Control: Israel's Pursuit of Arab Water Resources*. Washington, DC: American Educational Trust. A political history from 1919 to 1976 of Israel and the water uses it incurred over the years. Its tone sounds a little conspiratorial. It moves topic by topic through issues such as the National Water Carrier, The Johnston Plan, the West Bank, the Golan Heights, and Legal Issues.

Scudder, T. 1989. The African Experience with River Basin Development. *Natural Resources Forum* (May): 139–48. The purpose of this paper is to examine the record to date of river basin development in tropical Africa and in more detail the role and performance of institutions in this record of African river basin development. There is an assessment made of the experience of institutions with river basin development against the extent to which the resources of Africa's river basins have been developed in recent years and the role that institutions have played in that development.

Secretariat of the United Nations Commission for Europe. 1994. Protection and Use of Transboundary Watercourses and International Lakes in Europe. *Natural Resources Forum* 18(3): 171–80. This article analyses developments in national strategies and policies for the protection and use of transboundary water in Europe. In addition, it describes the progress made by European states in their cooperation on these issues. Particular attention is focused on the Convention on the Protection and Use of Transboundary Watercourses and International Lakes (Helsinki, 1992) pending its entry into force. (abstract)

Sellers, Jackie. 1993. Information Needs for Water Resources Decision-making. *Natural Resources Forum* (August): 228–34. Developing data and information systems is a never completed process, with continual updates and modifications. The purpose of this article is to identify the components of information needed to facilitate the planning process for resources utilization. This process is identified by Sellers as the proper development of Management Information Systems (MIS) or Decision Support Systems (DSS).

Shady, A., A. Adam, and K. Mohammed. 1994. The Nile 2002: the Vision Toward the Cooperation in the Nile Basin. *Water International* 19(2): 77–81. The Nile 2002 conference series started in February 1993 in Aswan, Egypt and was followed by the second in Khartoum, Sudan in January 1994. Both conferences dealt with the same general theme of "Comprehensive Water Resources Development of the Nile Basin." The Khartoum conference dealt with forty-two papers, including nine country papers from the Nile Basin, and several discussion series. The main feature of these papers is the emphasis on the need for cooperation among the co-basin countries and assistance from external support agencies and internal organizations. A comprehensive set of cooperation modalities, principles, and areas of potential support for external support agencies is identified. Institutional aspects are also examined emphasizing the complementary role of governmental, international support agencies, and nongovernmental organizations. (abstract)

Shah, R.B. 1994. Inter-state River Water Disputes: A Historical Review. *Water Resources Development* 10(2): 175–89. India has a federal set-up and about 80% of water resources are derived from interstate rivers. There have naturally been a number of disputes among the states regarding allocation and utilization of waters of interstate rivers. This article provides a historical review of the resolution of such disputes under the present constitutional provision and the experience gained. (abstract)

Sherk, G.W. 1989. Equitable Apportionment After Vermejo: The Demise of a Doctrine. *Natural Resources Journal* 29: 579–83. Historically, the Supreme Court has resolved interstate water conflicts under the doctrine of equitable apportionment. The decisions of the Court regarding a conflict between Colorado and New Mexico over the waters of the Vermejo river, however, have established burden of proof requirements that may eliminate equitable apportionment as a means of resolving interstate water conflicts. This article discusses the historical development of the doctrine and reviews the Vermejo decisions. Alternative means of resolving interstate water conflicts are discussed. Finally, given the inevitability of litigation regardless of which means of resolving interstate conflicts is selected, alternative litigation strategies are considered. (abstract)

Shrestha, Hari Man and Lekh Man Singh. 1995. The Ganges-Brahmaputra System: A Nepalese Perspective in the Context of Regional Cooperation. In *Asian Water Forum*. Asian Water Forum. United Nations University, 30 January–1 February. Discusses shortcomings of past efforts and negotiations. Includes a list of proposals not conducive to cooperation and water availability. Includes a proposal of an Indian-Nepalese water-sharing agreement.

Shuval, Hillel. 1992. Approaches to Resolving the Water Conflicts Between Israel and her Neighbors – a Regional Water-for-Peace Plan. *Water International* 17: 133–43. The disputes over surface and ground water resources between Israel, Jordan, and the Palestinians are particularly acute since all three are well below the Water Stress Level of 500 cubic metres/person/year as defined by Falkenmark. Based on currently accepted principles of international water law neither the upstream source country nor the downstream historic user country has absolute sovereign rights to control the use of international bodies of surface or groundwater. An accommodation, based on the principles of equitable apportionment and community of interest, considering the legitimate needs of each partner to the dispute, should be arrived at through direct negotiations. Since the total amount of water resources available is insufficient, it is a zero sum game and any attempt to find a solution by reallocation of the limited resources between the disputants is likely to lead to a deadlock. The approach proposed is to increase the size of the pie by developing a regional Water-for-Peace Plan. This would be funded by the major powers and import water from water-rich neighbouring countries such as Egypt, Lebanon, and Turkey, and/or desalinate seawater. However, it is proposed that baseline allocations be assured for "water security" for domestic, urban, industrial and fresh food supply needs of about 125 cubic metres/person/year. This should be abstracted directly within the territory of each side. (abstract)

Simms, R.A. 1989. Equitable Apportionment – Priorities and New Uses. *Natural Resources Journal* 29: 549–63. The doctrine of equitable apportionment was initially articulated as a conflict of laws doctrine designed to resolve the over-appropriation of an interstate stream resulting from the accumulation of rights perfected under differing regimes of water law. The doctrine also addresses such conflicts between states applying the same doctrine of water law. Historically, an equitable apportionment has been realized by the application of the doctrine of prior appropriation interstate between prior appropriation states, varying the doctrine only to protect economies predicated on junior appropriations. Recently, the Court appears to have changed the doctrine of equitable apportionment, suggesting that priority of appropriation might be varied to supplant existing uses with new uses of higher economic value. The new direction in equitable apportionment marks a radical departure from settled precedent. (abstract)

Smith, Scot E. and Hussam M. Al-Rawahy. 1990. The Blue Nile: Potential for Conflict and Alternatives for Meeting Future Demands. *Water International* 15(4): 217–22. Despite nearly a century of water regulation projects on the Nile basin, today Egypt faces a genuine crisis with respect to water supply. The crisis has been brought on by a combination of drought and greatly increased usage by all riparian nations along the Nile. It is highly probable that Egypt will need to search for new sources of water, reallocate existing water supply and reduce demand simultaneously in order to stave off major water shortfalls. It is unlikely that these measures can be accomplished without both internal and external conflict. This paper describes the current state of water supply and usage in Egypt. Further we examine potential sources of conflict over water allocation decisions. Finally, the paper presents a method for evaluating alternative strategies that could be implemented by Egypt to avoid future conflict. (abstract)

Sohn, Louis B. 1987. Peaceful Settlement of Disputes and International Security. *Negotiation Journal* (2, April): 155–66. A law professor offers his advice on dispute settlement methods, noting that the UN charter only says methods should be peaceable, but gives no framework. He therefore offers some settlement methods.

Solanes, Miguel. 1992. Legal and Institutional Aspects of River Basin Development. *Water International* 17: 116–23. Solanes discusses international law principles and procedural means for the effective application of those principles. He also touches on issues of international law such as joint development on rivers and equitable use.

Spalding, Mark J. 1995. Resolving International Environmental Disputes: Public Participation and the Right-to-Know. *Journal of Environment and Development* 4 (1, Winter): 141–54. The North American Free Trade Agreement is the first trade pact that provides considerable protections for a country's environmental standards. The environmental side agreement to the NAFTA, negotiated at the insistence of and with the participation of environmental groups, has promoted the important roles in environmental regulation and enforcement played by the principles of transparency (right-to-know) and public participation (right-to-sue). The side agreement provides for: citizen submissions to the Secretariat of the North American Commission for Environmental Cooperation; the US, Mexican, and Canadian governments' guarantees of citizens' right-to-know, as well as citizen suits and remedies regarding environmental harm and requests for enforcement of environmental laws. As procedures under the NAFTA and its environmental side agreement are developed, opportunities remain to incorporate the principles of transparency and public participation even more broadly. (abstract)

Sprinz, Detlef. 1995. Regulating the International Environment: A Conceptual Model and Policy Implications. In *Prepared for the 1995 Annual Meeting of the American Political Science Association*. 1995 Annual Meeting of the American Political Science Association. Chicago, IL, 31 August–3 September. The present scale of production and consumption is likely to generate adverse externalities (pollution). The dangers posed by pollution infringe on the welfare of countries, depending on the type of environmental problem they face. While research is under way which links environmental degradation to the onset of civil and international war, this article focuses on the non-military instruments which governments can use to improve the state of their environment. After briefly reviewing the literature on the relationship between environmental degradation and the onset of civil and international conflict, the basic model of environmental regulation in a closed economy will be developed; these assumptions will be relaxed step-by-step by introducing transboundary pollution, international trade, and global environmental problems. Furthermore, a brief sketch is provided for how to turn the static treatment into a simple dynamic perspective on defining and achieving environmental security. In the penultimate section, some suggestions for further refinement of the concept presented here are made before turning to a summary of the conclusions. (abstract)

Stahl, Michael. 1993. Land Degradation in East Africa. *Ambio* 22 (8, December): 505–08. The highlands of East Africa have a high agricultural potential and have historically supported kingdoms with stratified social structures. Today, traditional fertility practices cannot be maintained under conditions of mounting population growth and land scarcity. Land degradation is now threatening the very basis of the farming communities. This paper discusses land degradation in East Africa in the context of soil conservation. It describes technical and institutional responses to land degradation including regional cooperation and concludes that degradation is not yet irreversible. Relatively low-cost technologies exist that have the potential, given supportive institutions and incentives, to achieve widespread adoption among smallholders. (abstract)

Starr, Joyce R. and Daniel C. Stoll, eds. 1988. *The Politics of Scarcity: Water in the Middle East*. Boulder, CO: Westview Press. By the year 2000, the Middle East will confront yet another serious challenge to regional stability as Egypt, Jordan, Israel, and Syria cope with significant water shortages. The emerging crisis will involve actual water shortages, inefficient management practices, and deterioration of water quality. The authors review available water technologies such as desalination, water re-use processes, and large-scale engineering projects that might alleviate regional water shortages, with special consideration given to the recently proposed "Peace Pipeline" from Turkey. The book describes US government agency involvement with water issues and evaluates their respective programmes and development projects. (abstract)

—— and Daniel Stoll. 1987. *U.S. Foreign Policy on Water Resources in the Middle East*. Washington, DC: The Center for Strategic and International Studies. Overviews some water security issues, and the Nile basin as a case study in relations. The authors give a detailed account of United States policy structure on water resources, and also make some recommendations for change in the US policy.

Stein, Janice Gross. 1985. Structures, Strategies, and Tactics of Mediation: Kissinger and Carter in the Middle East. *Negotiation Journal* (October): 331–47. History and analysis of Kissinger's and Carter's different styles of negotiation, starting with Kissinger in 1973.

—— 1988. International Negotiation: A Multidisciplinary Perspective. *Negotiation Journal* (July): 221–31. Addresses negotiation in terms of game theory and history.

Stone, Paula J. 1980. A Systems Approach to Water Resource Allocation in International River Basin Development. *Water Resources Research* 16 (1, February): 1–13. A methodology is presented to assist the water resource systems analyst in providing and interpreting technical information to promote coordinated water resource use in international river basin development. The international river basin planning environment is examined to identify the analytical information which can address and satisfy comprehensive decision-making requirements. This information is organized wthin a multilevel decomposition framework that permits an analysis of the implications of alternative resource development patterns in terms of nations both achieving their objectives and pursuing development strategies that will lead to feasible basin-wide resource use. The proposed methodology is applied to an example of international river basin planning with the generation of resource transformation curves, and results are examined for their appropriateness as bases for coordinated development. (abstract)

Sun, Peter. 1994. *Multipurpose River Basin Development in China.* Economic Development Institute Seminar Series. Washington, DC: Economic Development Institute of the World Bank. This publication describes the discussions at a Senior Policy Seminar on Policies for Multipurpose River Basin Development in China held on 31 March–21 April 1990, in Nanjing, China, and jointly sponsored by the Science and Education and Foreign Affairs departments of the Ministry of Water Resources, China, and the Economic Development Institute of the World Bank. In 1989 China's Ministry of Water Resources had identified a strong need for such a seminar. The minister had indicated that following the seminar, which would re-examine policies in effect for water resource development in China, a report would be produced that would be and become the basis for future policy formulation. A set of lecture papers and discussion notes from the seminar was published in China in 1991 (in Chinese) in response to the minister's request. This publication is an abbreviated English language version that focuses on the seminar's conclusions and recommendations. (abstract)

Szekely, A. 1993. Emerging Boundary Environmental Challenges and Institutional Issues: Mexico and the United States. *Natural Resources Journal* 33 (1, Winter): 33–58. The purpose of this article is to examine the emerging transboundary resource and environmental issues that are impacting, or will be in the near future, the border region between Mexico and the United States. In addition, an analysis is also made on what type of institutions will be needed to deal with the issues in order to secure the necessary bilateral cooperation. Szekely also suggests that the relationship between the United States, Mexico, and Canada must also be explored for possible trilateral agreements.

—— 1993. How to Accommodate an Uncertain Future into Institutional Responsiveness and Planning: The Case of Mexico and the United States. *Natural Resources Journal* 33 (Spring): 397–403. This article is a short review of the history of bilateral agreements between Mexico and the United States and the International Boundary and Water Commission. The author suggests there is an immediate need to call attention to the uncertain future problems for more in-depth planning of response by Mexico and the United States. For a more detailed discussion by Szekely on this topic see the article, Szekely, Alberto. 1993. Emerging Boundary Environmental Challenges and Institutional Issues: Mexico and the United States. *Natural Resources Journal* 33 (Winter): 33–58.

Tahovnen, Olli, Veijo Kaitala, and Matti Pohjola. 1993. A Finnish-Soviet Acid Rain Game: Noncooperative Equilibria, Cost Efficiency, and Sulfur Agreements. *Journal of Environmental Economics and Management* 24: 87–100. This study analyses cost effectiveness in environmental cooperation between Finland and the Soviet Union. It is assumed that the aim of both countries is to attain a given target deposition level at minimum possible sulphur abatement costs. Cost-effective cooperation is compared to noncooperative equilibirum and to the agreement on sulphur emissions between these two countries. It is shown that the agreement is not cost-effective, implies higher abatement costs than under noncooperation, and is strategically unstable. However, the cost differences and the incentives to cheat are small. The computations reveal that the main source of potential cooperation benefits is not asymmetrical emission transportation or differences in abatement costs but rather different target depostition levels for Finland and the Soviet Union. (abstract)

Talbot, A.R. 1983. *Settling Things: Six Case Studies in Environmental Mediation*. Washington, DC: The Conservation Foundation and Ford Foundation. Documents six environmental disputes in the USA settled with the assistance of a mediator, and draws a number of conclusions. The disputes were: (1) the extension of Interstate 90 across Lake Washington into Seattle: some environmentalists were dissatisfied with the result; (2) resolution of river disputes on the River Hudson, including the dropping of plans to construct the Storm King power plant; (3) a dispute over the use of hydropower from a lake at Swanville, Maine: a local committee was established and min-max water levels agreed; (4) when access to a proposed park on Portage Island, Washington was affected by Indian rights, the Lummi tribe agreed to purchase the island and establish the park; (5) resolution of a waste disposal plan involving the implementation of the Wisconsin Environmental Policy Act: the towns of Eau Claire and Seymour (the proposed dump site) agreed conditions of dumping; and (6) when agreement was reached between local protestors and the developer at Port Townshend, Washington, over the siting of a ferry terminal. (R. Land)

Teclaff, L.A. and E. Teclaff. 1994. Restoring River and Lake Basin Eco-systems. *Natural Resources Journal* 34 (Fall): 905–32. Suggesting that water management policy is shifting from being development dominated to focusing on environmental values, this article examines examples of damage to ecosystems in the past, the restoration techniques currently being used, and relevant developments in domestic and international water law and policy. Strategies are suggested in the concluding section for the restoration of damaged aquatic ecosystems in a transboundary context.

Touval, Saadia. 1987. Frameworks for Arab-Israeli Negotiations – What Difference do They Make? *Negotiation Journal* (January): 37–52. Discusses the conditions for Arab-Israeli meetings, such as places for the meeting and the nature or nationality of the mediator. Israel's preferences for negotiation changed over the years, and the Arab preferences much more so. A discussion of the superpowers is also included.

Trevin, J.O. and J. Day. 1990. Risk Perception in International River Basin Management: The Plata Basin Example. *Natural Resources Journal* 30 (1, Winter): 87–105. Perception of the risk of multilateral cooperation has affected joint international action for the integrated development of the Plata river basin. The origins of sovereignty concerns among Argentina, Bolivia, Brazil, Paraguay, and Uruguay are explored in terms of their historical roots. The role of risk in determining the character of the Plata Basin Treaty, and the ways in which risk was managed in order to reach cooperative agreements, are analysed. The treaty incorporates a number of risk management devices that were necessary to achieve international cooperation. The institutional system implemented under the treaty produced few concrete results for almost two decades. Within the current favourable political environment in the basin, however, the structure already in place reopens the possibility of further rapid integrative steps. (abstract)

Turan, Ilter. 1993. Turkey and the Middle East: Problems and Solutions. *Water International* 18 (1, March): 23–29. Three types of solution have been proposed regarding the intensifying problems generated by insufficient water in the Middle East: (1) increasing the amount of available water; (2) redistributing the existing supply of water; and (3) using the existing water supply more efficiently. In the search for solutions, it is argued that the so-called water problem in the region should be disaggregated, and that each river system should be treated as a separate problem area since the needs and the political relations between the riparians are different in each case. Then, several plans to resolve water insufficiencies are discussed, such as transporting water from Turkey via a pipeline to the Gulf and via super tankers to any needy country in the region; reaching a regional agreement to pursue policies of water conservation, including the adoption of water-saving technologies and founding an international bank to support the adoption of these technologies; and, launching by riparian states of schemes to study land quality and patterns of water use as a preliminary step to further cooperation. It is concluded that the long-term solutions to water shortages necessitate an effort to control and reverse the trends that generate demands for larger and larger quantities of water. (abstract)

Ury, William L. 1987. Strengthening International Mediation. *Negotiation Journal*, July, 225–241. Ury focuses on six recurrent opportunities for an enlarged role of mediators: Start without being asked; Catch conflicts before they heat up; Getting parties to the table; Trying out innovative mediation techniques; Coordinating third party efforts; and following up after a mediator has left.

—— Jeanne M. Brett and Stephen B. Goldberg. 1988. Designing An Effective Dispute Resolution System. *Negotiation Journal* (October): 413. The authors provide a variety of ways to implement ADR techniques, from prevention to back-up and final offers. This article is a common reference and considered seminal.

Valencia, M.J. 1986. Taming Troubled Waters: Joint Development of Oil and Mineral Resources in Overlapping Claim Areas. *San Diego Law Review* 23: 661. Many offshore areas with mineral or petroleum potential are claimed by more than one nation. Joint development is an arrangement by which such nations can avoid questions of sovereignty through joint exploration and development of any resource in an agreed area. Frequently appearing elements in precedents for joint development include: the extent of the area; the contract type; financial arrangements; the process of selection of concessionaires or operators; the length of the agreement; and the nature and functions of the joint management body. Geology plays a fundamental role in the selection and evolution of joint development agreements. The sucess of joint development agreements is dependent on the given knowledge of actual deposits, good political relations, practical mindedness, and cooperative private companies. (abstract)

Van der Keij, W., R.H. Dekker, H. Kersten, and J.A. De Wit, W. 1991. Water Management of the River Rhine: Past, Present, and Future. *European Water Pollution Control* 1(1): 9–18. The present condition of the River Rhine is a result of one-sided management that was aimed only at security against floods and at the river's function as a shipping route. For some decades now, river management has also been aimed at amelioration of the water quality. This has led to a certain degree of ecological restoration in recent years, as compared to the all-time low water quality round 1970. The Sandoz calamity has made it clear that the protection of the drinking-water supply and of the river's ecosystem was still insufficient. A further restoration of the river ecosystem necessitates, next to water-quality improvement, morphological adjustments as well as consideration of the relations between the river and its floodplains. Only integral river management makes this possible, where all interests and functions of the river are taken into account from the beginning of every project, and optimum solutions are sought through cooperation of all disciplines concerned. A strengthening of international cooperation is necessary. In time we shall have to come to an international water authority for the management of the River Rhine. (abstract)

Van de Ven, F.H., D. Gutknecht, D.P. Loucks, and K.A. Salewicz. 1991. *Hydrology for the*

Water Management of Large River Basins. Proceedings of an International Symposium, Vienna, August 1991. International Association of Hydrological Sciences. Contains 37 papers presented at a symposium designed to improve understanding of the various hydrological processes and to investigate tools and methods that can be used to analyse hydrological impacts of anthropogenic activities. Case studies from many rivers, including the Zambezi, Mekong, and Nile are included. Papers are divided into four themes: water-management and cooperation in large and/or international river basins (6 papers); flow regimes and water management in relation to climate changes, river development and land use (19); water quality and sediment transport management (4); and operational flow and water quality forecasting (8). Fourteen papers are abstracted separately. (N. Davey)

Verghese, B.G. 1995. Towards An Eastern Himalayan Rivers Concord. In *Asian Water Forum*. Asian Water Forum. United Nations University, 30 January–1 February. History of the Indo-Gangetic plain, water difficulties, Farakka issue. Includes Nepal in the analysis, and even Bhutan. Verghese does not believe that the water issue is a zero sum game.

Vidal-Hall, J. 1989. Wellsprings of Conflict. *South*. The rapid rise of urbanization, population growth, industrial and agricultural development are depleting the once-abundant water resources of the Middle East. With three river systems shared by all the states, water bids fair to replace oil as the precious commodity of the future. (abstract)

Vlachos, Evan. 1986. *The Management of International River Basin Conflicts*. Washington DC: George Washington University Press. Another large conference results in a large body of aggregate material from it. Many people contributed to this work with their presentations. Each section offers approaches to transboundary river dispute resolution: Case Studies, Decision Support Systems, Conclusions/Recommendations, and Other River Basins of Interest.

Wagner, R. Harrison. 1983. The Theory of Games and the Problem of International Cooperation. *American Political Science Review* 77 (June): 330–46. Argues that the Prisoner's Dilemma and the Stag Hunt are inadequate models of international politics, and that a security dilemma does not necessarily have the implications people usually assign it.

Wakil, Mikhail. 1993. Analysis of Future Water Needs for Different Sectors in Syria. *Water International* 18 (1, March): 18–22. Water demands for different sectors in Syria have been increasing steadily and at a high rate, and it is expected that by the year 2010 the country will face a deficit in its water balance. To define the importance and the evolution of the projected water shortage, an analysis of the future water needs for the domestic, agricultural, and industrial sectors is performed. The actual water consumption in Syria for different sectors is estimated at about 12 billion m^3/yr, 90 per cent of which are consumed to irrigate about 700×10^3 ha. At the completion of the currently planned irrigation projects, which are expected to be finished by the year 2020, the irrigated areas will amount to 1350×10^3 ha, and will consume 20.5 billion m^3/yr. Sixty per cent of the new areas brought under irrigation will rely on the waters of the Euphrates river and its tributary, Al-Khabour. Within the same period, the Syrian population will reach approximately 32 million inhabitants who will consume 3.5 billion m^3/yr for domestic, municipal, and industrial uses. The available water resources in Syria are estimated at about 23.5 billion m^3, taking into account the Syrian share of the Euphrates river (13 billion m^3). Ten per cent of this amount is lost by evaporation. Therefore, it is expected that by the year 2010, Syria will reach an equilibrium in its water balance. However, starting from this date, it will experience an ever increasing water deficit. This situation will curtail future development plans and lead to decreased standards of living. It could further be a serious source of conflict in the Middle East if no agreement is reached regarding the allocation of Euphrates water. (abstract)

Waterbury, John. 1992. Three Rivers in Search of a Regime: The Jordan, the Euphrates and the Nile. William Stewart Tod Professor of Public and International Affairs, April. In this paper, we shall look at three Middle Eastern rivers that cross international boundaries and for which there are no binding, comprehensive accords governing their use. The three are the Jordan, the Euphrates, and the Nile. They will be treated in that order as it corresponds to a descending order of difficulty in obtaining cooperation among the riparian states in their basins in joint use of their water. The basic facts concerning the amounts of water available and current patterns of usage are fairly well know, and I shall make no attempt to go over them systematically here (inter alia, see Gischler, 1979; Naff and Matson, 1984; Starr, 1988; Kolars, 1992; Army Corps of Engineers, 1991; Allan Howell, 1990; Lowi, 1990; Waterbury, 1979 and 1983). (abstract)

White, Sally Blount and Margaret A. Neale. 1991. Reservation Prices, Resistance Points, and BATNAs: Determining the Parameters of Acceptable Negotiated Outcomes. *Negotiation Journal* 7 (4, October): 379–88. Defines terms of negotiations: BATNAs, Resistance Points, Reservation Prices, and the conjunctive use of two or more of the above. Empirical findings regarding these ideas are presented.

Whittington, D. and Kingsley Haynes. 1985. Nile Water for Whom? Emerging Conflicts in Water Allocation for Agricultural Expansion in Egypt and Sudan. In *Agricultural Development in the Middle East*, ed. Peter Beaumont and K. McLachlan. New York: John Wiley and Sons. The paper discusses agriculture and water use in Egypt and Sudan. It also reviews the 1959 Nile Waters Agreement. It offers a plan for long-term efficient water use.

—— and E. McClelland. 1992. Opportunities for Regional and International Cooperation in the Nile Basin. *Water International* 17(3): 144–54. Prospects for the Upper Nile basin states of Egypt, Sudan, and Ethiopia point to increasing competition for Nile water resources as population and development pressures intensify for all involved. This paper argues that it is in no one's best interest to maintain the lack of coordination in river basin development that persists in the Nile valley. Decisions made today about investment in water development projects, new irrigation schemes, and industrial projects will have consequences far into the future when water resources are in much greater demand. Furthermore, the unanticipated environmental and climatic changes of the 1980s have accelerated the need to make economic, political, and legal adjustments in the existing Nile management and allocation regime. Three possibilities are examined for cooperation in river basin development, with an emphasis on the importance of taking a basin-wide perspective on future water planning and investment. (abstract)

Wolf, A.T. 1993. The Jordan Watershed: Past Attempts at Cooperation and Lessons for the Future. *Water International* 18 (1, March): 5–17. Scarcity of water resources in the Mideast has led to bitter, occasionally armed, conflict; to generally undiscussed complications in political negotiations; and to unique opportunities for dialogue among the hostile actors of the Arab-Israeli conflict. This paper suggests a process of conflict resolution by which the dispute over the waters of the Jordan River watershed might be approached. Focus is on the "hydro-political" relations among Israelis, Jordanians, and Palestinians of the West Bank and Gaza. The process is designed for a hypothetical mediator, and the disciplinary tools he or she might use are described. In conjunction with regional peace talks and development plans, a four-step process for conflict resolution is developed, with political, management, and technical options explored. In addition, a large-scale project for the creation of power and water that may help ease regional political tensions is proposed. The paper is organized into three major sections: (1) background to Mideast "hydro-politics;" (2) a conceptual model for conflict resolution; and (3) four steps to water conflict resolution. (abstract)

—— and M. Murakami. 1995. Techno-political Decision Making for Water Resources Development: The Jordan River Watershed. *Water Resources Development* 11(2): 147–62. Discussions on water resources development generally focus on a variety of technical options, often without considering the potential repercussions of each option. This paper incorporates both technical and political considerations in a "techno-political" decision-making framework. Water resources development alternatives are then examined to evaluate their respective priorities for development in Israel, Palestine, and Jordan, which are the major riparians of the Jordan river. Particular account is taken of the Middle East peace negotiations, and consequent political changes. Each proposal is designed to provide incentives for sharing resources and benefits among the riparian states. (abstract)

——— and T. Ross. 1992. The Impact of Scarce Water Resources on the Arab-Israeli Conflict. *Natural Resources Journal* 32: 921–58. The Jordan river watershed is included within the borders of countries and territories each of whose water consumption is currently approaching or surpassing annual recharge. The region is also particularly volatile politically, with five Arab-Israeli wars since 1948, and many tenacious issues yet unresolved. This paper suggests that scarce water resources are actually inextricably related to regional conflict, having led historically to intense, and sometimes armed, competition, but also to occasional instances of cooperation between otherwise hostile players. The focus on water as a strategic resource has particular relevance to policy options given, for example, Israel's reliance on the West Bank for a share of its water resources, and the water needs of at least one impending wave of immigration to the region. Included in the paper are sections describing the natural hydrography and water consumption patterns of the region, a brief history of political events affected by 'hydro-strategic' considerations, and a survey of some resource strategy alternatives for the future. (abstract)

Wood, W.B. 1992. Ecopolitics: Domestic and International Linkages. *Geographica* 27(1): 49–54. Czechoslovakia, landlocked in the heart of Central Europe, suffers the consequences of being both an exporter and importer of air and water pollutants. Bilateral and multilateral efforts to curb transboundary environmental degradation require an understanding of the linkages between domestic and international ecopolitics. This brief paper explores some of the economic and ecological ties that can be a source of conflict between countries as well as a means for closer environmental cooperation. Countries that share several common ecological concerns, such as Czechoslovakia and its neighbours, are at the forefront of testing international environmental institutions and political commitments to environmental quality. The success of their mutual efforts to resolve local, cross-border problems will determine whether the international community is ready to tackle the global ecological problems. (abstract)

World Bank/UNDP/UNEP. 1994. Aral Sea Program – Phase 1. Proposed Donors Meeting, 23–24. The main objective of this paper is to provide a briefing to the donor countries and international agencies on the Aral Sea Program – Phase 1 which was formulated by the Executive Committee (EC) of the Interstate Council (ICAS) with the assistance of a World Bank Mission. It will be discussed at the donors' meeting scheduled for 23–24 June 1994 in Paris. The paper is organized in seven sections: Background, Aral Sea Program – Phase 1, Program Prospects and Issues, Risks, Program Financing, Recommendations, and Next Steps. The paper discusses the Program and its issues in sufficient detail to enable donors to indicate their support. The paper is based on the World Bank Mission's Aide Memoire (Vols. 1 and 2), which provides more detailed information on specific projects. It includes, however, additional information. For example, it explains the implementation plan in greater detail, updates the cost estimates, discusses the Program's prospects, issues, risks and financing options, and proposes the next steps for follow-up actions. (abstract)

Yaron, Dan. 1994. An approach to the problem of water allocation to Israel and the Palestinian Entity. *Resource and Energy Economics* 16 (4, November): 271–86. This article reviews the water scarcity problem in Israel/Palestine, including discussion of conservation and supply augmentation. It also offers a method for allocation through price mechanisms.

Young, Ralph. 1991. The Economic Significance of Environmental Resources: A Review of the Evidence. *Review of Marketing and Agricultural Economics* 59 (3, December): 229–54. The products and services of many environmental resources do not enter commercial markets and remain unpriced. The absence of market values presents a major difficulty for environmental projects in competing for ever-tightening budgets. In response to the need for assessing costs and benefits, a number of methods have been devised and applied to generate estimates of the value of unpriced resources. This paper reviews briefly the approaches employed for generating value estimates for unpriced environmental resources, with particular attention being paid to the contingent valuation method. Estimates of value for environmental products and services for both the United States and Australia are presented. This evidence clearly demonstrates that a wide range of unpriced environmental resources have significant economic value to the community. The paper concludes with a discussion of the implications for project development, funding, and policy in Australia. (abstract)

Zaman, Munir. 1982. The Ganges Basin and the Water Dispute. *Water Supply and Management* 6(4): 321–8. History of negotiations 1951–1977 and hydrologic data. Zaman discusses augmentation schemes and the Indian and Nepalese views of the Ganges dispute.

Zarour, H. and J. Isaac. 1993. Nature's Apportionment and the Open Market: A Promising Solution to the Arab-Israeli Water Conflict. *Water International* 18(1): 40–53. The populations and consumption levels of countries throughout the Middle East are steadily increasing and, correspondingly, their water demands. Water resources available to most of these countries remain more or less the same, if not being reduced by abusive utilization or pollution. Many now speculate that the region's next war will be fought over water. Today's international legislative structure is incapable of solving complex water disputes such as that of the Middle East. However, principles of international legitimacy provide the basis for solutions. This paper presents a pragmatic, practical, and dispassionate formula compatible with the principles of international law and legitimacy for dealing with international water resources allocation problems. The presented formula is built around resolving the Middle East's water conflict according to natural apportionment of resources and the open market approach. (abstract)

Zartman, I. William. 1978. *The Negotiation Process.* London: Sage Publications. Talks of negotiation as a social process and discusses decision-making in society. Covers game theory and some modifications of the Nash model.

—— 1992. International Environmental Negotiation: Challenges for Analysis and Peace. *Negotiation Journal* 8 (2, April): 113–24. Discusses the creation and analysis of the varied outcomes of international environmental negotiation. Identifies the challenge for all negotiators as the creation of the greatest Pareto-optimal outcome (the biggest pie possible with the largest shares possible for each party in the negotiations).

10

Bibliography

It was our hope to include as many materials as possible relevant to the study of water disputes and water treaties. We are pleased with the result but realize that we may have omitted important items. In an attempt to keep our records as up to date as possible, because this document is part of an ongoing study, if you know of citations you feel are relevant that we have missed, please forward them to us for inclusion of future documents. Thank you.

Send to: Aaron Wolf
              Department of Geosciences
              104 Wilkinson Hall
              Oregon State University
              Corvallis, OR 97330-5506
              email: wolfa@geo.orst.edu

Abbas, A.T. 1982. *The Ganges Water Dispute*. New Delhi: Vikas Publishing House.

Abbott, P.C. and S.L. Haley. 1988. International Trade Theory and Natural Resource Concepts. In *Agricultural Trade and Natural Resources: Discovering the Critical Linkages*, ed. John D. Sutton, 35–62. Boulder, CO: Lynne Rienner.

Abduhadi, Rami. 1992. Water Resources for the State of Palestine. Council on Foreign Relations, June.

Abebe, Mesfin. 1995. Nile: source of regional cooperation or conflict? *Water International* 20 (March): 32–35.

Adams, Gregory D., Gordon C. Rausser, and Leo K. Simon. 1993. *Modelling Natural Resource Negotiations: An Application to California Water Policy*. Berkeley, CA: UC-Berkeley.

Adler, A. 1995. Positioning and Mapping International Land Boundaries. *Boundary and Territory Briefing*, Vol. 2.1. Durham: International Boundaries Research Unit.

Adler, E. 1993. Cognitive Evolution: A Dynamic Approach for the Study of International Relations and Their Progress. In *Progress in International Relations*, ed. E. Adler and B. Crawford. New York, NY: Columbia University Press.

Adler, Peter S. 1990. Casting Sunshine on Negotiated Settlements. *Negotiation Journal* 6 (4, October): 305–9.

af Ornas, A.H. and S. Lodgaard. 1992. *The Environment and International Security*. Oslo, Norway: Uppsala University, Department of Human Geography/International Peace Research Institute.

Agnew, C.T. 1992. *Water Resources in the Arid Realm*. London: Routledge.

Aiken, J.D. 1980. *Legal Aspects of Conflicts Between Users of Surface and Ground Water*. Lincoln, NE: University of Nebraska, Department of Agricultural Economics.

Akatsuka, Yuzo and Takashi Asaeda. 1995. Econo-Political Environments of the Mekong Basin: Developments and Related Transport Infrastructures with Particular Emphasis on the Mekong River Navigation. Asian Water Forum. Bangkok, Thailand: United Nations University, 30 January–1 February.

Akdogan, Haluk. 1992. An Economic Analysis of Inter-Country Water Transfer through Pipelines. Sharing Fresh Water Resources. Crete, unpublished mimeo.

Al-Jarwan, Saud Ali. 1993. Water Regime Formation: An Empirical, Explanatory, Quantitative and Comparative Study of the Nile and Ganges River. diss. Carbondale, IL: Southern Illinois University.

Al-jayyousi, Odeh Rashed. 1993. Evaluating Potential Water Conflict in the Middle East: Strategies for Cooperation. diss. UMI. Chicago: University of Illinois.

Al-Loub, B. and T.T. Al-Shemmeri. 1995. Sustainable Development of Water Resources and Possible Enhancement. *Water International* 20 (2, June): 106–9.

Alam, Undala. 1997. The Indus Water Treaty: Peace Amidst War. IXth World Water Congress. Montreal, Canada, 1–6 September.

Alemu, Senai. 1995. The Nile Basin: Data Review and Riparian Issues. Final Draft Report. Washington, DC: AGRPW, The World Bank.

—— 1995. Problem Definition and Stakeholder Analysis of the Nile River Basin. In *3rd Annual Nile 2002 Conference*. 3rd Annual Nile 2002 Conference. Arusha, Tanzania, 13–15 February.

Alheritiere, Dominique. 1985. Settlement of Public International Disputes on Shared Resources. *Natural Resources Journal* 25(3): 701–11.

Ali, Mohammed. 1965. *The River Jordan and the Zionist Conspiracy*. Cairo: Information Department.

—— 1987. *Water Resources Policy for Asia*. Boston, MA: A.A. Balkema.

Ali, Salamat. 1988. Stemming the Flood: Few Results from Bangladesh-Indian Talks on Water Control. *Far Eastern Economic Review*, 13 October, 24–6.

Allan, J.A. 1992. Substitutes for Water Are Being Found in the Middle East and North Africa. *GeoJournal* 28(3): 375–85.

—— and Chibli Mallat. 1995. *Water in the Middle East: Legal, Political, and Commercial Implications*. London and New York: Tauris Academic Studies.

Allan, T. 1994. Economic and Political Adjustments to Scarce Water in the Middle East. *Studies in Environmental Science* 58: 375–87.

Allee, D.J. 1993. *Managing Transboundary Water Conflicts: The United States and its Boundary Commissions*. Ithaca, New York: Cornell University.

—— 1993. Subnational Governance and the International Joint Commission: Local Management of United States and Canadian Boundary Waters. *Natural Resources Journal* 33 (Winter): 133–51.

Allen, T. 1995. A Transition of the Political Economy of Water and Environment in Israel-Palestine. In *Joint Management of Shared Aquifers: The Second Workshop November 27–December 1, 1994*, ed. M. Haddad and E. Feitelson. Jerusalem, Israel: The Harry S. Truman Research Institute and the Palestine Consultancy Group.

Altinbilek, H. Dogan. 1997. Water and Land Resources Development in Southeastern Turkey. *International Journal of Water Resources Development 1997* 13 (3, September): 311–32.

American Academy of Arts and Sciences. 1994. Introduction. *Bulletin of the American Academy of Arts and Sciences* XLVIII (2, November).

Amery, Hussein. 1993. The Litani River of Lebanon. *The Geographical Review* 83 (3, July): 229–37.

Amy, Douglas J. 1987. *The Politics of Environmental Mediation*. New York: Columbia University Press.

Andah, K. and F. Siccardi. 1991. Prediction of Hydrometeorlogical Extremes in the Sudanese Nile Region: A Need for International Co-operation. In *Hydrology for the Water Management of Large River Basins*, ed. F.H.M. Van de Ven et al., 3–12. International Association of Hydrological Sciences.

Anderson, E.W. 1991. Hydropolitics. *Geographical Magazine* (February): 10–14.

—— 1991. Making Waves on the Nile. *Geographical Magazine* (April): 31–4.

—— 1991. The Middle East and Hydropolitics. *World Energy Council Journal* (December): 35–8.

—— 1991. The Source of Power. *Geographical Magazine* (March): 12–15.

—— 1992. Water Conflict in the Middle East – a New Initiative. *Jane's Intelligence Review* 4(5): 227–30.

Anderson, Kirsten Ewers. 1995. Institutional Flaws of Collective Forest Management. *Ambio* 24 (6, September): 349–53.

Anderson, T. 1982. The New Resource Economics: Old Ideas and New Applications. *American Journal of Agricultural Economics* 64 (5, December): 928–34.

Anderson, Terry Lee. 1983. *Water Rights: Scarce Resource Allocation, Bureaucracy, and the Environment*. San Francisco, CA: Pacific Institute for Public Policy Research.

Andras, Eva and Mercedes Lentz. 1993. *Enhancement of Middle East Water Supply: A Literature Review of Technologies and Applications*. Ottawa: International Water Engineering Centre.

Anon. 1994. Protection and Use of Transboundary Watercourses and International Lakes in Europe. *Natural Resources Forum* 18(3): 171–80.

Anonymous. 1975. Whither the Mekong Committee. *Bangkok Bank Monthly Review* 16 (10, October): 589 ff.

—— 1977. The Future of the Mekong Basin. *Bangkok Bank Monthly Review* 18 (3, 4, March, April): 123 ff.

—— 1987. Where Dams Can Cause Wars. *The Economist* 18 (July): 37–8.

Antle, J.M. and G. Heidebrink. 1995. Environment and Development: Theory and International Evidence. *Economic Development and Cultural Change* 43(3): 603–25.

Appelgren, Bo and Wulf Klohn. 1997. Management of Transboundary Water Resources for Water Security; Principles, Approaches and State Practice. *Natural Resources Journal* 21 (2, May): 91–100.

Apprey, Victor. 1988. Multiagent Mathematical Programming with an Application to Conflict Resolution. diss. Charlottesville, VA: University of Virginia, System Science Engineering.

Arbatov, Alexander A. 1986. Oil as a Factor in Strategic Policy and Action: Past and Present. In *Global Resources and International Conflict: Environmental Factors in Strategic Policy and Action*, ed. Arthur H. Westing, 21–37. New York, NY: Oxford University Press.

Arlosoroff, S. 1995. Water Resources Management Within Regional Cooperation in the Middle East. In *Joint Management of Shared Aquifers: The Second Workshop November 27–December 1, 1994*, ed. M. Haddad and E. Feitelson. Jerusalem, Israel: The Harry S. Truman Research Institute and the Palestine Consultancy Group.

Arnoux, Rosemary, Richard Dawson, and Martin O'Connor. 1993. Logics of Death and Sacrifice in the Resource Management Law Reforms of Aotearoa, New Zealand. *Journal of Economic Issues* 27 (December): 1059–96.

Aryal, Manisha. 1995. Dams: The Vocabulary of Protest. *Himal* (July/August): 11–21.

Ashworth, John and Ivy Papps. 1991. Equity in European Community Pollution Control. *Journal of Environmental Economics and Management* 20 (1, January): 46–54.

Associated Press. 1995. Water Crisis Looms, World Bank Says. *The Washington Post*, Monday, 7 August.

Ataov, Turkkaya. 1981. The Use of Palestinian Waters and International Law. In *Third United Nations Seminar on the Question of Palestine*. Third United Nations Seminar on the Question of Palestine. Colombo, Sri Lanka: United Nations, 10–14 August.

Auer, Brad and Joshua Foster. 1992. Water is Thicker Than Blood: A Developmental Construct Examining Water Scarcity in the Middle East as a Threat to International Security. Yale School of Forestry and Environmental Studies, 29 December.

Avnimelech, Yoram. 1994. Water Scarcity – Israel's Experience and Approach. In *Water as an Element of Cooperation and Development in the Middle East*, ed. Ali Ihsan Bagis. Ankara: Hacettepe University.

Axelrod, R. 1990. *The Evolution of Cooperation*. New York: Basic Books.

Azar, Christian and John Holmberg. 1995. Defining the Generational Environmental Debt. *Ecological Economics* 14 (1, July): 7–20.

Azar, Edward E. 1990. *The Management of Protracted Social Conflicts*. London: Aldershot.

Bacow, L.S. and M. Wheeler. 1984. *Environmental Dispute Resolution*. New York, NY: Plenum Press.

Bagis, Ali Ihsan. 1994. Water in the Region: Potential and Prospects. In *Water as an Element of Cooperation and Development in the Middle East*, ed. Ali Ihsan Bagis. Ankara: Hacettepe University.

Baker, G. 1987. Drought Renews Jordan Debate. *World Water* 10 (1, January, February): 20–22.

Bakour, Y. 1991. *Planning and Management of Water Resources in Syria*. Damascus, Syria: Arab Organization for Agricultural Development.

Balek, J. 1977. *Hydrology and Water Resources in Tropical Africa*. New York: Elsevier Press.

Bandhyopadhyay, Jayantha and Dipak Gyawali. 1994. Himalayan Water Resources: Ecological and Political Aspects of Management. *Mountain Research and Development* 14: 1–24.

Bar-Shira, Z. 1992. Nonparametric Test of the Expected Utility Hypothesis. *American Journal of Agricultural Economics* 74 (3, August): 523–33.

Bari, Zohurul. 1977. Syrian-Iraqi Dispute Over the Euphrates Waters. *International Studies* 16 (2, April–June): 227–44.

Barnard, William S. 1994. From Obscurity to Resurrection: The Lower Oranger River as International Boundary. In *Political Boundaries and Coexistence: Proceedings of the IGU-Symposium, Basle/Switzerland, 24–27 May*, ed. Werner A. Galluser, 125–34. New York, NY: Peter Lang.

Barret, S. 1994. *Conflict and Cooperation in Managing International Water Resources*. Centre for Social and Economic Research on the Global Environment, Vol. Wm 94094.

—— 1994. Conflict and Cooperation in Managing International Water Resources. *Policy Research Working Paper, The World Bank, Policy Research Department* 1303 (May).

Bartlett, Richard A. 1995. *Troubled Waters: Champion International and the Pigeon River Controversy*. Knoxville, TN: University of Tennessee Press.

Bath, C.R. 1981. Resolving Water Disputes. *Proceedings of the Academy of Political Science* 34: 181–8.

Bau, Jao. 1991. *Integrated Approaches to Water Pollution Problems: Proceedings of the International Symposium (SISIPPA), Lisbon, Portugal 19–23 June 1989*. New York, NY: Elsevier Applied Science.

—— 1995. Cooperation Among Water Research and Development Institutions of Europe. *Water International* 20: 129–35.

Baumann, Duane D., John J. Boland, and John H. Sims. 1984. Water Conservation: The Struggle Over Definition. *Water Resources Research* 20 (4, April): 428–34.

Baumol, William J. and Wallace E. Oates. 1992. *The Theory of Environmental Policy*. Cambridge, MA: Cambridge University Press.

Baxter, R.R. 1967. The Indus Basin. In *The Law of International Drainage Basins*, ed. A.H. Garretson, D.R. Hayton, and C.J. Olmstead, 443–85. New York, NY: Oceana Books.

Beaumont, Peter. 1991. Transboundary Water Disputes in the Middle East. Transboundary Waters in the Middle East. Ankara, September.

—— 1994. The Myth of Water Wars and the Future of Irrigated Agriculture in the Middle East. *Water Resources Development* 10(1): 9–22.

—— 1997. Dividing the Waters of the River Jordan: An Analysis of the 1994 Israel-Jordan Peace Treaty. *International Journal of Water Resources Development* 13 (3, September): 415–24.

Beck, J. 1994. An Experimental Test of Preferences for the Distribution of Income and Individual Risk Aversion. *Eastern Economic Journal* 20 (2, Spring): 131–46.

Beck, Robert E. 1991. *Waters and Water Rights*. Charlottesville, VA: Michie.

Becker, Mimi Larsen. 1993. The International Joint Commission and Public Participation: Past Experiences, Present Challenges, Future Tasks. *Natural Resources Journal* 33 (2, Spring): 235–74.

Becker, Nir and K. William Easter. 1994. Dynamic Games on the Great Lakes. In *Resolution of Water Quantity and Quality Conflicts*, ed. Ariel Dinar and Edna Loehman. New York, NY: Praeger.

Beckerman, Wilfred. 1992. Economic Growth and the Environment: Whose Growth? Whose Environment? *World Development* 20(4): 481–96.

Begum, Khurshida. 1988. *Tension Over the Farakka Barrage*. Stuttgart: Steiner Verlag.

Behera, N.C., Paul Evans, and C. Rizvi. 1997. *Beyond Boundaries: A Report of the State of Non-official Dialogues on Peace, Security and Cooperation in South Asia*. New York: Joint Center for Asia Pacific Affiars.

Bell, F.C. 1988. The Sharing of Scarce Water Resources. *Geoforum* 19(3): 353–66.

Ben-Zvi, Abraham. 1989. *Between Lausanne and Geneva: International Conferences and The Arab-Israeli Conflict*. London: Westview Press.

Bendor, Jonathan and Piotr Swistak. 1995. Types of Evolutionary Stability and the Problem of Cooperation. *Proceedings of the National Academy of Sciences, USA* 92: 3596–600.

Benhabib, Jess and Aldo Rustichini. 1991. *Social Conflict, Growth and Income Distribution*. Chicago, IL: Northwestern University.

Bennett, P.G. 1987. *Analyzing Conflict and its Resolution: Some Mathematical Contributions*. Oxford: Clarendon Press.

Benshu, Z. 1995. South to North Transfer Project. *International Water Power and Dam Construction* 47 (January): 42–5.

Benvenisti, E. and H. Gvirtzman. 1993. Harnessing International Law to Determine Israeli-Palestinian Water Rights: The Mountain Aquifer. *Natural Resources Journal* 33 (Summer): 543–67.

Berber, Friedrich Joseph. 1959. *Rivers in International Law*. London: Oceana Publications.

Berck, P., S. Cecchetti, N. Gallini, and N. Williams. 1984. *The Theory of Exhaustible Resources*. Berkeley, CA: The University of California, Division of Agricultural and Natural Resources.

Bercovitch, Jacob. 1992. Mediators and Mediation Strategies in International Relations. *Negotiation Journal* 8 (2, April): 99–112.

Berger, Thomas R. 1988. Conflict in Alaska. *Natural Resources Journal* 28 (1, Winter): 37–62.

Bergstrom, Margareta. 1990. The Release in War of Dangerous Forces from Hydrological Facilities. In *Environmental Hazards of War: Releasing Dangerous Forces in an Industrialized World*, ed. Arthur H. Westing, 38–47. London, England: Sage Publications.

Berkoff, Jeremy. 1994. *A Strategy for Managing Water in the Middle East*. Washington, DC: World Bank.

Berthelot, R. 1989. The Multidonor Approach in Large River and Lake Basin Development in Africa. *Natural Resources Forum* (August): 209–15.

Beschorner, Natasha. 1992. *Water and Instability in the Middle East*. Adelphi Paper, Vol. 273.

Bhatti, Neeloo, David G. Streets, and Wesley K. Foell. 1992. Acid Rain in Asia. *Environmental Management* 16(4): 541–62.

Biddle, G.C. and R. Steinberg. 1984. Allocation of Joint and Common Costs. *Journal of Accounting Literature* 3: 1–45.

Bilen, Ozden. 1994. Prospects for Technical Cooperation in the Euphrates-Tigris Basin. In *International Waters of the Middle East: From Euphrates-Tigris to Nile*, ed. Asit K. Biswas, 5–43. Bombay, Delhi, Calcutta, Madras: Oxford University Press.

Bingham, Gail. 1986. *Resolving Environmental Disputes. A Decade of Experience*. Washington, DC: The Conservation Foundation.

—— and Suzanne Goulet Orenstein. 1991. The Role of Negotiation in Managing Water Conflicts, submitted to American Society of Civil Engineers for publication.

——, A.T. Wolf, and Tim Wohlgenant. 1994. *Resolving Water Disputes: Conflict and Cooperation in the United States, the Near East, and Asia*. Washington, DC: U.S. Agency for International Development.

Biot, Yvan, Piers M. Blaikie, Cecile Jackson, and Richard Palmer-Jones. 1995. *Rethinking Research on Land Degradation in Developing Countries*. World Bank Discussion Papers, Vol. 289. Washington, DC: World Bank.

Biswas, Asit K. 1981. Integrated Water Management: Some International Dimensions. *Journal of Hydrology* 51 (1/4, May): 369–80.

—— 1983. *Long Distance Water Transfer: A Chinese Case Study and International Experiences*. Dublin: Tycooly International Publishing.

—— 1983. Some Major Issues in River Basin Management for Developing Countries. In *River Basin Development. Proceedings of the National Symposium 1981, Dacca*, ed. M. Zaman et al., 17–27. Dublin: Tycooly International Publishing.

—— 1991. Water for Sustainable Development, a Global Perspective. *Development and Co-operation* 5: 20.

—— 1992. Indus Water Treaty: The Negotiating Process. *Water International* 17: 201–09.

—— 1992. Water for Third World Development. A Perspective from the South. *Water Resources Development* 8(1): 3–9.

—— 1993. Management of International Waters: Problems and Perspective. *Water Resources Development* 9(2): 167–88.

—— 1994. Management of International Water Resources: Some Recent Developments. In *International Waters of the Middle East: From Euphrates-Tigris to Nile*, ed. Asit K. Biswas, 185–214. Bombay, Delhi, Calcutta, Madras: Oxford University Press.

—— 1994. Sustainable Water Resources Development: Some Personal Thoughts. *Water Resources Development* 10(2): 109–16.

—— 1995. Institutional Arrangements for International Cooperation in Water Resources. *Water Resources Development* 11(2): 139–45.

—— 1997. Editorial: Management of International Waters. *International Journal of Water Resources Development* 13 (3, September): 277–8.

——, and ed. 1994. *International Waters of the Middle East: From Euphrates-Tigris to Nile*. Water Resources Management Series, Vol. 2. Bombay, Delhi, Calcutta, Madras: Oxford University Press.

Biswas, Asit K., T.N. Khoshoo, and Ashok Khosla. 1990. *Environmental Modelling for Developing Countries*. New York, NY: Tycooly Publishing.

——, J. Kolars, M. Murakami, J. Waterbury, and A.T. Wolf. 1996 (forthcoming). *Core and Periphery: A Comprehensive Approach to Middle Eastern Water*. Oxford: Oxford University Press.

Blainey, Geoffrey. 1988. *The Causes of War*, 3rd edn. London: Macmillan.

Blake, Gerald. 1994. International Transboundary Collaborative Ventures. In *Political Boundaries and Coexistence: Proceedings of the IGU-Symposium, Basle/Switzerland, 24–27 May*, ed. Werner A. Galluser, 359–71. New York, NY: Peter Lang.

——, William J. Hildesley, Martin A. Pratt, Rebecca J. Ridley, and Clive H. Schofield, eds. 1995. *The Peaceful Management of Transboundary Resources*. International Environmental Law and Policy Series. Boston, MA: Graham & Trotman/Martinus Nijhoff.

Bloomfield, L.M. and Gerald F. Fitzgerald. 1958. *Boundary Waters Problems of Canada and the United States*. Toronto: Carswell Company.

Blum, Charlotte. 1994. Oman: Fast Track to Private Infrastructure. *Middle East Business Weekly* 38 (29 July): 13.

Bochniarz, Zbigniew. 1992. Water Management Problems in Economies in Transition. *Natural Resources Forum*, February, 55–63.

Boisvert, R.N. and N.L. Bills. 1976. *A Non-Survey Technique for Regional I-O Models: Application to River Basin Planning*. Ithaca, NY: New York Cornell Agricultural Experiment Station, New York State College of Agriculture and Life Sciences Department of Agricultural Economics.

Bolin, I. 1990. Upsetting the Power Balance: Cooperation, Competition, and Conflict Along an Andean Irrigation System. *Human Organization* 49(2): 140–8.

Bolukbasi, Suha. 1993. Turkey Challenges Iraq and Syria: The Euphrates Dispute. *Journal of South Asian and Middle Eastern Studies* XVI (4, Summer): 9–32.

Bond, Andrew R. et al. 1993. Geography of Human Resources in the Post-Soviet Realm. *Post-Soviet Geography* 34: 219–80.

Born, Stephen M. 1989. *Redefining National Water Policy: New Roles and Directions*. AWRA Special Publication No. 89-1. Bethesda, MD: American Water Resources Association.

Bourne, Charles B. 1992. The International Law Commission's Draft Articles on the Law of International Watercourses: Principles and Planned Measures. *Colorado Journal of Environmental Law and Policy* 3(1): 65–92.

—— 1996. The International Law Association's Contributions to Internaitonal Water Resources Law. *Natural Resources Journal* 36: 155–216.

Braden, John B. and C.D. Kolstad, eds. 1991. *Measuring the Demand for Environmental Quality*. New York, NY: North Holland.

—— and Stephan B. Lovejoy. 1989. *Agriculture and Water Quality: International Perspectives*. Boulder, CO: Lynne Rienner.

Brams, Stephen J. 1990. *Negotiation Games: Applying Game Theory to Bargaining and Arbitration*. New York: Routledge.

Brecher, Michael. 1974. *Decisions in Israel's Foreign Policy*. London: Oxford University Press.

Brooks, David B. 1993. Adjusting the Flow: Two Comments on the Middle East Water Crisis. *Water International* 18 (1, March): 35–39.

—— 1994. Economics, Ecology, and Equity: Lessons From the Energy Crisis in Managing Water Shared by Israelis and Palestinians. *Studies in Environmental Science* 58: 441–50.

—— 1995. Planning for a Different Future: Soft Water Paths. In *Joint Management of Shared Aquifers: The Second Workshop November 27–December 1, 1994*, ed. M. Haddad and E. Feitelson. Jerusalem, Israel: The Harry S. Truman Research Institute and the Palestine Consultancy Group.

—— 1997. Between the Great Rivers: Water in the Heart of the Middle East. *International Journal of Water Resources Development* 13 (3, September): 291–310.

Brown, C.A. 1984. The Central Arizona Water Control Study: A Case for Multiobjective Planning and Public Involvement Water Storage and Flood Control Alternatives, Conflict Resolution. *Water Resources Bulletin* 20 (3, June): 331–7.

Brown, H.J. 1993. *ADR Principles and Practice*. London, England: Sweet and Maxwell.

Brown, Jennifer. 1989. *Environmental Threats: Perception, Analysis, and Management*. New York, NY: Belhaven Press.

Brown, Lester R. 1977. *Redefining National Security*. Worldwatch Paper, Vol. 14. Washington, DC: Worldwatch Institute.

—— 1978. *Human Needs and the Security of Nations*. Headline series, Vol. 238. New York, NY: The Foreign Policy Association.

Brown, Seyom. 1987. *The Causes and Prevention of War*. New York: St. Martin's.

Bruce, J.P. 1976. Water Management in International Basins. *Proceedings of the Soil Conservation Society of America* 31: 152–60.

Bruhacs, J. 1992. Evaluation of the Legal Aspects of Projects in International Rivers. *European Water Pollution Control* 2(3): 10–13.

—— 1993. *The Law of Non-Navigational Uses of International Watercourses*. Dordrecht: Martinus Nijhoff.

Buck, S.J., G.W. Gleason, and M.S. Jofuku. 1993. "The Institutional Imperative": Resolving Transboundary Water Conflict in Arid Agricultural Regions of the United States and the Commonwealth of Independent States. *Natural Resources Journal* 33: 595–628.

Buckle, L.G. and S.T. Buckle. 1986. Placing Environmental Mediation in Context: Lessons from "Failed" Mediations. *Environmental Impact Assessment Review* 6: 50–70.

Bulloch, John and Adel Darwish. 1993. *Water Wars: Coming Conflicts in the Middle East*. London: Victor Gollancz.

Burchi, Stefano, ed. 1993. *Treaties Concerning the Non-Navigational Uses of International Watercourses – Europe*. Rome: Food and Agriculture Organization of the United Nations.

Burton, Ian, Robert W. Kates, and Gilbert F. White. 1993. *The Environment as Hazard*, 2nd edn. New York, NY: The Guilford Press.

Burton, L. 1987. The American Indian Water Rights Dilemma: Historical Perspective and Dispute-Settling Policy Recommendations. *Journal of Environmental Law* 7(1): 1–31.

Butler, William Elliott. 1991. *Control Over Compliance with International Law*. Dordrecht: Martinus Nijhoff.

Butterworth, Robert L. 1976. *Managing Interstate Conflict, 1945–1974*. New Haven, CT: Yale University Press.

Butts, Kent Hughes. 1993. *Environmental Security: What is DOD's Role?* Carlisle, PA: Strategic Studies Institute, U.S. Army War College.

—— 1994. *Environmental Security: A DOD Partnership for Peace*. Carlisle, PA: Strategic Studies Institute, U.S. Army War College.

—— 1996. The Environment: A Geographical Component of Conflict. In *Association of American Geographers Annual Meeting*. Association of American Geographers Annual Meeting. Charlotte, NC, March.

Buzan, B. 1991. *People, States and Fear: An Agenda for International Security Studies in the Post-Cold War Era*. London, England: Harvester Wheatsheaf.

Calleros, J. Roman. 1991. The Impact on Mexico of the Lining of the All-American Canal. *Natural Resources Journal* 31 (Fall): 829–38.

Cameron, T.A. 1992. Nonuser Resource Values. *American Journal of Agricultural Economics* 74 (5, December): 1133–7.

Cano, Guillermo. 1982. Laws of Nature and Water Laws. *Water International* 7: 81–3.

—— 1989. The Development of the Law in International Water Resources and the Work of the International Law Commission. *Water International* 14: 167–71.

Capistrano, Ana Doris and Clyde F. Kiker. 1995. Macro-scale Economic Influences on Tropical Forest Depletion. *Ecological Economics* 14 (1, July): 21–9.

Caponera, Dante A. 1978. *Water Laws in Moslem Countries*. Rome: Food and Agriculture Organization of the United Nations.

—— 1980. *The Law of International Water Resources*. Rome: Food and Agriculture Organization of the United Nations.

—— 1985. Patterns of Cooperation in International Water Law: Principles and Institutions. *Natural Resources Journal* 25 (3, July): 563–87.

—— 1987. International Water Resources Law in the Indus Basin. In *Water Resources Policy for Asia*, ed. Mohammed Ali and et al. Boston, MA: A.A. Balkema.

—— 1991. Legal and Transboundary Concepts of Cooperation. Transboundary Waters in the Middle East. Ankara, September.

—— 1992. *Principles of Water Law and Administration: National and International*. Rotterdam: A.A. Balkema.

—— 1993. Legal Aspects of Transboundary River Basins in The Middle East: The Al Asi (Orontes), The Jordan, and The Nile. *Natural Resources Journal* 33: 629–63.

—— 1994. The Legal-Institutional Issues Involved in the Solution of Water Conflicts in the Middle East: The Jordan. *Studies in Environmental Science* 58: 163–80.

—— and Dominique Alheritiere. 1979. *Water Law in Selected African Countries (Benin, Burundi, Ethiopia, Gabon, Kenya, Mauritius, Sierra Leone, Swaziland, Upper Volta, Zambia)*. Rome: Food and Agriculture Organization of the United Nations.

Carlson, Lisa J. 1995. A Theory of Escalation and International Conflict. *Journal of Conflict Resolution* 39 (3, September): 511–34.

Carlton, Dennis W. 1991. *The Theory of Allocation and Its Implications for Marketing and Industrial Structure*. Cambridge, MA: National Bureau of Economic Research.

Carpenter, Susan L. and W.J. Kennedy. 1988. The Denver Metropolitan Water Roundtable: A Case Study in Researching Agreements. *Natural Resources Journal* 28 (1, Winter): 21–36.

Carr, M.K.V., P.B. Leeds-Harrison, and R.C. Carter. 1990. Water Management. *Outlook On Agriculture* 19 (4, December): 229–35.

Carroll, John Edward. 1986. Water Resource Management as an Issue in Environmental Diplomacy. *Natural Resources Journal* 26 (2, Spring): 207–20.

Carson, R. 1980. A Theory of Co-operatives. *Canadian Journal of Economics* 10(4): 563–89.

Cashman, Greg. 1993. *What Causes War? An Introduction to Theories of International Conflict*. New York: Lexington Books.

Changnon, Stanley. 1987. An Assessment of Climate Change, Water Resources, and Policy Research. *Water International* 12: 69–76.

Chapman, Duane. 1973. Economic Aspects of a Nuclear Desalination Agro-Industrial Project in the United Arab Republic. *American Journal of Agricultural Economics* 55(3): 433–40.

Charles T. Main Inc. 1966. *Program for Development of Surface Storage in the Indus Basin and Elsewhere within West Pakistan*. Boston: World Bank.

Chaube, U.C. 1990. Water Conflict Resolution in the Ganga-Brahmaputra Basin. *Water Resources Development* 6(2): 79–85.

—— 1992. Multilevel Hierarchical Modelling of an International Basin. In *Proceedings of the International Conference on Protection and Development of the Nile and Other Major Rivers*, Vol. 2/2. International Conference on Protection and Development of the Nile and Other Major Rivers. Cairo, Egypt (February) 3–5.

Chaudhry, Mohammed Talib. 1973. Conjunctive Use of Indus Basin Waters-Pakistan. diss. Fort Collins: Colorado State University.

Chauhan, Babu Ram. 1981. *Settlement of International Water Law Disputes in International Drainage Basins*. Berlin: Erich Schmidt.

—— 1992. *Settlement of International and Inter-State Water Disputes in India*. Bombay: N.M. Tripathi.

Chitale, M.A. 1995. Institutional Characteristics for International Cooperation in Water Resources. *Water Resources Development* 11(2): 113–23.

Choucri, Nazli. 1991. Resource Scarcity and National Security in the Middle East. In *New Perspectives for a Changing World Order*, ed. Eric H. Arnett, 99–107. Washington, DC: Science and International Security, American Association for the Advancement of Science.

Christy, Francis T., ed. 1980. The Law of the Sea: Problems of Conflict and Management of Fisheries in Southeast Asia. Conference Proceedings, ed. Francis T. Christy. ICLARM/ISEAS Workshop on the Law of the Sea. Manila: International Center for Living Aquatic Resources Management, 26–9 November, 1978.

Cioffi-Revilla, Claudio. 1990. *The Scientific Measurement of International Conflict: Handbook of Datasets on Crises and War, 1495–1988 A.D.* Boulder, CO: Lynne Rienner.

Clark, Edwin H., II, Gail Bingham, and Suzanne Goulet Orenstein. 1991. Resolving Water Disputes: Obstacles and Opportunities. *Resolve* 23: 1–10.

Clarke, R. 1991. *Water: The International Crisis*. London: Earthscan.

Clay, Daniel C., Mark Guizlo, and Sally Wallace. 1994. *Population and Land Degradation*. The Environmental and Natural Resources Policy and Training Project Working Paper, Vol. 14. Madison, WI: EPAT/MUCIA Research and Training.

Cline, William R. 1992. *The Economics of Global Warming*. Washington, DC: Institute for International Economics.

Cloete, Fanie. 1990. Prospects for a Negotiated Settlement in South Africa. *Negotiation Journal* 6 (2, April): 119–34.

Cobban, Helena. 1991. Defining Interests and Managing Change in the Middle East: The Challenge for the United States in the 1990s. In *New Perspectives for a Changing World Order*, ed. Eric H. Arnett, 109–21. Washington, DC: Science and International Security, American Association for the Advancement of Science.

Cohen, Saul B. 1986. *The Geopolitics of Israel's Border Question*. Tel Aviv, Israel: Tel Aviv University.

—— 1986. *The Geopolitics of Israel's Border Question*. Boulder, CO: Westview.

—— 1992. Middle East Geopolitical Transformation: The Disappearance of a Shatterbelt. *Journal of Geography* 91 (1, January/February): 2–10.

Committee on Western Water Management. 1992. *Water Transfers in the West: Efficiency, Equity, and the Environment*. Washington, DC: National Academy Press.

Conant, M.A. 1990. The Middle East Agenda. *Geopolitics of Energy* 12(6): 3–7.

Conca, Ken. 1994. In the Name of Sustainability: Peace Studies and Environmental Discourse. *Peace and Change* 19(2): 91–113.

Connah, G. 1981. *Three Thousand Years in Africa*. New York: Cambridge University Press.

Cooley, John. 1973. *Green March, Black September*. London: Frank Cass.

—— 1984. The War Over Water. *Foreign Policy* (Spring): 3–26.

Cooper, Jerrold. 1983. *Reconstructing History from Ancient Inscriptions: The Lagash-Umma Border Conflict*. Malibu, CA: Undena.

Copeland, Miles. 1969. *The Game of Nations*. New York: College Notes and Texts.

Coplin, W. and M. O'Leary. 1974. Activities of the PRINCE Project. *Policy Studies Journal*, Summer.

—— and M. O'Leary. 1976. *Everyman's PRINCE: A Guide to Understanding Your Political Problems*. New York: Dusbury Press.

—— and M. O'Leary. 1983. *Political Analysis Through the PRINCE System*. Policy Studies Association.

Cowell, Alan. 1989. Next Flashpoint in the Middle East: Water. *New York Times*, 16 April.

Crane, M. 1991. Diminishing Water Resources and International Law: U.S.-Mexico, A Case Study. *Cornell International Law Journal* 24(2): 299–323.

Crow, B. and A. Lindquist. 1990. Development of the Rivers Ganges and Brahmaputra: The Difficulty of Negotiating a New Line. *DPP Working Paper – Open University* 19: 1–46.

Cummings, R.G. and V. Nercissiantz. 1992. The Use of Water Pricing as a Means for Enhancing Water Use Efficiency in Irrigation: Case Studies in Mexico and the United States. *Natural Resources Journal* 32: 731–55.

Curtuis, M. 1995. More Contentious Than West Bank is Resource Below: Water. *Los Angeles Times*, Saturday, July 15.

Dabelko, Geoffrey D. and David D. Dabelko. 1995. Environmental Security: Issues of Conflict and Redefinition. *Environmental Change and Security Project Report* (1, Spring): 3–13.

Dahal, Rajendra. 1995. Gunning for Kosi High. *Himal* (July/August): 24–28.

Dalby, Simon. 1992. Ecopolitical Discourse: "Environmental Security" and Political Geography. *Progress in Human Geography* 16: 303–22.

—— 1992. Security, Modernity, Ecology: The Dilemmas of Post-Cold War Security Discourse. *Alternatives* 17(1): 95–134.

—— 1994. Environmental Security. In *Dictionary of Geopolitics*, ed. John V. O'Loughlin. Westport, CT: Greenwood Press.

Danielson, P. 1995. With Fire: The Ethical Aftermath of Incentive Compatible Devices. Computes and Philosophy Conference. New York, NY: Carnegie Mellon University, August.

Davis, J.R., P. Whigham, and I.W. Grant. 1988. Representing and Applying Knowledge About Spatial Processes in Environmental Management. *AI Applications in Natural Resource Management* 2(4): 17–25.

Davis, Ray Jay. 1991. Atmospheric Water Resources Development and International Law. *Natural Resources Journal* 31 (Winter): 11–44.

Davis, Uri, Antonia Maks, and John Richardson. 1980. Israel's Water Policies. *Journal of Palestine Studies* 9(2): 3–32.

Day, J. 1972. International Management of the Rio Grande Basin the United States and Mexico – Water. *American Water Resources Association Bulletin* 8 (5, October): 935–47.

Day, R. and J.V. Day. 1977. A Review of the Current State of Negotiation Order Theory; and Appreciation and a Critique. *The Sociological Quarterly* 18 (Winter): 126–42.

Delft Declaration. 1991. *Reprinted in A Strategy for Water Sector Capacity Building*. Delft: International Institute for Hydraulic and Environmental Engineering and United Nations Development Programme.

Dellapenna, J. (forthcoming). Treaties as Instruments for Managing Internationally-Shared Water Resources: Restricted Sovereignty vs. Community of Property. In *Middle East Water: The Potential and Limits of Law*.

—— 1995. Building International Water Management Institutions: The Role of Treaties and Other Legal Arrangements. In *Water in the Middle East: Legal, Political, and Commercial Implications*, ed. J.A. Allan and Chibli Mallat, 55–89. London and New York: Tauris Academic Studies.

Delli Priscoli, Jerome. 1988. Conflict Resolution in Water Resources: Two 404 General Permits. *Journal of Water Resources Planning and Management* 114 (1, January): 66–77.

—— 1989. Public Involvement, Conflict Management: Means to EQ and Social Objectives. *Journal of Water Resources Planning and Management* 115 (1, January): 31–42.

—— 1992. Collaboration, Participation, and Alternative Dispute Resolution: Process Concepts for the World Bank's Role in Water Resources. Draft.

—— 1994. *Participation in Conflict Resolution*. The Participation Forum. Washington, DC: AID.

Derman, Bill and Anne Ferguson. 1995. Human Rights, Environment, and Development: The Dispossession of Fishing Communities on Lake Malawi. *Human Ecology* 23 (2, June): 125–42.

Deshan, T. 1995. Optimal Allocation of Water Resources in Large River Basins: II, Application to Yellow River Basin. *Water Resources Management* 9: 53–66.

—— 1995. Optimal Allocation of Water Resources in Large River Basins: I, Theory. *Water Resources Management* 9: 39–51.

Deudney, Daniel. 1990. The Case Against Linking Environmental Degradation and National Security. *Millennium* 19: 461–76.

—— 1991. Environment and Security: Muddled Thinking. *The Bulletin of Atomic Scientists* (April): 23–28.

Devine, Timothy. 1990. A Preemptive Approach in Multilateral Negotiation: Charles Evans Hughes and the Washington Naval Conference. *Negotiation Journal* 6 (4, October): 369–82.

Devlin, John F. 1992. Effects of Leadership Style on Oil Policy. *Energy Policy*, November, 1048–54.

Dewitt, David. 1994. Common, Comprehensive, and Cooperative Security. *The Pacific Review* 7(1): 1–15.

De Bono, E. 1970. *Lateral Thinking*. Harmondsworth, England: Penguin Books.

De Silva, K.M. 1994. Conflict Resolution in South Asia. *International Journal on Group Rights* 1(4): 247–67.

Diamond, Louise and John McDonald. 1992. *Multi-Track Diplomacy: A System Guide and Analysis*. Grinnell, IA: Iowa Peace Institute.

Diehl, Paul. 1991. Geography and War: A Review and Assessment of the Empirical Literature. *International Interactions* 17: 11–27.

Dillman, Jeffrey. 1989. Water Rights in the Occupied Territories. *Journal of Palestine Studies* 19(1): 46–71.

Dinar, Ariel and Edna Tusak Loehman. 1995. *Water Quantity/Quality Management and Conflict Resolution: Institutions, Processes, and Economic Analyses*. Westport, CT: Praeger.

—— and A.T. Wolf. 1994. Economic Potential and Political Considerations of Regional Water Trade: The Western Middle East Example. *Resources and Energy Economics* 16 (4, Winter): 335–56.

—— and A.T. Wolf. 1994. International Markets for Water and the Potential for Regional Cooperation: Economic and Political Perspectives in the Western Middle East. *Economic Development and Cultural Change* 43 (1, October): 43–66.

—— and D. Zilberman. 1991. The Economics of Resource-Conservation, Pollution-Reduction Technology Section: The Case of Irrigation Water. *Resources and Energy* 13: 323–48.

——, P. Seidl, H. Olem, V. Jorden, A. Duda, and Johnson R. 1995. *Restoring and Protecting the World's Lakes and Reservoirs*, Technical Paper No. 289. Washington, DC: World Bank.

Dixit, Ajaya. 1995. Mapping Nepal's Water Resource. *Himal* (July/August): 23.

Donkers, Henk. 1995. Water as a Factor in the Israelisch-Arabisch Conflict. *Internationale Spectator* 49 (January): 13–18.

Doran, Charles F. and Stephen W. Buck. 1991. *The Gulf, Energy, and Global Security – Political and Economic Issues*. Boulder, CO: Lynne Rienner.

Dorman, S. 1991. Who Will Save the Aral Sea? *Environmental Policy Review* 5(2): 45–54.

Downing, Paul B. 1984. *Environments, Economics, and Policy*. Boston, MA: Little, Brown and Company.

Drezon-Tepler, Marcia. 1990. *Interest Groups and Political Change in Israel*. New York: State University of New York Press.

Druckman, D. 1993. The Situational Levers of Negotiating Flexibility. *Journal of Conflict Resolution* 37(2): 236–76.

Drysdale, A. 1992. Syria and Iraq – the Geopathology of a Relationship. *GeoJournal* 28(3): 347–55.

Dryzek, John S. 1983. *Conflict and Choice in Resource Management: The Case of Alaska*. Boulder, CO: Westview Press.

—— and Susan Hunter. 1987. Environmental Mediation for International Problems. *International Studies Quarterly* 31: 87–102.

Duda, Alfred M. and David La Roche. 1997. Joint Institutional Arrangements for Addressing Transboundary Water Resources Issues – Lessons for the GEF. *Natural Resources Journal* 21 (2, May): 127–38.

—— and David La Roche. 1997. Sustainable Development of International Waters and their Basins: Implementing the GEF Operational Strategy. *International Journal of Water Resources Development* 13 (3, September): 383–402.

Dudley, N.J. 1992. Water Allocation by Markets, Common Property and Capacity Sharing: Companions or Competitors? *Natural Resources Journal* 32 (Fall): 766–78.

—— 1994. An Innovative Institutional Arrangement With Potential for Improving the Management of International Water Resources. *Studies in Environmental Science* 58: 469–80.

Dudley, R.L. 1990. A Framework for Natural Resource Management. *Natural Resources Journal* 30 (Winter): 107–22.

Dufournaud, Christian. 1982. On the Mutually Beneficial Cooperative Scheme: Dynamic Change in the Payoff Matrix of International River Basin Schemes. *Water Resources Research* 18(4): 764–72.

Duloy, John H. 1979. *Preliminary Report on Validation of and Some Water Allocation Experiments with the Indus Basin Model*. Washington, DC: World Bank.

—— and Gerald T. O'Mara. 1984. *Issues of Efficiency and Interdependence in Water Resource Investments: Lessons from the Indus Basin of Pakistan*. Washington, DC: World Bank.

——, Gerald T. O'Mara, and Anthony Brooke. 1980. *The Indus Basin Model: Validation Revisited and a Look at the Short Run Impact of the Tarbela Dam*. Washington, DC: World Bank.

——, Gerald T. O'Mara, and Shen-Shing Ting. 1978. *Programming and Designing Investment in the Indus Basin*. Washington, DC: World Bank.

Dworsky, L.B. et al. 1993. The North American Experience: Managing International

Transboundary Water Resources. *Natural Resources Journal* 33 (Winter, Spring): 1–228, 233–459.

Dworsky, L.B. and A.E. Utton. 1993. Assessing North America's Management of Its Transboundary Waters. *Natural Resources Journal* 33 (Spring): 413–59.

Eagleson, Peter S. 1986. The Emergence of Global-Scale Hydrology. *Water Resources Research* 22 (9, August): 6S–14S.

Eaton, David, and J. Eaton. 1995. Joint Management of Aquifers Between the Jordan River Basin and the Mediterranean Sea by Israelis and Palestinians: An International Perspective. In *Joint Management of Shared Aquifers: The Second Workshop November 27–December 1, 1994*, ed. M. Haddad and E. Feitelson. Jerusalem, Israel: The Harry S. Truman Research Institute and the Palestine Consultancy Group.

—— and ed. 1992. *The Ganges-Brahmaputra Basin: Water Resources Cooperation Between Nepal, India, and Bangladesh*. Austin, TX: Lyndon B. Johnson School of Public Affairs.

—— 1993. *Water Resource Challenges in The Ganges-Brahmaputra River Basin*. Austin, TX: Lyndon B. Johnson School of Public Affairs.

Eden, S. 1988. Negotiation and the Resolution of Water Allocation Disputes. M.Sc. thesis. Tucson: University of Arizona.

Ehrmann, John R. 1982. *A Bibliography on Natural Resources and Environmental Conflict: Management Strategies and Processes*. Chicago, IL: Council of Planning Libraries.

El-Ashry, Mohamed T. 1991. International Cooperation, the Environment, and Global Security. *Population* 18(3): 16–26.

El-Hindi, Jamal L. 1990. The West Bank Aquifer and Conventions Regarding Laws of Belligerent Occupations. *Michigan Journal of International Law* 11(4): 1400–23.

El-Khatib, Monib. 1994. The Syria Tabqa Dam: Its Development and Impact. *The Geographical Bulletin* 26: 19–28.

El-Yussif, Faruk. 1983. Condensed History of Water Resources Development in Mesopotamia. *Water International* 8: 19–22.

Elliott, Michael. 1991. Water Wars. *Geographical Magazine* (May): 28–30.

Elmusa, S.S. 1993. Dividing the Common Palestinian-Israeli Water: An International Water Law Approach. *Journal of Palestine Studies* (Spring): 57–78.

—— 1994. Towards an Equitable Distribution of the Common Palestinian-Israeli Waters: An International Water Law Framework. *Studies in Environmental Science* 58: 451–67.

—— 1995. Dividing Common Water Resources According to International Water Law: The Case of the Palestinian-Israeli Waters. *Natural Resources Journal* 35 (Spring): 223–73.

—— 1995. The Jordan-Israel Water Agreement: A Model or an Exception? *Journal of Palestine Studies* 24 (3, Spring): 63–73.

Ember, Carol R. and Melvin Ember. 1992. Resource Unpredictability, Mistrust, and War: A Cross-Cultural Study. *Journal of Conflict Resolution* 36 (June): 242–62.

Emel, J.L. 1987. Groundwater Rights Definition and Transfer. *Natural Resources Journal* 27 (Summer): 653–73.

—— and Elizabeth Brooks. 1988. Changes in Form and Function of Property Rights Institutions Under Threatened Resource Scarcity. *Annals of the Association of American Geographers* 78(2): 241–52.

Enell, M. 1989. Impact of Utilization in River Basin Management. *Aqua Fennica* 19(2): 153–7.

Engle, M. 1992. A Global View. *California Grower* 16 (11, November): 34.

Esco Corporation. 1947. *Palestine: A Study of Jewish, Arab, and British Policies*. New Haven: Yale University Press.

Eva, Fabrizio. 1994. From Boundaries to Borderlands: New Horizons for Geopolitics. In *Political Boundaries and Coexistence: Proceedings of the IGU-Symposium, Basle/Switzerland, 24–27 May*, ed. Werner A. Galluser, 372–80. New York, NY: Peter Lang.

Evans, Norman A. 1969. *Water and Western Destiny: From Conflict to Cooperation. Proceedings.* Fort Collins: Western Interstate Water Conference, 3D, Colorado State University.

Fairclough, A.J. 1991. Policy Challenges of the Environment. *The Washington Quarterly* 14 (1, Winter).

Falkenmark, Malin. 1986. Fresh Waters as a Factor in Strategic Policy and Action. In *Global Resources and International Conflict: Environmental Factors in Strategic Policy and Action*, ed. Arthur H. Westing, 85–182. New York, NY: Oxford University Press.

—— 1987. Water-Related Limitations to Local Development. *Ambio* 16(4).

—— 1989a. The Massive Water Shortage in Africa – Why Isn't It Being Addressed? *Ambio* 18(2): 112–18.

—— 1989b. Middle East Hydropolitics: Water Scarcity and Conflicts in the Middle East. *Ambio* 18(6): 350–52.

—— 1990. Global Water Issues Confronting Humanity. *Journal of Peace Research* 27(2): 177–90.

—— and J. Lundqvist. 1995. Looming Water Crisis: New Approaches are Inevitable. In *Hydropolitics*, ed. L. Ohlsson. London, England: Zed Books.

—— and C. Widstrand. 1992. Population and Water Resources: A Delicate Balance. *Population Bulletin* 47(3).

——, J. Lundqvist, and C. Widstrand. 1989. Macro-scale Water Scarcity Requires Micro-scale Approaches: Aspects of Vulnerability in Semi-Arid Development. *Natural Resources Forum* (November): 258–67.

Fano, E. 1977. The Role of International Agencies. In *Water in a Developing World – The Management of a Critical Resource*, ed. A.E. Utton and L. Teclaff, 219–30. Boulder, CO: Westview Press.

Fashchevsky, B. 1992. Ecological Approach to Management of International River Basins. *European Water Pollution Control* 2(3): 28–31.

Fearnside, Philip M. 1993. Deforestation in Brazilian Amazonia: The Effect of Population and Land Tenure. *Ambio* 22 (8, December): 537–45.

Feitelson, E. and M. Haddad. 1995. *Joint Management of Shared Aquifers: Final Report.* Jerusalem, Israel: The Palestine Consultancy Group and the Harry S. Truman Research Institute.

—— and S.D. Sylvain. 1995. Possible Institutional Models for a Joint Aquifer Management Body. In *Joint Management of Shared Aquifers: The Second Workshop November 27–December 1, 1994*, ed. M. Haddad and E. Feitelson. Jerusalem, Israel: The Harry S. Truman Research Institute and the Palestine Consultancy Group.

Feiveson, Harold A., Frank W. Sinden, and Robert H. Socolow. 1976. *Boundaries of Analysis: an Inquiry into the Tocks Island Dam Controversy.* Cambridge, MA: Ballinger Publishing Company.

Fejerdy-Dobolyi, H. 1993. A Description of Water Quality in Hungary. In *Two Essays on Water Quality in Central and Eastern Europe, ENR 93-20*, 26–30. Washington, DC: Resources for the Future.

Feldman, David Lewis. 1991. *Water Resources Management.* Baltimore, MD: The Johns Hopkins University Press.

Firor, John. 1990. *The Changing Atmosphere: A Global Challenge.* New Haven, CT: Yale University Press.

Fishel, Walter L., ed. 1971. *Resource Allocation in Agricultural Research.* Minneapolis, MN: University of Minnesota Press.

Fishelson, Gideon. 1989. *Economic Cooperation in the Middle East.* Boulder, CO: Westview Press.

—— 1990. Water and the Middle East. *Israel Scene* (October): 4–5.

—— 1992. Solutions for the Scarcity of Water in the Middle East in Times of Peace. In *Kfar*

Blum Conference. Kfar Blum Conference. Tel Aviv: Armand Hammer Fund for Economic Cooperation in the Middle East, December.

Fisher, Franklin M. 1993. *An Economic Framework for Water Negotiation and Management.* Massachusetts Institute of Technology (November): 7.

Fisher, Roger and William L. Ury. 1981. *Getting to Yes: Negotiating Agreement Without Giving In.* New York: Penguin.

Flack, J.E. and D.A. Summers. 1971. Computer-Aided Conflict Resolution in Water Resource Planning: An Illustration. *Water Resources Research* 7(6): 1410–14.

Flatters, Frank and Theodore Horbulyk, M. 1995. Water and Resource Conflicts in Thailand: An Economic Perspective. Prepared for Natural Resource and Environment Program, Thailand Development Research Institute, April.

Flint, C.G. 1995. Recent Developments of the International Law Commission Regarding International Watercourses and Their Implications for the Nile River. *Water International* 20(4): 197–204.

Fogel, M.M. 1985. International and Educational Aspects of Watershed Management. In *Watershed Management in the Eighties: Proceedings of a Symposium, Denver, Colorado, April 30–May 1, 1985,* ed. E. Bruce Jones and Timothy J. Ward, 7–14. New York, NY: ASCE.

Folk-Williams, J.A. 1982. Negotiation Becomes More Important in Settling Indian Water Rights Disputes in the West. *Resolve* (Summer): 1–5.

—— 1988. The Use of Negotiated Agreements to Resolve Water Disputes Involving Indian Rights. *Natural Resources Journal* 28 (Winter): 63–103.

Food and Agriculture Organization of the United Nations. 1980. *The Law of International Resources: Some General Conventions.* Rome: Food and Agriculture Organization of the United Nations.

—— 1984. *Systematic Index of International Water Resource Treaties, Declarations, Acts and Cases by Basin,* Vols. I and II. Rome: Food and Agriculture Organization of the United Nations.

Food Policy Research Institute. 1995. *Review of Water Resource Issues and Irrigation Strategy in Sub-Saharan Africa.* Monograph. Wageningen, Netherlands: World Bank.

Forest Resources Division of FAO. 1994. *Forest Resources Assessment 1990: Non-tropical Developing Countries Mediterranean Region.* Rome: Food and Agriculture Organization of the United Nations.

Forster, Bruce. 1989. The Acid Rain Games: Incentives to Exaggerate Control Costs and Economic Disruption. *Journal of Environmental Management* 28 (4, June): 349–60.

Foster, Charles H.W. and Peter P. Rogers. 1988. *Federal Water Policy: Toward an Agenda for Action.* Cambridge, MA: Harvard University Press.

Fox, Irving K. and David LeMarquand. 1979. International River Basin Co-operation: The Lessons from Experience. *Water Supply and Management* 3: 9–27.

Fox, Jefferson. 1992. The Problem of Scale in Community Resource Management. *Environmental Management* 16 (3, May/June): 289–97.

——, John Krummel, Sanay Yarnasarn, Methi Ekasingh, and Nancy Podger. 1995. Land Use and Landscape Dynamics in Northern Thailand: Assessing Change in Three Upland Watersheds. *Ambio* 24 (6, September): 328–34.

Fox, W.M. 1990. *Effective Group Problem Solving: How to Broaden Participation, Improve Decision Making, and Increase Commitment to Action.* San Francisco, CA: Jossey–Bass.

Frankel, N. 1991. Water and Turkish Foreign Policy. *Political Communication and Persuasion* 8: 257–311.

Fraser, N.M. and K.W. Hipel. 1984. *Conflict Analysis: Models and Resolutions,* Series Vol. 11. New York, NY: North-Holland.

——, K.W. Hipel, J. Jaworsky, and R. Zuljan. 1990. A Conflict Analysis of the Armenian-Azerbaijani Dispute. *Journal of Conflict Resolution* 34 (3, December): 652–77.

Frederick, K.D. 1993. *Balancing Water Demands with Supplies, The Role of Management in a World of Increasing Scarcity*, Technical Paper Number 189. Washington, DC: World Bank.

Frederiksen, H.D. 1992. *Water Resources Institutions: Some Principles and Practices*, Technical Paper Number 191, Washington, DC: World Bank.

—— 1996. Water Crisis in the Developing World: Misconceptions About Solutions. *Journal of Water Resources Planning and Management* (March/April): 79–87.

Freeman, A. Myrick. 1993. *The Measurement of Environmental and Resource Values: Theory and Methods*. Washington, DC: Resources for the Future.

Frey, Fred W. 1993. The Political Context of Conflict and Cooperation Over International River Basins. *Water International* 18(1): 54–68.

—— 1993. Power, Conflict, and Cooperation. *Research and Exploration* 9 (November): 18–37.

—— and T. Naff. 1985. Water: An Emerging Issue in the Middle East? *ANNALS, AAPSS* 482: 65–84.

Friday, Laurie and Ronald Laskey, eds. 1989. *The Fragile Environment: The Darwin College Lectures*. Cambridge, MA: Cambridge University Press.

Frisvold, George B., and Margriet F. Caswell. 1994. Transboundary Water Agreements and Development Assistance. International Conference on Coordination and Decentralization in Water Resources Management. Rehovot, Israel, 3–6 October.

Frohlich, N. 1992. An Impartial Reasoning Solution to the Prisoner's Dilemma. *Public Choice Studies* 74 (4, December): 447–60.

—— 1995. The Incompatibility of Incentive Compatible Devices and Ethical Behavior: Some Experimental Results and Insights. *Public Choice Studies* 25: 24–51.

—— and J.A. Oppenheimer. 1992. *Choosing Justice: An Experimental Approach to Ethical Theory*. Berkeley, CA: California University Press.

—— and J.A. Oppenheimer. 1993. Structuring Observation in Moral Realist and Cognitive Theories. Annual Meetings of the American Political Science Association. Washington, DC, September.

—— and J.A. Oppenheimer. 1994. Preferences for Income Distribution and Distributive Justice: A Window on the Problems of Using Experimental Data in Economics and Ethics. *Eastern Economics Journal* 20 (2, Spring): 149–55.

—— and J.A. Oppenheimer. 1996. Experiencing Impartiality to Invoke Fairness in the n-PD: Some Experimental Results. *Public Choice Studies* 86: 117–35.

Gale, R.P. 1986. Sociological Theory and the Differential Distribution of Impacts in Rural Resource Development Projects. In *Differential Social Impacts of Rural Resource Development*, ed. Pamela D. Elkind-Savatsky and Judith D. Kaufman, 37–59. Boulder, CO: Westview Press.

Gallusser, Werner A. 1994. Borders as Necessities and Challenges – Human Geographic Insights Paving the Wat to a New Border Awareness. In *Political Boundaries and Coexistence: Proceedings of the IGU-Symposium, Basle/Switzerland, 24–27 May*, ed. Werner A. Galluser, 381–9. New York, NY: Peter Lang.

Galnoor, Itzhak. 1978. Water Policymaking in Israel. *Policy Analysis* 4: 339–67.

Galtung, Johan. 1994. Coexistence in Spite of Borders: On the Borders in the Mind. In *Political Boundaries and Coexistence: Proceedings of the IGU-Symposium, Basle/Switzerland, 24–27 May*, ed. Werner A. Galluser, 5–14. New York, NY: Peter Lang.

Garbell, Maurice. 1965. The Jordan Valley Plan. *Scientific American* 212(3): 23–31.

Garber, Andra and Elia Salameh. 1992. *Jordan's Water Resources and their Future Potential*. Amman: Friedrich Ebert Stiftung.

Gardner, Roy, Elinor Ostrom, and James Walker. 1991. The Nature of Common-Pool Resource Problems. *Rationality and Society* 2 (3, July): 335–58.

Garfinkel, M.R. 1994. Domestic Politics and International Conflict. *American Economic Review* 84(5): 1294–309.

Garfinkle, Adam. 1994. *War, Water, and Negotiation in the Middle East: The Case of the Palestine-Syria Border, 1916–1923.* Tel Aviv, Israel: Tel Aviv University Press.

Garnham, David. 1976. Dyadic International War, 1816–1965: The Role of Power Parity and Geographical Proximity. *Western Political Quarterly* 29: 231–42.

Garretson, A.H. et al. 1967. *The Law of International Drainage Basins.* Dobbs Ferry, New York: Oceana Publications, Inc.

Gautam, S. 1976. Interstate Water Disputes: A Case Study of India Politics. *Water Resource Bulletin* 12 (5, October): 1061–70.

Gazdar, Muhammad Nasir. 1990. *An Assessment of the Kalabagh Dam Project on the River Indus, Pakistan.* Karachi: Pakistan Environmental Network.

Gedeon, Raja. 1994. Water Sector of Jordan in Perspective. In *Water as an Element of Cooperation and Development in the Middle East,* ed. Ali Ihsan Bagis. Water as an Element of Cooperation and Development in the Middle East. Ankara: Hacettepe University.

Gehring, T. 1990. International Environmental Regimes: Dynamic Sectoral Legal Systems. *Yearbook of International Environmental Law* 1: 35–56.

Giannias, Dimitrios and Joseph N. Lekakis. 1994. Inter-Country River Water Allocation: The Case of Nestos in the Balkans. In *Sharing Scarce Fresh Water Resources in the Mediterranean Basin – An Economic Perspective,* ed. M. Shechter. Boston, MA: Kluwer Academic Publishers.

Gibb, Sir Alexander. 1966. *The Tarbela Project: Evaluation of the Tarbela Project within the Development Progamme of the Indus Basin.* London: A. Gibb and Partners.

Glasbergen, Pieter. 1992. Comprehensive Policy Planning for Water Systems. *Water Resources Development* 3(1): 45–52.

—— 1992. Seven Steps Towards an Instrumentation Theory for Environmental Policy. *Policy and Politics* 20(3): 191–200.

—— 1993. Managing Environmental Conflict in an International Context. In *International Issues in Environmental Sciences,* ed. P.B. Sloep. Heerlen: Open Universiteit.

—— 1995. Environmental Dispute Resolution as a Management Issue: Towards New Forms of Decision Making. In *Managing Environmental Disputes: Network Management as an Alternative,* ed. Pieter Glasbergen, 1–17. Boston, MA: Kluwer Academic Publishers.

—— 1995. Project Management for Water Conflicts: An Analysis of Factors Conducive to Success or Failure. In *Managing Environmental Disputes: Network Management as an Alternative,* ed. Pieter Glasbergen, 105–18. Boston, MA: Kluwer Academic Publishers.

—— 1995. *Managing Environmental Disputes: Network Management as an Alternative.* Boston, MA: Kluwer Academic Publishers.

Glassner, Martin, ed. 1983. *Global Resources: Challenges of Interdependence.* New York: Praeger.

Gleick, Peter. 1988. The Effects of Future Climatic Changes on International Water Resources: The Colorado River, the United States, and Mexico. *Policy Sciences* 21(1): 23–39.

—— 1990. Climate Changes, International Rivers, and International Security: The Nile and the Colorado. In *Greenhouse Glasnost,* ed. Robert Redford and Terill J. Minger, 147–65. New York, NY: The Ecco Press.

—— 1990. Environment, Resources, and International Security and Politics. In *Science and International Security: Responding to a Changing World,* ed. Eric Arnett, 501–23. Washington, DC: American Association for the Advancement of Science.

—— 1991. Environment and Security: The Clear Connections. *The Bulletin of the Atomic Scientists* 47: 17–21.

—— 1992. Effects of Climate Change on Shared Fresh Water Resources. In *Confronting Climate Change*, ed. I.M. Mintzer, 127–40. Cambridge University Press, for Stockholm Environment Institute.

—— 1992. *Water and Conflict*. Environmental Change and Acute Conflict, Vol. 1. Toronto, Canada: University of Toronto and the American Academy of Arts and Sciences.

—— 1993. Water and Conflict: Fresh Water Resources and International Security. *International Security* 18(1): 79–112.

—— 1993. Water and War in the Middle East. Briefing for U.S. Congress. Washington, DC: Energy and Environmental Study Institute, 5 November.

—— 1993. *Water in Crisis: A Guide to the World's Fresh Water Resources*. New York, NY: Oxford University Press.

—— 1994. Reducing the Risks of Conflict Over Fresh Water Resources in the Middle East. *Studies in Environmental Science* 58: 41–54.

—— 1994. Water, War and Peace in the Middle East. *Environment* 36 (3, April): 6–42.

—— and Miriam R. Lowi. 1992. *Environmental Change and Acute Conflict*. Toronto: Peace and Conflict Studies Program.

Gochman, Charles S. 1990. The Geography of Conflict: Militarized Interstate Disputes Since 1816. Annual Meeting of the International Studies Association. Washington, DC.

—— and Zeev Maoz. 1984. Militarized Interstate Disputes 1816–1976. *Journal of Conflict Resolution* 28: 585–616.

Godana, Bonaya Adhi. 1985. *Africa's Shared Water Resources: Legal and Institutional Aspects of the Nile, Niger, and Senegal River Systems*. London, England: F. Pinter.

Goldfarb, William. 1988. *Water Law*. Michigan: Lewis Publishers, Inc.

Goldie, L.F. 1985. Equity and the International Management of Transboundary Resources. *Natural Resources Journal* 25 (3, July).

Golubev, Genady N. and Asit K. Biswas, eds. 1979. *Interregional Water Transfers*. Oxford: Pergamon Press.

Gonzalez, Arturo and Santiago Rubio. 1992. Optimal Interbasin Water Transfers in Spain. In *Sharing Scarce Fresh Water Resources in the Mediterranean Basin: An Economic Perspective*. Sharing Scarce Fresh Water Resources in the Mediterranean Basin: An Economic Perspective. Padova, Italy, 23–24 April.

Goodman, A. and K. Edwards. 1992. Integrated Water Resources Planning. *Natural Resources Forum* (February): 65–70.

Goslin, I.V. 1977. International River Compacts: Impact on Colorado. In *Water Needs for the Future – Political, Economic, Legal, and Technological Issues in a National and International Framework*. Boulder, CO: Westview Press.

Gottleib, P.D. 1995. The "Golden Egg" as a Natural Resource: Toward a Normative Theory of Growth Management. *Society and Natural Resources* 8 (1, Jan/Feb): 49–56.

Government of the People's Republic of Bangladesh. 1976. *Deadlock on the Ganges*. Dacca: Government of the People's Republic of Bangladesh.

—— 1976. *White Paper on the Ganges Water Dispute*. Dacca: Government of the People's Republic of Bangladesh.

Gradus, Y. 1994. The Israel-Jordan Rift Valley: A Border of Cooperation and Productive Coexistence. In *Political Boundaries and Coexistence: Proceedings of the IGU-Symposium, Basle/Switzerland, 24–27 May*, ed. Werner A. Galluser. New York, NY: Peter Lang.

Gramann, J.H. and R.J. Burdge. 1981. The Effect of Recreation Goals on Conflict Perception: the Case of Water Skiers and Fisherman. *Journal of Leisure Research* 13 (1, 1st quarter): 15–27.

Grant, R. 1994. The Geography of International Trade. *Progress in Human Geography* 18(3): 298–312.

Gray, B. 1989. *Collaborating, Finding Common Ground for Multiparty Problems*. San Francisco: Jossey-Bass Publishers.

Green, Peter. 1992. Damming the Danube: A Political Saga. *U.S. News and World Report*, 9 November, 18.

Grether, D.M., R.M. Isaac, and C.R. Plott. 1989. *The Allocation of Scarce Resources: Experimental Economics and the Problem of Allocating Airport Slots*. Boulder, CO: Westview Press.

Grey, D. 1994. Joint Management of Aquifers Under Conditions of Scarcity. In *Joint Management of Shared Aquifers: The First Workshop*, ed. M. Haddad and E. Feitelson. Jerusalem, Israel: The Harry S. Truman Research Institute and the Palestine Consultancy Group.

Griffin, R.C. 1995. On the Meaning of Economic Efficiency in Policy Analysis. *Land Economics* 71 (1, February): 1–15.

Grigg, Neil S. 1978. *Transfer of Water Resources Knowledge: An Assessment*. Raleigh, NC: University of North Carolina.

—— 1993. Western Water: An Issue Dripping with Conflict. *Forum of Applied Research and Public Policy* 8 (Winter): 71–7.

Grover, B. and M. Jefferson. 1995. A World Water Council: One Possible Model. *Water Resources Development* 11(2): 125–38.

Gruen, George. 1991. *The Water Crisis: The Next Middle East Crisis?* Los Angeles: Wiesenthal Center.

—— 1993. Recent Negotiations over the Waters of the Euphrates and Tigris, *Proceedings of the International Symposium on Water Resources in the Middle East: Policy and Institutional Aspects*. Urbana, IL, 24–7 October.

Guariso, Giorgio and Dale Whittington. 1987. Implications of Ethiopian Water Development for Egypt and Sudan. *International Journal of Water Resources Development* 3(2): 105–14.

——, Dale Whittington, Baligh Shindi Zikri, and Khalil Hosny Mancy. 1981. Nile Water for Sinai: Framework for Analysis. *Water Resources Research* 17 (6, December): 1585–93.

Gulhati, Niranjan Das. 1973. *The Indus Waters Treaty*. New York: Allied Publishers.

Gur, Schlomo. 1991. *The Jordan Rift Valley: A Challenge for Development*. Jerusalem: Office of the Prime Minister.

Haas, Peter M. 1990. *Saving the Mediterranean: The Politics of International Environmental Cooperation*. New York, NY: Columbia University Press.

—— and E.B. Haas. 1995. Learning to Learn: Improving International Governance. *Global Governance* 1 (3, September–December): 255–84.

Habeeb, William Mark. 1988. *Power and Tactics in International Negotiation*. Baltimore, MD: Johns Hopkins University Press.

Haddad, Marwan. 1994. An Approach for Regional Management of Water Shortages in the Middle East. In *Water as an Element of Cooperation and Development in the Middle East*, ed. Ali Ihsan Bagis. Ankara: Hacettepe University.

Haddad, Sweilum. 1967. Principles and Procedures used in Planning and Executing the East Ghor Irrigation Project. In *International Conference on Water for Peace*. International Conference on Water for Peace. Washington, DC, 23–31 May.

Hafner, Gerhard. 1991. The Application of the Optimum Utilization Principle to the Euphrates and Tigris Drainage Basin. In *International Conference on Transboundary Waters in the Middle East: Prospects for Regional Cooperation*. International Conference on Transboundary Waters in the Middle East: Prospects for Regional Cooperation, 2–3 September.

Hakim, Bahzad. 1994. The Water Question in Lebanon: Needs and Resources. In *Water as an Element of Cooperation and Development in the Middle East*, ed. Ali Ihsan Bagis. Ankara: Hacettepe University.

Haley, Stephen L. 1993. *Environmental and Agricultural Policy Linkages in the European Community: The Nitrate Problem and CAP Reform*. Washington, DC: IATRC.

Hamner, J. and A. Wolf. 1998. Patterns in International Water Resource Treaties: The Transboundary Freshwater Dispute Database. *Colorado Journal of International Environmental Law and Policy 1997 Yearbook.*

Hamdy, Atef, Mahmoud Abu-Zeid, and C. Lacirignola. 1995. Water Crisis in the Mediterranean: Agricultural Water Demand Management. *Water International* 20 (4, December): 176–87.

Handl, G. 1990. Environmental Security and Global Change: The Challenge to International Law. *Yearbook of International Environmental Law* 1: 3–33.

—— 1992. The International Law Commission's Draft Articles on the Law of International Watercourses (General Principles and Planned Measures): Progressive or Retrogressive Development of International Law? *Colorado Journal of Environmental Law and Policy* 3(1): 123–43.

Handley, Paul. 1993. River of Promise. *Far Eastern Economic Review*, 16 September, 68–72.

Handmer, John W. and Anthony H. Dorcey, J. 1991. *Negotiating Water: Conflict Resolution in Australian Water Management.* Canberra, Australia: Centre for Resource and Environmental Studies, Australian National University.

Hanqin, Xue. 1992. Relativity in International Water Law. *Colorado Journal of Environmental Law and Policy* 3(1): 45–57.

Harashina, S. 1988. Environmental Dispute Resolution in Road Construction Projects in Japan. *Environmental Impact Assessment Review* 1: 29–41.

Harza Engineering Company. 1968. *Water Resource Development in the Indus Basin.* Lahore, Pakistan: Harza Engineering Company.

Hashimoto, Tsuyoshi. 1995. Regional Cooperative Development for the Salween River. In *Asian Water Forum.* Asian Water Forum. United Nations University, 30 January–1 February.

Hawdon, David and Peter Pearson. 1993. *Acid Deposition and Global Warming: Simulating Energy, Environment, Economy Interaction in the UK with ENDAM2.* Surrey Energy Economics Discussion Paper Series (SEEDS), Vol. 65. Guilford: SEEC, Economics Department, University of Surrey.

Hayes, Douglas L. 1991. The All-American Canal Lining Project: A Catalyst for Rational and Comprehensive Groundwater Management on the United States–Mexico Border. *Natural Resources Journal* 31 (Fall): 803–27.

Hayes, Peter and Lyuba Zarsky. 1994. Environmental Issues and Regimes in Northeast Asia. *International Environmental Affairs* 6 (4, Fall): 283–319.

Haynes, Kingsley and Dale Whittington. 1981. International Management of the Nile – Stage Three? Water Supply. *Geographical Review* 71 (1, January): 17–32.

Hayton, R.D. 1982. Co-operation in the Development of Shared Water Resources. *Natural Resources Forum* 6(2): 167–81.

—— 1983. Report on the Dakar Meeting of International River Commissions. *Natural Resources Journal* 23 (2, April).

—— 1992. Observations on the International Law Commission's Draft Rules on the Non-Navigational Uses of International Watercourses. *Colorado Journal of Environmental Law and Policy* 3(1): 31–44.

—— 1993. The Matter of Public Participation. *Natural Resources Journal* 33 (2, Spring): 275–81.

—— and A.E. Utton. 1989. Transboundary Groundwaters: The Bellagio Draft Treaty. *Natural Resources Journal* 29 (Summer): 663–722.

—— and A.E. Utton. 1992. *Transboundary Groundwaters: The Bellagio Draft Treaty.* Albuquerque, NM: University of New Mexico School of Law.

Hayton, Robert D. 1988. *River and Lake Basin Development.* UN Department of Technical Cooperation and Development Natural Resources Water Series, Vol. 20. New York, NY: United Nations.

—— 1991. Reflections on the Estuarian Zone. *Natural Resources Journal* 31.

Hefny, M.A. 1995. International Water Issues and Conflict Resolution: Some Reflections. *African Journal of International and Comparative Law* 17 (2, June): 360–79.

Hellier, Chris. 1990. Draining the Rivers Dry. *Geographical Magazine* 62 (7, July): 32–5.

Hennessy, J. and N. Widgery. 1995. River Basin Development – the Holistic Approach. *International Water Power and Dam Construction* 47(5): 24–6.

Heraclides, Alexis. 1989. Conflict Resolution, Ethnonationalism and the Middle East Impasse. *Journal of Peace Research* 26(2): 197–212.

Herzog, A.J. 1981. The 'Gatekeeper' Hypothesis and the International Transfer of Scientific Knowledge. *The Journal of Technology Transfer* 6 (1, Fall): 57–72.

Herzog, Lawrence A. and Stephen P. Mumme. 1994. Where North Meets South: Cities, Space, and Politics on the U.S.-Mexican Border. *Studies in Comparative International Development* 29(3).

Hill, Barbara J. 1982. An Analysis of Conflict Resolution Techniques: From Problem-Solving Workshops to Theory. *Journal of Conflict Resolution* 26 (1, March): 109–38.

Hill, Christopher V. 1990. Water and Power: Riparian Legislation and Agrarian Control in Colonial Bengal. *Environmental History Review* (Winter).

Hillel, Daniel. 1987. *The Efficient Use of Water in Irrigation*. World Bank Technical Paper, Vol. 64. Washington, DC: World Bank.

Hindley, Angus. 1992. Water: The Source of Regional Conflict. *Middle East Business Weekly* 36 (31 January): 12.

Hindley, Martin. 1995. Undercurrents of Conflict. *International Water Power and Dam Construction* 47 (January): 30–31.

Hinojosa, J. IV. 1987. Water from the Rio Grande: The Only Game in Town. *Journal of the Rio Grande Valley Horticultural Society* 40: 55–6.

Hipel, K.W. 1983. *Operational Research Techniques in River Basin Management*. University of Waterloo, Department of Geography Publication Series, Vol. 20. University of Waterloo, Department of Geography.

—— and Niall M. Fraser. 1980. Metagame Analysis of Garrison Conflict. *Water Resources Research* 16(4): 629–37.

——, R.K. Ragade, and T.E. Unny. 1976. Metagame Theory and Its Applications to Water Resources. *Water Resources Research* 12 (3, June): 331–9.

——, R.K. Ragade, and T.E. Unny. 1976. Political Resolution of Environmental Conflicts. *Water Resources Bulletin* 12(4): 813–27.

Hirsch, M. and D. Housen-Couriel. 1993. Aspects of the Law of International Water Resources. *Water Science and Technology: A Journal of the International Association on Water Pollution Research.* 27 (7/8): 213–21.

Hoch, Gary. 1993. The Politics of Water in the Middle East. *Middle East Insight* 9 (March/April): 17–21.

Hodren, John P. 1992. *Energy and International Security*. American Association for the Advancement of Science, Vol. 92–33S. Washington, DC: Committee on Science and International Security, American Association for the Advancement of Science.

Hof, Frederick. 1985. *Galilee Divided: The Israel-Lebanon Frontier 1916–1984*. Boulder, CO: Westview.

—— 1995. The Yarmouk and Jordan Rivers in the Israel-Jordan Peace Treaty. *Middle East Policy* 3 (4, April): 47–56.

Hofius, K. 1991. Co-operation in Hydrology of the Danube Basin Countries. In *Hydrology for the Water Management of Large River Basins*, ed. F.H.M. Van de Ven et al., 37–43. International Association of Hydrological Sciences.

—— 1991. Co-operation in Hydrology of the Rhine Basin Countries. In *Hydrology for the Water Management of Large River Basins*, ed. F.H.M. Van de Ven et al., 25–35. International Association of Hydrological Sciences.

Hollander, Jack M., ed. 1992. *The Energy-Environment Connection*. Washington, DC: Island Press.

Holling, C. 1978. Model Invalidation and Belief. In *Adapting Environmental Assessment and Management*. New York: John Wiley and Sons.

Holst, Johan Jorgen. 1989. Security and the Environment: A Preliminary Exploration. *Bulletin of Peace Proposals* 20(2): 123–8.

Homer-Dixon, Thomas F. 1991. On the Threshold: Environmental Changes as Causes of Acute Conflict. *International Security* 16 (2, Fall): 76–116.

—— 1992. *Population Growth and Conflict*. American Association for the Advancement of Science, Vol. 92–33S. Washington, DC: Committee on Science and International Security, American Association for the Advancement of Science.

—— 1994. Environmental Scarcities and Violent Conflict: Evidence From Cases. *International Security* 19(1): 4–40.

—— 1995. Strategies for Studying Causation in Complex Ecological Political Systems. *The Project on Environment, Population, and Security*.

—— 1995/96. Correspondence: Environment and Security. *International Security* 20 (3, Winter): 189–94.

——, Jeffrey H. Boutwell, and George W. Rathjens. 1993. Environmental Change and Violent Conflict. *Scientific American* 268(2): 38–45.

—— and Marc A. Levy. 1995. Correspondence: Environment and Security. *International Security* 20 (3, Winter): 189–98.

——, Marc A. Levy, Gareth Porter, and Jack Goldstone. 1996. Environmental Scarcity and Violent Conflict: A Debate. *Environmental Change and Security Project* (2, Spring): 49–71.

Hori, Hiroshi. 1993. Development of the Mekong River Basin, Its Problems and Future Prospects. *Water International* 18: 110–15.

Hosh, Leonardo and Jad Isaac. 1992. Roots of the Water Conflict in the Middle East. In *The Middle East Water Crisis*. The Middle East Water Crisis. University of Waterloo, 7–9 May.

Housen-Couriel, Deborah. 1992. Aspects of the Law of International Water Resources. Draft.

—— 1994. *Some Examples of Cooperation in the Management and Use of International Water Resources*. Tel Aviv: Tel Aviv University.

Howarth, R.B. and R.B. Norgaard. 1990. Intergenerational Resource Rights, Efficiency, and Social Optimality. *Land Economics* 66 (1, February): 1–11.

Howe, C.W. 1987. On the Theory of Optimal Regional Development Based on an Exhaustible Resource. *Growth and Change* 18 (2, Spring): 53–68.

—— and K. Easter. 1971. *Interbasin Transfers of Water: Economic Issues and Impacts*. Baltimore, MD: Johns Hopkins Press.

Howell, P.P. and J.A. Allan. 1994. *The Nile: Sharing a Scarce Resource. An Historical and Technical Review of Water Management and of Economic and Legal Issues*. Cambridge, MA: Cambridge University Press.

——, Michael Lock, and Stephen Cobb. 1988. *The Jonglei Canal: Impact and Opportunity*. Cambridge, MA: Cambridge University Press.

Hsiao, J.C. 1975. The Theory of Share Tenancy Revisited. *Journal of Political Economy* 83 (5, October): 1023–32.

Huddle, Franklin P. 1972. *The Mekong Project: Opportunities and Problems of Regionalism*. Report to House Committee on Foreign Affairs. Washington, DC: U.S. Government Printing Office.

Hufschmidt, M.M. 1982. New Approaches to Economic Analysis of Natural Resources and Environmental Quality. In *Economic Approaches to Natural Resource and Environmental*

Quality Analysis, ed. M.M. Hufschmidt and E.L. Hyman, 2–30. Dublin: Tycooly International Publishing.

—— and Eric L. Hyman, eds. 1982. *Economic Approaches to Natural Resource and Environmental Quality Analysis*. Dublin, Ireland: Tycooly International Publishing Limited.

Hulme, Mike and Mick Kelly. 1993. Exploring the Links Between Desertification and Climate Change. *Environment* 35 (6, July/August): 4–11.

Hultin, J. 1992. Source of Life, Source of Conflict: Fear and Expectations Along the Nile. In *Regional Case Studies of Water Conflicts*, ed. L. Ohlsson, 20–45. Goteborg: Padrigu Papers.

Hundley, Norris. 1966. *Dividing the Waters: A Century of Controversy Between the U.S. and Mexico*. Berkeley, CA: University of California Press.

—— 1992. *The Great Thirst: Californians and Water, 1770s–1990s*. Berkeley, CA: University of California Press.

Hunter, David. 1993. *Interbasin Water Transfers after NAFTA*. Washington, DC: Center for International Environmental Law.

Hunter, J.M., L. Rey, K.Y. Chu, E.O. Adekolu-John, and K.E. Mott. 1993. *Parasitic Diseases in Water Resources Development: The Need for Intersectoral Negotiation*. Geneva: World Health Organization.

Hurrell, Andrew and Benedict Kingsbury, eds. 1992. *The International Politics of the Environment: Actors, Interests, and Institutions*. New York, NY: Oxford University Press.

Hurst, Christopher and Peter Rogers. 1985. *Tale of Two Simulations – Integrating Water Sector Planning into Economy-wide Planning Models*. Discussion Paper No. 142-D. Harvard University: John F. Kennedy School of Government.

Husarska, Anna. 1992. Dam Check. *The New Republic* (21 December): 15.

Hussain, M. Basheer. 1972. *The Cauvery Water Dispute: An Analysis of Mysore's Case*. Mysore: Rao and Raghavan.

Hutchinson, C.F. et al. 1992. Development in Arid Lands: Lessons from Lake Chad. *Environment* 34: 18.

Inbar, Moshe and Jacob Maos. 1984. Water Resources Planning and Development in the Northern Jordan Valley. *Water International* 9: 18–25.

Ingram, H. 1990. *Water Politics: Continuity and Change*. Albuquerque, NM: University of New Mexico Press.

—— 1994. Managing Transboundary Resources: Lessons from Ambos Nogales. *Environment* 36 (4, May): 6 ff.

—— and D.R. White. 1993. International Boundary and Water Commission: An Institutional Mismatch for Resolving Transboundary Water Problems. *Natural Resources Journal* 33 (Winter): 153–200.

——, Dean E. Mann, Gary D. Weatherford, and Hanna J. Cortner. 1984. Guidelines for Improved Institutional Analysis in Water Resources Planning. *Water Resources Research* 20 (3, March): 323–34.

——, Robert G. Varady, and Lenard Milich. 1995. Enhancing Environmental Policy: Some Principles for the New Border Environment Cooperation Commission (BECC). *Transboundary Resources Report* 7 (2, Summer): 4.

Ingram, Helen, Nancy Laney, David Gilliam, and Stephen P. Mumme. 1996. Divided Waters: Bridging the U.S.-Mexico Border. *The American Political Science Review* 90(2).

International Association on Water Pollution Research and Control. 1987. *Water Quality International*. London, England: IAWPRC.

International Association on Water Quality. 1994. *Water Quality International 1994. Part 3, Chemical and Petrochemical Waste Management; Industrial Waste Treatment: Selected Proceedings of the 17th Biennial Conference of the International Association on Water Quality, held in Budapest, Hungary, 24–30 July 1994*. Tarrytown, NY: Pergamon.

International Boundary and Water Commission. 1981. *Joint Projects of the U.S. and Mexico through the International Boundary and Water Commission.* El Paso, Texas: International Boundary and Water Commission.

International Joint Commission. 1988. *International Joint Commission Activities.* Washington, DC: The Commission.

International Transboundary Resources Center. 1992. *Borders and Water: North American Water Issues.* Albuquerque, NM: International Transboundary Resources Center.

International Water Resources Association. 1986. IWRA Seminar on Interbasin Water Transfer. Beijing, China: International Water Resources Association, 15–19 June.

Islam, M. Rafqul. 1987. The Ganges Water Dispute: An Appraisal of a Third Party Settlement. *Asian Survey* 27: 918–35.

Islam, N. 1992. Indo-Bangladesh Common Rivers: The Impact on Bangladesh. *Contemporary South Asia* 1(2): 203–25.

Izac, A.M. 1986. Resource Policies, Property Rights, and Conflicts of Interest. *The Australian Journal of Agricultural Economics* 30 (1, April): 23–37.

Jacobs, Harvey M. and Richard G. Rubino. 1988. *Predicting the Utility of Environmental Mediation: Natural Resource and Conflict Typologies as a Guide to Environmental Assessment.* Madison, Wisconsin: Institute for Legal Studies.

Jacobs, J.W. and James L. Wescoat, Jr. 1994. Flood Hazard Problems and Programmes in Asia's Large River Basins. *Asian Journal of Environmental Management* 2: 91–104.

Jacobs, Jeffrey W. 1994. Toward Sustainability in Lower Mekong River Basin Development. *Water International* 19: 43–51.

—— 1995. Mekong Committee History and Lessons for River Basin Development. *The Geographical Journal* 161 (2, July): 135–48.

Jaffee Center for Strategic Studies. 1989. *The West Bank and Gaza: Israel's Options for Peace.* Tel Aviv, Israel: Tel Aviv University.

Jamail, M.H. and Stephen Mumme. 1982. The International Boundary and Water Commission as a Conflict Management Agency in the U.S.-Mexico Borderlands. *The Social Science Journal* 19(1): 45–60.

Johnson, G.L. 1983. Ethical Issues in Resource Economics: Discussion. *American Journal of Agricultural Economics* 65 (5, December): 1033–4.

Jordan, Jeffrey L. 1992. *Resolving Intergovernmental Water Disputes Through Negotiation.* Athens, GA: University of Georgia.

—— 1994. The Effectiveness of Pricing as a Stand-Alone Water Conservation Program. *Water Resources Bulletin* 30 (5, September/October): 871–8.

Jovanovic, D. 1985. Ethiopian Interests in the Division of the Nile River Waters. *Water International* 10(2): 82–5.

—— 1986. Response to Discussion by M.M.A. Shahin of Paper "Ethiopian Interests in the Division of the Nile River Waters." *Water International* 11(1): 20–2.

—— 1986. Response to Discussion of Paper "Ethiopian Interests in the Division of the Nile River Waters." *Water International* 11(2): 89.

Joyce, Charles L. 1978. *Comparative Evaluations of the Regional Institutions for the Development of the Senegal River, Niger River, and Lake Chad.* Contract AID/AFR-C-1407. Washington, DC: U.S. Agency for International Development.

Judge, Shana. 1994. The Nile: River of Hope or Conflict? *Transboundary Resources Report* 8 (2, Summer): 1–3.

Just, Richard E., John K. Horowitz, and Sinaia Netanyahu. 1994. Problems and Prospects in the Political Economy of Trans-Boundary Water Issues. University of Maryland, College Park, MD, September.

Kahan, David. 1987. *Agriculture and Water Resources in the West Bank and Gaza.* Jerusalem: *Jerusalem Post.*

Kahhaleh, Subhi. 1981. *The Water Problem in Israel and its Repercussion on the Arab-Israeli Conflict*. Beirut: Institute for Palestine Studies.

Kakonen, Jyrki, ed. 1992. *Perspectives on Environmental Conflict and International Politics*. London: Pinter Publishers.

Kally, Elisha. 1989. The Potential for Cooperation in Water Projects in the Middle East at Peace. In *Economic Cooperation in the Middle East*, ed. Gideon Fishelson, 303–25. Boulder, CO: Westview Press.

—— 1989. *Water In Peo :*. Tel Aviv: Sifriat Poalim. Hebrew.

—— and Gideor :ishelson. 1993. *Water and Peace: Water Resources and the Arab-Israeli Peace Pr* Westport, CT: Praeger.

Karan, Pradyumna P. 1961. Dividing the Water: A Problem in Political Geography. *The Professional Geographer* 13 (1, January): 6–10.

Karp, L.S. 1994. *Monopoly Power Can Be Disadvantageous in the Extraction of a Durable Nonrenewable Resource*. Working Paper 732. Berkeley, CA: California Agricultural Experiment Station.

Karsh, Efraim. 1991. Neutralization: The Key to an Arab-Israeli Peace. *Bulletin of Peace Proposals* 22(1): 11–24.

Kassem, Atef M. 1992. The Water Use Analysis Model (WUAM): A River Basin Planning Model. In *Proceedings of the International Conference on Protection and Development of the Nile and Other Major Rivers*, Vol. 2/2. International Conference on Protection and Development of the Nile and Other Major Rivers. Cairo, Egypt, 3–5 February.

Kattelmann, Richard. 1990. Conflicts and Cooperation over Floods in the Himalaya-Ganges Region. *Water International* 15(4): 189–94.

Kaufman, E. 1994. *The Relevance of the International Protection of Human Rights to Democratization and Peace*. The J.B. Kroc Institute for International and Peace Studies, University of Notre Dame.

—— 1996. *Innovative Problem Solving Workshops: A Comprehensive and Practical Approach*. Monograph, Vol. 8. College Park, MD: CIDCM, University of Maryland at College Park.

——, J. Oppenheimer, A. Wolf, and A. Dinar. 1997. Transboundary Fresh Water Disputes and Conflict Resolution: Planning an Integrated Approach. *Water International* 22 (1, February).

Kaye, Lincoln. 1989. Buyer's Market. *Far Eastern Economic Review* 2 (February): 22.

—— 1989. Resources and Rights: Rivalries Hamper Indo-Bangladesh Water Sharing. *Far Eastern Economic Review* 2 (February): 19–22.

—— 1989. The Wasted Waters. *Far Eastern Economic Review* 2 (February): 16–18.

Keeney, R.L. and E.F. Wood. 1977. An Illustrative Example of the Use of Multiattribute Utility Theory for Water Resource Planning. *Water Resource Res* 13 (4, August): 705–12.

Kelman, Herbert C. 1982. Creating the Conditions for Israeli-Palestinian Negotiations. *Journal of Conflict Resolution* 26 (1, March): 39–75.

—— 1990. Interactive Problem Solving: A Social-Psychological Approach to Conflict Resolution. In *Readings in Conflict Management and Resolution*, ed. J. Burton and F. Dukes, 199–215. New York: St Martin's Press.

Kelsey, Robert. 1994. Come Hell or High Water. *Project & Trade Finance* (132, April): 28–30.

Kemp, Geoffrey. 1991. Conditions for a Stable and Lasting Peace in the Middle East. *Perspectives on War & Peace* 8 (2, Spring): 1–6.

Keohane, R., M. McGinnis, and E. Ostrom. 1992. *Proceedings of a Conference on Linking Local to Global Common*. Cambridge, MA: Harvard University.

Kershner, Isabel. 1990. Talking Water: Secret U.S.-Mediated Negotiations Could Herald Regional Cooperation in the Middle East. *The Jerusalem Report*, 25 October, 44–5.

Khan, M. Yunus. 1990. Boundary Water Conflict Between India and Pakistan. *Water International* 15 (4, December): 195–9.

Khan, Tauhidul Anwar. 1994. Challenges Facing the Management and Sharing of the Ganges. *Transboundary Resources Report* 8 (1, Spring): 1–4.

Khan, Z.A. 1976. *Basic Documents on the Farakka Conspiracy From 1951 to 1976*. Dacca: Khoshroz Kitab Mahal.

Khassawneh, Awn. 1995. The International Law Commission and Middle East Waters. In *Water in the Middle East: Legal, Political, and Commercial Implications*, ed. J.A. Allan and Chibli Mallat, 21–8. London and New York: Tauris Academic Studies.

Kilgour, D. Marc and Ariel Dinar. 1995. *Are Stable Agreements for Sharing International River Waters Now Possible?* Policy Research Working Paper, Vol. 1474. Washington, DC: World Bank.

———, Norio Okada, and A. Nishikori. 1988. Load Control Regulation of Water Pollution: An Analysis Using Game Theory. *Journal of Environmental Management* 27: 179–94.

Kim, C.S., M.R. Moore, and J.J. Hanchar. 1989. A Dynamic Model of Adaptation to Resource Depletion: Theory and an Application to Groundwater Mining. *Journal of Environmental Economics and Management* 17 (1, July): 66–82.

Kira, T. 1988. Towards the Environmentally Sound Management of Lake/Watershed Systems: the Aims and Activities of ILEC. *International Journal of Water Resources Development* 4 (2, June): 124–6.

Kirk, Elizabeth J. 1991. The Greening of Security: Environmental Dimensions of National, International, and Global Security After the Cold War. In *New Perspectives for a Changing World Order*, ed. Eric H. Arnett, 47–71. Washington, DC: Science and International Security, American Association for the Advancement of Science.

Kirmani, S. 1990. Water, Peace and Conflict Management: The Experience of the Indus and Mekong River Basins. *Water International* 15 (December): 200–5.

——— 1993. Pakistan: Managing Indus Water: a Whole Basin Approach. *Water Resources Journal* (176, March): 109–11.

——— and Guy LeMoigne. 1997. *Fostering Riparian Cooperation in International River Basins: The World Bank at its Best in Development Diplomacy*. Technical Paper, Vol. 335. Washington, DC: The World Bank.

——— and R. Rangeley. 1994. *International Inland Waters: Concepts for a More Active World Bank Role*. Technical Paper, Vol. 239. Washington, DC: World Bank.

Kishel, J. 1993. Lining the All-American Canal: Legal Problems and Physical Solutions. *Natural Resources Journal* 33 (Summer): 697–726.

Kliot, Nurit. 1994. *Water Resources and Conflict in the Middle East*. New York, NY: Routledge.

——— 1995. Building a Legal Regime for the Jordan-Yarmouk River System – Lessons from Other International Rivers. *Transboundary Resources Report* 9(1): 1–3.

——— and Yoel Mansfeld. 1994. The Dual Landscape of a Partitioned City: Nicosia. In *Political Boundaries and Coexistence: Proceedings of the IGU-Symposium, Basle/Switzerland, 24–27 May*, ed. Werner A. Galluser, 151–61. New York, NY: Peter Lang.

——— and Stanley Waterman. 1991. *The Political Geography of Conflict and Peace*. London: Belhaven Press.

Kliot, Nuritl, Deborah Shmueli, and Uri Shamir. 1997. *Institutional Frameworks for the Management of Transboundary Water Resources*. Haifa, Israel: Water Research Institute, two vols.

Kolars, John. 1991. The Role of Geographic Information Systems (GIS) Technology in the Future Management of Middle Eastern Rivers. 1991 MESA Annual Meeting. Washington, DC, 12–26 November.

——— 1992. Fine Tuning the Future Euphrates-Tigris System. In *Presented to the Council on Foreign Relations*, 1–15. Presented to the Council on Foreign Relations, 17 June.

—— 1992. Trickle of Hope: Negotiating Water Rights is Critical to Peace in the Middle East. *The Sciences*, November/December.

—— 1992. Water Resources of the Middle East. *Canadian Journal of Development Studies*.

—— 1994. Problems of International River Management: The Case of the Euphrates. In *International Waters of the Middle East: From Euphrates-Tigris to Nile*, ed. Asit K. Biswas, 44–94. Bombay, Delhi, Calcutta, Madras: Oxford University Press.

—— and William A. Mitchell. 1991. *The Euphrates River and the Southeast Anatolia Development Project*. Carbondale and Edwardsville, IL: Southern Illinois University Press.

Kolb, Deborah M. and Susan S. Silbey. 1990. Enhancing the Capacity of Organizations to Deal with Disputes. *Negotiation Journal* 6 (4, October): 297–304.

Kolberg, W.C. 1992. Approach Paths to the Steady State: A Performance Test of Current Period Decision Rule Solution Methods for Models of Renewable Resource Management. *Land Economics* 68 (1, February): 11–27.

Konow, J. 1994. *A Positive Theory of Economic Fairness*. Working Paper, Vol. 94081. Los Angeles, CA: Loyola Marymount University.

Koopman, Cheryl, Jack Snyder, and Robert Jervis. 1990. Theory-driven Versus Data-driven Assessment in a Crisis. *Journal of Conflict Resolution* 34 (4, December).

Kost, I.L. 1990. *Rights and Duties of Riparian States ... Bibliography*. The Hague: Peace Palace Library.

Kovar, Karel and Hans-Peter Nachtnebel. 1993. *Application of Geographic Information Systems in Hydrology and Water Resources Management: Proceedings of an International Conference Held in Vienna, Austria, from 19 to 22 April 1993: the Conference was Jointly Organized by the International Commission on Groundwater of the International Association of Hydrological Sciences (IAHS), the United Nations Educational, Scientific and Cultural Organization (UNESCO)*. Wallingford, England: International Association of Hydrological Sciences.

Kowalok, Michael E. 1993. Common Threads: Research Lessons from Acid Rain, Ozone Depletion, and Global Warming. *Environment* 35 (6, July/August): 12–20.

Krasner, S.D. 1983. *International Regimes*. Ithica, NY: Cornell University Press.

—— 1985. *Structural Conflict: The Third World Against Global Liberalism*. Berkeley, CA: Berkeley University Press.

Kremenyuk, Victor, ed. 1991. *International Negotiation: Analysis, Approaches, Issues*. San Francisco: Jossey-Bass.

Kriesberg, Louis. 1988. Strategies of Negotiating Agreements: Arab-Israeli and American-Soviet Cases. *Negotiation Journal* (January): 19.

—— 1992. *International Conflict Resolution: the U.S.-USSR and Middle East Cases*. New Haven, CT: Yale University Press.

Krishna, Raj. 1995. International Watercourses: World Bank Experience and Policy. In *Water in the Middle East: Legal, Political, and Commercial Implications*, ed. J.A. Allan and Chibli Mallat, 29–54. London and New York: Tauris Academic Studies.

Krutilla, John V. 1969. *The Columbia River Treaty – The Economics of an International River Basin Development*. Baltimore, MD: Johns Hopkins Press.

Kuffner, Ulrich. 1993. Water Transfer and Distribution Schemes. *Water International* 18 (1, March): 30–4.

Kumar, Chetan. 1995. Environmental Degradation and Security in South Asia. In *South Asia Approaches the Millennium: Reexamining National Security*, ed. Marvin G. Weinbaum and Chetan Kuman, 145–66. Boulder, CO: Westview Press.

Kuttab, J. and J. Ishaq. 1994. Approaches to the Legal Aspects of the Conflict on Water Rights in Palestine/Israel. *Studies in Environmental Science* 58: 239–49.

Lake, L.M., ed. 1980. *Environmental Mediation: the Search for Consensus*. Boulder, CO: Westview Press.

Lancaster, J. 1995. Critical Need, Crucial Divide: Water Issue Stalls Talks in West Bank. *The Washington Post*, Monday, 24 July, A14.

Landau, George D. 1980. The Treaty for Amazonian Cooperation: A Bold New Instrument for Development. *Georgia Journal of International and Comparative Law* 10 (3, Fall): 463–89.

Land Degradation in South Asia: Its Severity, Causes and Effects Upon the People. 1994. World Soil Resources Report, Vol. 78. Rome: Food and Agriculture Organization of the United Nations.

Laqueur, Walter. 1967. *The Road to Jerusalem: The Origins of the Arab-Israeli Conflict 1967.* New York: Macmillan.

Lashkar, M.I. 1990. Water Distribution: Calling Provincial Chief Minister's Meeting. *Pakistan & Gulf Economist* 9 (13–19 January): 16–17.

Lassker, A.J. 1994. Legal Aspects of International Cooperation on Transboundary Water Resources. *Studies in Environmental Science* 58: 515–21.

Laurent, Pierre-Joseph. 1994. Authority and Conflict in Management of Natural Resources: a Story about Trees and Immigrants in Southern Burkina Faso. *Forests, Trees and People Newsletter* (25, October): 37–44.

Lax, David A. and James K. Sebenius. 1986. *The Manager as Negotiator: Bargaining for Cooperation and Competitive Gain.* New York: Macmillan.

Laylin, John G. and Rinaldo L. Bianchi. 1959. The Role of Adjudication in International River Disputes. *Americal Journal of International Law* 53: 30–49.

Lee, D.C. 1990. *The Use of Production Indices in Planning and Evaluating Fisheries Management Programs. Discussion Paper QE90–23.* Washington, DC: Resources for the Future.

Lee, D.J. and A. Dinar. 1995. *Review of Integrated Approaches to River Basin Planning, Development and Management.* Policy Research Working Paper, Vol. 1446. Washington, DC: The World Bank.

Lee, Kai N. 1982. Defining Success in Environmental Dispute Resolution. *Resolve* (Spring): 1–3.

Lee, T. 1992. Water Management Since the Adoption of the Mar del Plata Action Plan: Lessons for the 1990s. *Natural Resources Forum* (August): 202–11.

—— 1995. The Management of Shared Water Resources in Latin America. *Natural Resources Journal* 35 (3, Summer): 541–53.

Lekakis, Joseph and Dimitrios Giannias. 1991. Optimal Freshwater Allocation: The Case of Nestos. Draft.

LeMarquand, David. 1976. Politics of International River Basin Cooperation and Management. *Natural Resources Journal* 16 (October): 883–901.

—— 1977. *International Rivers: The Politics of Cooperation.* British Columbia: Westwater Research Centre, University of British Columbia.

—— 1981. International Action for International Rivers. *Water International* 6: 147–51.

—— 1986. Preconditions to Cooperation in Canada-United States Boundary Waters. *Natural Resources Journal* 26 (2, Spring): 221–42.

—— 1989. Developing River and Lake Basins for Sustained Economic Growth and Social Progress. *Natural Resources Forum* (May): 127–38.

—— 1993. The International Joint Commission and Changing Canada-United States Boundary Relations. *Natural Resources Journal* 33 (1, Winter): 59–91.

LeMoigne, G.L., Shawki Barghouti, Gershon Feder, Lisa Garbus, and Mei Xie. 1992. *Country Experiences with Water Resources Management: Economic, Institutional, Technological and Environmental Issues.* Technical Paper, Vol. 175. Washington, DC: The World Bank.

—— and J. Delli Priscoli. 1992. Waters, Rivers, and Mankind: Challenges for the Coming Millennium. *Water International* 17: 81–7.

——, A. Subramanian, M. Xie, and S. Giltner. 1994. *A Guide to the Formulation of Water Resources Strategy*. Technical Paper, Vol. 263. Washington, DC: The World Bank.

Lepawsky, A. 1963. International Development of River Resources. *International Affairs* 39: 533–55.

Lesnick, Michael T. 1981. *A Bibliography for the Study of Natural Resources and Environmental Conflict*. Chicago: Council of Planning Libraries.

Lesotho Highlands Development Authority. 1987. *Lesotho Highlands Water Project*. Monograph. Maseru: Lesotho Highlands Development Authority.

Lesser, J.A. 1990. Resale of the Columbia River Treaty Downstream Power Benefits: One Road from Here to There. *Natural Resources Journal* 30 (Summer): 609–28.

Levy, Marc A. 1995. Is the Environment a National Security Issue? *International Security* 20 (2, Fall): 35–62.

—— 1995. Time for a Third Wave of Environment and Security Scholarship? *Environmental Change and Security Project* (1, Spring): 44–6.

—— 1995/96. Correspondence: The Author Replies. *International Security* 20 (3, Winter).

Lewicki, Roy J. and Joseph A. Litterer. 1985. *Negotiation: Readings, Exercises, and Cases*. Homewood, IL: Irwin.

Libby, Lawrence W. 1994. Conflict on the Commons: Natural Resource Entitlements, The Public Interest, and Agricultural Economics. *American Journal of Agricultural Economics* 76 (December): 997–1009.

Libiszewski, Stephan. 1994. Sources of Life, Sources of Strife. *Swiss Review of World Affairs* 6: 8–10.

Lieftinck, Pieter and Robert Sadove. 1967. *Study of the Water and Power Resources of West Pakistan*. Washington, DC: World Bank.

Linnerooth, J. 1990. The Danube River Basin: Negotiating Settlements to Transboundary Environmental Issues. *Natural Resources Journal* 30: 629–58.

Linnerooth-Bayer, Joanne. 1993. Current Danube River Events and Issues. *Transboundary Resources Report* (Winter): 7.

Lintner, S., S. Arif and M. Hatziolos. 1996. *The Experience of the World Bank in the Legal Institutional, and Financial Aspects of Regional Environmental Programs. Potential Applications of Lessons Learned for the RODME and PERSGA Programs*. Washington, DC: The World Bank.

Lippitt, Gordon L., Peter Langseth, and Jack Mossop. 1985. *Implementing Organizational Change*, 1st edn. San Francisco, CA: Jossey-Bass.

Lipschutz, Ronnie D. 1992. *What Resources Will Matter? Environmental Degradation as a Security Issue*. American Association for the Advancement of Science, Vol. 92-33S. Washington, DC: Committee on Science and International Security, American Association for the Advancement of Science.

—— and John Holdren. 1990. Crossing Borders: Resource Flows, the Global Environment, and International Security. *Bulletin of Peace Proposals* 21(2): 121–33.

List, Martin and Volker Rittberger. 1992. Regime Theory and International Environmental Management. In *The International Politics of the Environment: Actors, Interests, and Institutions*, ed. Andrew Hurrell and Benedict Kingsbury, 85–109. New York, NY: Oxford University Press.

Loehman, Edna Tusak and Ariel Dinar. 1995. Introduction. In *Water Quantity/Quality Management and Conflict Resolution: Institutions, Processes, and Economic Analyses*, ed. Ariel Dinar and Edna Tusak Loehman, xxi–xxx. Westport, CT: Praeger.

Lohmann, L. 1991. Engineers Move in on the Mekong. *New Scientist* 1777: 44–7.

—— 1990. Remaking the Mekong. *The Ecologist* 20(2): 61–6.

London, James B. and Harry W. Miley. 1990. Interbasin Transfer of Water: An Issue of Efficiency and Equity. *Water International* 15 (December): 231–5.

Lonergan, Stephen C. and David B. Brooks. 1994. *Watershed: The Role of Fresh Water in the Israeli-Arab Conflict.* Ottawa: International Development and Research Centre.

—— and Barb Kavanagh. 1991. Climate Change, Water Resources and Security in the Middle East. *Global Environmental Change* (September): 272–90.

Lord, W. 1980. Water Resource Planning: Conflict Management. *Water Spectrum* 12: 2–10.

—— et al. 1979. *Conflict Management in Federal Water Resource Planning.* Boulder, CO: University of Colorado Press.

Loucks, Daniel P. 1995. Developing and Implementing Decision Support Systems: A Critique and a Challenge. *Water Resources Bulletin* 31 (4, August): 571–82.

Lovins, A. 1977. *Soft Energy Paths: Toward a Durable Peace.* Cambridge, MA: Ballinger and Friends of the Earth.

Lowi, Miriam R. 1985. *The Politics of Water: The Jordan River and the Riparian States.* McGill Studies in International Development, Vol. 35. Montreal: McGill Studies in International Development.

—— 1990. The Politics of Water Under Conditions of Scarcity and Conflict: The Jordan River and Riparian States. diss. Princeton, NJ: Princeton.

—— 1993. Bridging the Divide: Transboundary Resource Disputes and the Case of West Bank Water. *International Security* 18(1): 113–38.

—— 1993. *Water and Power: The Politics of a Scarce Resource in the Jordan River Basin.* New York, NY: Cambridge University Press.

—— 1995. Rivers of Conflict, Rivers of Peace. *Journal of International Affairs* 49 (1, Summer).

Lubell, Harold and Charbel Zarour. 1990. Resilience Amidst Crisis: The Informal Sector of Dakar. *International Labour Review* 129(3): 387–96.

Lynne, Gary D., J. Walter Milon, and Michael E. Wilson. 1990. Identifying and Measuring Potential Conflict in Water Institutions. *Water Resources Bulletin* 26 (4, August): 669–76.

Maass, Arthur and Raymond Lloyd Anderson. 1978. *And the Desert Shall Rejoice: Conflict, Growth, and Justice in Arid Environments.* Cambridge, MA: MIT Press.

MacAvoy, Peter V. 1986. The Great Lakes Charter: Toward a Basinwide Strategy for Managing the Great Lakes. *Case Western Reserve Journal of International Law* 18(49): 49–65.

MacChesney, Brunson. 1959. Judicial Decisions. *Americal Journal of International Law* 53: 156–71.

MacDonnell, L.J. 1988. Natural Resources Dispute Resolution: An Overview. *Natural Resources Journal* 28 (Winter): 5–19.

—— 1989. Federal Interests in Western Water Resources: Conflict and Accommodation. *Natural Resources Journal* 29 (Spring): 389–411.

Madsen, Robert A. 1995. Of Oil and Rainforests: Using Commodity Cartels to Conserve Depletable Natural Resources. *International Environmental Affairs* 7 (3, Summer): 207–34.

Mageed, Y.A. 1994. The Nile Basin: Lessons from the Past. In *International Waters of the Middle East: From Euphrates-Tigris to Nile,* ed. Asit K. Biswas, 156–84. Bombay, Delhi, Calcutta, Madras: Oxford University Press.

—— and G.F. White. 1995. Critical Analysis of Existing Institutional Arrangements. *Water Resources Development* 11(2): 103–11.

Mahendrarajah, S. and P.G. Warr. 1991. Water Management and Technological Change: Village Dams in Sri Lanka. *Journal of Agricultural Economics* 42(3): 309–24.

Maitreyi, K.R. 1993. Ecological Movements: Strategies for Managing Conflicts Around Natural Resources: A Case Study of Forests. *Forests, Trees and People Newsletter* (20, April): 21–2.

Maluwa, T. 1988. Legal Aspects of the Niger River Under the Niamey Treaties. *Natural Resources Journal* 28 (Fall): 671–97.

Management Science Center. 1967. *A Model Study of the Escalation and De-escalation of Conflict*. PA: University of Pennsylvania, Management Science Center.

Mandel, R. 1991. Sources of International River Basins Disputes. Annual Meeting of the International Studies Association. Vancouver, Canada.

Manig, W. 1994. Situation-specific Management in Appropriate Irrigation Organizations: Areas of Conflict in Water Distribution. *Quarterly Journal of International Agriculture* 33(3): 243–57.

Manner, Eero J. 1987. *Diversion of Waters and The Principle of Equitable Utilization*. Oxford: Clarendon Press.

—— 1988. *The Work of the International Law Association*. Helsinki: Association Committee on International Water Rights Law.

Maoz, Zeev and Nasrin Abdolali. 1989. Regime Types and International Conflict, 1816–976. *Journal of Conflict Resolution* 33: 3–35.

Marantz, Paul and Janice Gross Stein, eds. 1985. *Peace-Making in the Middle East: Problems and Prospects*. London: Croom Helm.

Margat, Jean. 1989. The Sharing of Common Water Resources in the European Community (EEC). *Water International* 14: 59–61.

Marr, P. and W. Lewis, eds. 1993. *Riding the Tiger: The Middle East Challenge After the Cold War*. Boulder, CO: Westview Press.

Marseille, Claudia. 1983. *Conflict Management: Negotiating [American] Indian Water Rights*. Cambridge, MA: Lincoln Institute of Land Policy.

Maruyama, M. 1963. The Second Cybernetics: Deviation-Amplifying Mutual Causal Processes. *American Scientist* 51: 164–79.

Marwell, Gerald and David R. Schmitt. 1975. *Cooperation: An Experimental Analysis*. New York: Academic Press.

Maslow, Abraham H. 1990. *Motivation and Personality*. New York: Columbia University Press.

Mateo, R.M. 1992. Administration of Water Resources: Institutional Aspects and Management Modalities. *Natural Resources Forum* (May): 117–25.

Mather, T. 1989. The Planning and Management of African River and Lake Basin Development and Conservation. *Natural Resources Forum* (February): 59–70.

Mathews, Jessica Tuchman. 1989. Redefining Security. *Foreign Affairs* 68 (Spring): 162–77.

Matsuura, Shigenori. 1995. China's Air Pollution and Japan's Response to It. *International Environmental Affairs* 7 (3, Summer): 235–48.

Mattair, Thomas R. 1992. The Peace Process: Can it Bridge Water This Troubled? *Middle East Policy* (Spring): 57–85.

Matthew, Richard A. 1995. Environmental Security: Demystifying the Concept, Clarifying the Stakes. *Environmental Change and Security Project* (1, Spring): 14–23.

Matthews, Olen Paul. 1984. *Water Resources: Geography and Law*. Washington, DC: Association of American Geographers.

—— 1994. Judicial Resolution of Transboundary Water Conflicts. *Water Resources Bulletin* 30 (3, May/June): 375–84.

McAuslan, Patrick. 1987. The Lesotho Highlands Water Project and Environmental Law: A Case Study. *Lesotho Law Journal* 3(2): 41–66.

McBean, Edward A. and Norio Okada. 1988. Use of Metagame Analysis in Acid Rain Conflict Resolution. *Journal of Environmental Management* 27: 153–62.

McCaffrey, S. 1989. The Law of International Watercourses: Some Recent Developments and Unanswered Questions. *Denver Journal of International Law and Policy* 17: 505–26.

—— 1991. International Organizations and the Holistic Approach to Water Problems. *Natural Resources Journal* 31 (1, Winter): 139–65.

—— 1992. Background and Overview of the International Law Commission's Study of the Non-Navigational Uses of International Watercourses. *Colorado Journal of Environmental Law and Policy* 3(1): 17–30.

—— 1992–3. A Human Right to Water: Domestic and International Implications. *Georgetown International Environmental Law Review* 5: 1–24.

—— 1993. The Evolution of the Law of International Watercourses. *Australian Journal of Public and International Law* 45.

McCool, Daniel. 1993. Indian Water Settlements: The Prerequisites of Successful Negotiation. *Policy Studies Journal* 21 (Summer): 227–42.

—— 1993. Intergovernmental Conflict and Indian Water Rights: An Assessment of Negotiated Settlements. *Publius: The Journal of Federalism* 32(1): 85–102.

McCormick, John. 1985. *Acid Earth: The Global Threat of Acid Pollution*. Washington, DC: International Institute for Environment and Development.

McGinnis, Michael D. 1986. Issue Linkage and the Evolution of International Cooperation. *Journal of Conflict Resolution* 30: 141–70.

McKinney, Matthew. 1992. Designing a Dispute Resolution System for Water Policy and Management. *Negotiation Journal* 8 (2, April): 153–64.

Meagher, Robert F. 1995. Himalayan Power: The World Bank as a Third Party in India-Nepal Riparian Negotiations. diss. Fletcher School of Law and Diplomacy.

Medzini, Meron, ed. 1976. *Israel's Foreign Relations*. Jerusalem: Ministry for Foreign Affairs.

Megahan, Walter F. and Peter N. King. 1985. Identification of Critical Areas on Forest Lands for Control of Nonpoint Sources of Pollution. *Environmental Management* 9 (1, January): 7–18.

Mehta, J.S. 1988. The Indus Water Treaty: A Case Study in the Resolution of an International River Basin Conflict. *Natural Resources Forum* 12(1): 69–77.

Meier, Richard L. 1991. A Global Role for the Palestine Arabs: The Integrated Agro-Industrial Complex. University of California, Berkeley, April.

Mesfin, Abebe. 1995. The Nile – Source of Regional Cooperation or Conflict? *Water International* 20 (1, March): 32–4.

Michel, Aloys Arthur. 1967. *The Indus Rivers: A Study of the Effects of Partition*. New Haven, CT: Yale University Press.

Mikhail, Wakil. 1992. Analysis of Future Water Needs for Different Sectors in Syria. Conference on the Mideast Water Crisis. Waterloo, April.

Miller, Morris. 1995. Transformation of a River Basin – Case of the Mekong Committee. Asian Water Forum. Bangkok, Thailand: United Nations University, 30 January–1 February.

Miller, P. 1982. Value as Richness: Toward a Value Theory for an Expanded Naturalism in Environmental Ethics. *Environmental Ethics* 4 (2, Summer): 101–14.

Milon, J.W. 1990. *Uncertainty and the Allocation of Renewable and Nonrenewable Resources*. SRDC Series, Vol. 137. Mississipi State, MS: Southern Rural Development Center.

Ministry of Agriculture and Irrigation. 1979. *Agreements on Development of Inter-State and International Rivers*. New Delhi, India: Government of India, Ministry of Agriculture and Irrigation, Central Water Commission.

Ministry of Foreign Affairs. 1995. *Water Issues Between Turkey, Syria, and Iraq*. Department of Regional and Transboundary Water, Republic of Turkey, Ministry of Foreign Affairs.

Mitchell, Bruce. 1990. *Integrated Water Management: International Experiences and Perspectives*. New York, NY: Belhaven Press.

—— 1991. *Resource Management and Development: Addressing Conflict and Uncertainty.* Toronto, Canada: Oxford University Press.

—— 1995. *Resources and Environmental Management in Canada.* Don Mills, Ontario: Oxford University Press.

Mitchell, John E. 1967. Planning of Water Development in Hashemite Kingdom of Jordan. In *International Conference on Water for Peace.* International Conference on Water for Peace. Washington, DC, 23–31 May.

Mitchell, Ronald B. 1994. Regime Design Matters: Intentional Oil Pollution and Treaty Compliance. *International Organization* 48 (Summer): 425–58.

—— 1995. Compliance with International Treaties: Lessons from Intentional Oil Pollution. *Environment* 37 (May): 10–15, 36–41.

Mitrany, David. 1975. *The Functional Theory of Politics.* New York: St Martin's Press.

Mizutani, Masakazu. 1989. Trends and Characteristics of Water Transfer Projects. *Journal of Irrigation Engineering and Rural Planning* (16, July): 71–86.

Moermond, James O., III and Shirley Erickson. 1987. A Survey of the International Law of Rivers. *Denver Journal of International Law and Policy* 16(1): 139–59.

Moffett, George D. III. 1990. Four-part Series on Middle East Water Problems. *Christian Science Monitor*, 8, 13, 14, 16 March.

—— 1990. Tigris-Euphrates Basin: Downstream Fears Feed Tensions. *Transboundary Resources Report* 4 (3, Winter): 4–5.

—— 1991. Nile River Rights Run Deep. *Transboundary Resources Report* 5 (1, Spring 1991): 3.

Montagnon, Peter. 1989. Lesotho Looks to $2 Billion Dam for Economic Lift. *Financial Times*, 13 April.

Montville, Joseph V. 1987. The Arrow and the Olive Branch: A Case for Track Two Diplomacy. In *Conflict Resolution: Track Two Diplomacy*, ed. John McDonald and D.B. Bendahmane, 5–20. Washington, DC: Foreign Service Institute.

Moore, C.W. 1986. *The Mediation Process: Practical Strategies for Resolving Conflict.* San Francisco: Jossey-Bass Publishers.

Moore, J. 1992. *Water-Sharing Regimes in Israel and the Occupied Territories.* Ottawa: Department of National Defence, Canada.

—— 1994. An Israeli-Palestinian Water-Sharing Regime. *Studies in Environmental Science* 58: 181–92.

Morris, Mary E. 1991. Poisoned Wells: The Politics of Water in the Middle East. *Middle East Insight* 8(2): 35–9.

—— 1993. *Dividing the Waters: Reaching Equitable Water Solutions in the Middle East.* Santa Monica, CA: RAND.

Morrow, James D. 1986. A Spatial Model of International Conflict. *American Political Science Review* 80: 1131–50.

—— 1992. Signaling Difficulties with Linkage in Crisis Bargaining. *International Studies Quarterly* 36: 153–72.

Mousa, M. 1994. A General View of the Water Situation in the Occupied Palestinian Territories (OPT). *Studies in Environmental Science* 58: 523–8.

Mudgal, Vipul. 1988. Troubled Waters. *India Today*, 31 October, 112–13.

Mumme, Stephen. 1988. *Apportioning Groundwater Beneath the United States-Mexico Border: Obstacles and Alternatives.* San Diego: University of California Press.

—— 1988. *Continuity and Change in United States-Mexico land and Water Relations: The Politics of the International Boundary and Water Commission.* Washington, DC: The Wilson Center.

—— 1992. New Directions in United States – Mexican Transboundary Environmental Management: A Critique of Current Proposals. *Natural Resources Journal* 32: 539–62.

—— 1992. System Maintenance and Environmental Reform in Mexico: Salinas's Preemptive Strategy. *Latin American Perspectives.* 72(19): 123–43.

—— 1995. Western Times and Water Wars: State, Culture, and Rebellion in California by John Walton. *The American Journal of Sociology* 98 (6, May).

—— 1993. Innovation and Reform in Transboundary Resource Management: A Critical Look at the International Boundary and Water Commission, United States and Mexico. *Natural Resources Journal* 33 (Winter): 93–132.

—— 1995. The North American Commission for Environmental Cooperation and the United States-Mexico Border Region. *Transboundary Resources Report* 9 (2, Summer): 1.

—— 1995. The New Regime for Managing US-Mexican Water Resources. *Environmental Management* 19(6).

—— 1999. Managing Acute Water Scarcity on the US-Mexico Border: Institutional Issues Raised by the 1990's Drought. *Natural Resources Journal* 39(1).

—— and Pamela Duncan. 1997. The Commission for Environmental Cooperation and Environmental Management in the Americas. *Journal of Interamerican Studies and World Affairs* 39(4).

—— and Carl Grundy-Warr. 1998. Structuration Theory and the Analysis of International Territorial Disputes: Lessons from an Application to the El Chamizal Controversy. *Political Research Quarterly* 51(4).

—— and S.T. Moore. 1990. Agency Autonomy in Transboundary Resource Management: The United States Section of the International Boundary and Water Commission, United States, and Mexico. *Natural Resources Journal* 30 (3, Summer): 661–84.

—— and Roberto A. Sanchez. 1992. New Directions in Mexican Environmental Policy. *Environmental Management.* 16(4): 465–74.

Murakami, M. 1995. *Managing Water for Peace in the Middle East: Alternative Strategies.* New York, NY: United Nations University Press.

—— and Katsumi Musiake. 1994. The Jordan River and the Litani. In *International Waters of the Middle East: From Euphrates-Tigris to Nile*, ed. Asit K. Biswas, 117–55. Bombay, Delhi, Calcutta, Madras: Oxford University Press.

—— and Katsumi Musiake. 1997. Eco-political Decision-making and Confidence-building Measures in the Development of International Rivers. *International Journal of Water Resources Development* 13 (3, September): 403–14.

—— and A.T. Wolf. 1995. Techno-political Water and Energy Development Alternatives in the Dead Sea and Aqaba Regions. *Water Resources Development* 11(2): 163–83.

——, U. El-Hanbali, and A.T. Wolf. 1995. Technopolitical Alternative Strategies in the Interstate Regional Development of the Jordan Rift Valley at Peace. *Water International* 20 (4, December).

Murphy, I.L. 1990. Resolving Conflicts on the Danube: The Gabcikovo-Nagymaros Power Dam Project. *International and Transboundary Water Resource Issues*, April, 146.

—— and J.E. Sabadell. 1986. International River Basins: A Policy Model for Conflict Resolution. *Resources Policy* 12 (June): 133–44.

Murray, John S. 1990. Dispute Systems: Design, Power, and Prevention. *Negotiation Journal* 6 (2, April): 105–9.

Musallam, Ramzi. 1990. *Water: Source of Conflict in the Middle East in the 1990s.* London: Gulf Centre for Strategic Studies.

Mustafa, I. 1994. The Arab-Israeli Conflict Over Water Resources. *Studies in Environmental Science* 58: 123–33.

Myers, Norman. 1987. Linking Environment and Security. *Bulletin of Atomic Scientists* 43 (June): 46–7.

—— 1993. *Ultimate Security: The Environmental Basis of Political Stability.* New York, NY: W.W. Norton and Company.

Nachmani, A. 1994. The Politics of Water in the Middle East. In *The Southeast European Yearbook 1993*, ed. T.A. Couloumbis, T.M. Veremis, and T. Dokos, 147–66. Athens, Greece: Hellenic Foundation for European and Foreign Policy.

Naff, Thomas. 1987. *Euphrates River Basin: Hydrological Analysis*. Philadelphia, PA: Associates for Middle East Research (AMER), Inc.

—— 1987. *Israel: Political, Economic and Strategic Analysis*. Philadelphia, PA: Associates for Middle East Research (AMER), Inc.

—— 1987. *Jordan: Political, Economic and Strategic Analysis*. Philadelphia, PA: Associates for Middle East Research (AMER), Inc.

—— 1987. *The Litani River: Hydrology and Management*. Philadelphia, PA: Associates for Middle East Research (AMER), Inc.

—— 1987. *The Potential and Limitations of Technology*. Philadelphia, PA: Associates for Middle East Research (AMER), Inc.

—— 1987. *Syria: Political, Economic and Strategic Analysis*. Philadelphia, PA: Associates for Middle East Research (AMER), Inc.

—— 1987. *Turkey: Political, Economic and Strategic Analysis*. Philadelphia, PA: Associates for Middle East Research (AMER), Inc.

—— 1990. *Water: The Middle East Imperative. A Summary of the Middle East Water Project*. Philadelphia, PA: Associates for Middle East Research (AMER), Inc.

—— 1992. *Water Scarcity, Resource Management, and Conflict in the Middle East*. American Association for the Advancement of Science, Vol. 92–33S. Washington, DC: Committee on Science and International Security, American Association for the Advancement of Science.

—— and Ruth C. Matson. 1984. *Water in the Middle East: Conflict or Cooperation?* Boulder, CO: Westview Press.

—— and Marvin E. Wolfgang, eds. 1985. *Changing Patterns of Power in the Middle East*. Beverly Hills, CA: Sage Publications.

Nakayama, Mikiyasu. 1997. Successes and Failures of International Organizations in Dealing with International Waters. *International Journal of Water Resources Development* 13 (3, September): 367–82.

Nanda, Ved P. 1977. Emerging Trends in the Use of International Law and Institutions for the Management of International Water Resources. In *Water Needs for the Future: Political, Economic, Legal, and Technological Issues in a National and International Framework*, ed. Ved P. Nanda, 15–38. Boulder, CO: Westview Press.

Narby, Jeremy and Shelton H. Davis, eds. 1983. *Resource Development and Indigenous Peoples: A Comparative Bibliography*. Boston, Massachusetts: Anthropology Resource Center.

Nazem, Nurul Islam. 1994. The Impact of River Control on an International Boundary: The Case of the Bangladesh-India Border. In *Eurasia, World Boundaries*, Vol. 3, ed. Carl Grundy-Warr, 101–10. London, England: Routledge.

—— and Mohammad Humayun Kabir. 1986. *Indo-Bangladesh Common Rivers and Water Diplomacy*. Dacca: The Bangladesh Institute of International and Strategic Studies.

Neiland, Arthur E. and I. Verinumbe. 1991. Fisheries Development and Resource-Usage Conflict: A Case Study of Deforestation Associated with the Lake Chad Fishery in Nigeria. *Environmental Conservation* 18 (Summer): 111–17.

Netherlands. Directoraat-Generaal Internationale Samenwerking. 1992. *Policy Plan for 1992–1995: Nile and Red Sea Region*. The Hague: Directorate General for International Cooperation of the Netherlands Ministry for Foreign Affairs.

Newham, Mark. 1988. Nile Decline a Threat to Egypt. *Pakistan and Gulf Economist* 7 (April): 40–1.

Newson, M. 1992. Water and Sustainable Development: the "Turn Around Decade?" *Journal of Environmental Planning and Management* 35(2): 175–83.

Nickum, J.E. and K.W. Easter. 1990. Institutional Arrangements for Managing Water Conflicts in Lake Basins. *Natural Resources Forum* 14 (3, August): 210–21.

Nishat, Aminun. 1995. Impact of Ganges Water Dispute on Bangladesh. Asian Water Forum. United Nations University, 30 January–1 February.

Nollkaemper, Andre. 1993. *The Legal Regime for Transboundary Water Pollution: Between Discretion and Constraint.* Dordrecht: Martinus Nijhoff.

Nomas, H.B. 1988. The Water Resources of Iraq: An Assessment. diss, 486 pp. University of Durham.

North, Ronald M. 1993. Application of Multiple Objective Models to Water Resources Planning and Management. *Natural Resources Forum,* August, 216–27.

North American Institute for Borders and Water. 1992. *North American Water Issues.* Albuquerque, NM: International Transboundary Resources Center.

Nutt, P.C. and R.W. Backoff. 1987. A Strategic Management Process for Public and Third-Sector Organizations. *APA Journal* 44 (Winter).

Ohlsson, L., ed. 1992. *Regional Case Studies of Water Conflicts.* Goteborg: Padrigu Papers.

—— 1995. *Hydropolitics.* London, England: Zed Books.

Okada, Norio, Keith W. Hipel, and Yoshiharu Oka. 1985. Hypergame Analysis of the Lake Biwa Conflict. *Water Resources Research* 21 (7, July): 917–26.

Okidi, C.O. 1982. Review of Treaties on Consumptive Utilization of Waters of Lake Victoria and Nile Drainage System. *Natural Resources Journal* 22 (January): 161–99.

—— 1988. The State and the Management of International Drainage Basins in Africa. *Natural Resources Journal* 28 (Fall): 645–69.

Okolicsanyi, Karoly. 1992. Slovak-Hungarian Tension: Bratislava Diverts the Danube. *RFE-RL Research Report* 11 (December): 49–55.

Olson, Kent W. 1983. Economics of Transferring Water to the High Plains. *Quarterly Journal of Business and Economics* 22 (Autumn): 63–80.

Olson, W.J. 1995. Small Wars. *The Annals of the American Academy of Political and Social Science* 541 (September).

Onorato, W. 1985. A Case Study on Joint Development: the Saudi Arabia/Kuwait Partitioned Neutral Zone. *Energy* 10: 539.

Onta, Pushpa Raj, Ashim Das Gupta and Rainer Loof. 1995. Potential Water Resources Development in the Salawin River Basin. In *Asian Water Forum.* Asian Water Forum. United Nations University, 30 January–1 February.

Ophuls, William. 1977. *Ecology and the Politics of Scarcity: A Prologue to a Political Theory of the Steady State.* San Francisco, CA: Freeman.

Oppenheimer, J.A. and C. Russell. 1983. A Tempest in a Teapot: The Analysis and Evaluation of Environmentalist Trading in Markets for Pollution Permits. In *Buying a Better Environment: Cost-Effective Regulation Through Permit Trading,* ed. David Joeres. Madison, WI: University of Wisconsin Press.

Opschoor, Johannes B. 1989. North-South Trade, Resource Degradation and Economic Security. *Bulletin of Peace Proposals* 20(2): 135–42.

Organization for Economic Cooperation and Development. 1993. Hydropower, Energy and the Environment: Options for Increasing Output and Enhancing Benefits. Conference Proceedings. Stockholm, Sweden: Organization for Economic Cooperation and Development, 14–16 June.

Ostrom, Elinor. 1982. Crafting Institutions: Self-Governing Irrigation Systems. In *The Art and Science of Negotiation,* ed. Howard Raiffa. Cambridge, MA: Harvard University Press.

—— 1992. *Crafting Institutions for Self-Governing Irrigation Systems.* San Francisco, CA: Institute for Contemporary Studies.

Ozawa, Connie P. and Lawrence Susskind. 1985. Mediating Science-Intensive Policy Disputes. *Journal of Policy Analysis and Management* 5(1): 23–39.

O'Connor, David. 1992. The Design of Self-Supporting Dispute Resolution Programs. *Negotiation Journal* 8 (2, April): 85–91.

O'Loughlin, John V., ed. 1994. *Dictionary of Geopolitics*. Westport, CT: Greenwood Press.

Painter, An. 1988. The Future of Environmental Dispute Resolution. *Natural Resources Journal* 28 (1, Winter): 145.

—— 1995. Resolving Environmental Conflicts Through Mediation. In *Water Quantity/ Quality Management and Conflict Resolution: Institutions, Processes, and Economic Analyses*, ed. Ariel Dinar and Edna Tusak Loehman, 249–58. Westport, CT: Praeger.

Paisley, Richard Kyle and Timothy L. McDaniels. 1995. International Water Law, Acceptable Pollution Risk and The Tatshenshini River. *Natural Resources Journal* 25 (Winter): 111–32.

Palo, Matti and Jyrki Salmi. 1988. *Deforestation or Development in the Third World?* Vol. II. Division of Social Economics of Forestry, Vol. 309. Helsinki: The Finnish Forest Research Institute.

Pandey, Bikash. 1995. Because It Is There: Foreign Money, Foreign Advice, and Arun III. *Himal*, July/August, 29–35.

Parry, Clive, ed. 1969. *The Consolidated Treaty Series*. New York: Oceana Publications.

Paul, T.V. 1994. *Asymmetric Conflicts: War Initiation by Weaker Powers*. Cambridge, MA: Cambridge University Press.

Paulsen, C.M. 1993. Policies for Water Quality Management in Central and Eastern Europe. In *Two Essays on Water Quality in Central and Eastern Europe. ENR93–20*, 15–25. Washington, DC: Resources for the Future.

Payne, M. 1986. Agricultural Polution – the Farmer's View. In *Effects of Land Use on Fresh Waters: Agriculture, Forestry, Mineral Exploitation, Urbanisation*, ed. J.F. Solbe, 329–40. West Sussex, England: Ellis Horwood Limited.

Pearce, Fred. 1991. Wells of Conflict on the West Bank. *New Scientist*, 1 June, 36–40.

Pearce, Stephen. 1993. Water and Politics: Lesotho's Highlands Water Scheme May Deprive that Country of Not Only its Water but Its Culture as Well. *Cultural Survival Quarterly*, Summer, 48–50.

Perritt, R. 1989. African River Basin Development: Achievements, the Role of Institutions, and Strategies for the Future. *Natural Resources Forum*, August, 204–8.

Peters, Charles M. 1996. *The Ecology and Management of Non-Timber Forest Resources*. Technical Paper, Vol. 322. Washington, DC: World Bank.

Petts, G.E. 1988. Water Management: the Case of Lake Biwa, Japan. *Geographical Journal* 154(3): 367–76.

Piddington, Kenneth. 1992. The Role of the World Bank. In *The International Politics of the Environment: Actors, Interests, and Institutions*, ed. Andrew Hurrell and Benedict Kingsbury, 212–27. New York, NY: Oxford University Press.

Pike, David. 1991. Gulf Fishing in Troubled Waters. *MEED*, 23 August, 4–6.

Platter, Adele G. and Thomas F. Mayer. 1989. A Unified Analysis of International Conflict and Cooperation. *Journal of Peace Research* 26(4): 367–83.

Plott, Charles R. 1978. Rawls's Theory of Justice: An Impossible Result. In *Decision Theory and Social Ethics, Issues in Social Choice*, ed. Hans W. Gottinger and Werner Leinfellner, 207. Dordrecht: Reidel.

Porter, Gareth and Janet Welsh Brown. 1991. *Global Environmental Politics*. Boulder, CO: Westview Press.

Postel, Sandra. 1984. *Air Pollution, Acid Rain, and the Future of Forests*. Worldwatch Paper, Vol. 58. Washington, DC: Worldwatch Institute.

—— 1984. *Water: Rethinking Management in an Age of Scarcity*. Worldwatch Paper, Vol. 62. Washington, DC: Worldwatch Institute.

—— 1985. *Conserving Water: The Untapped Alternative*. Worldwatch Paper, Vol. 67. Washington, DC: Worldwatch Institute.

—— 1989. *Water for Agriculture: Facing the Limits*. Worldwatch Paper, Vol. 93. Washington, DC: Worldwatch Institute.

—— 1992, 1997. *Last Oasis: Facing Water Scarcity*. New York and London: W.W. Norton and Company.

—— 1993. The Politics of Water. *World Watch* 6 (4, July/August): 10–18.

—— 1996. *Dividing the Waters: Food Security, Ecosystem Health, and the New Politics of Scarcity*. Worldwatch Paper, Vol. 132. Washington, DC: Worldwatch Institute.

—— 1997. Changing the Course of Transboundary Water Management. *Natural Resources Forum* 21 (2, May): 85–90.

Precoda, Norman. 1991. Requiem for the Aral Sea. *Ambio* 20 (3–4, May): 109–14.

Priest, J.E. 1992. International Competition for Water and Motivations for Dispute Resolution. *Agricultural Water Management* 21(1–2): 3–11.

Prizker, David M. and Deborah S. Dalton. 1990. *Negotiated Rulemaking Sourcebook*. Washington, DC: US Goverment Printing Office.

Quigg, Phillip W. 1977. A Water Agenda to the Year 2000. *Common Ground* 3(4): 11–16.

Quinn, J.T. and J.J. Harrington. 1992. Generating Alternative Designs for Interjurisdictional Natural Resource Development Schemes in the Greater Ganges River Basin. *Papers in Regional Science* 71(4): 373–91.

Rabe, B.G. 1988. The Politics of Environmental Dispute Resolution. *Policy Studies Journal* 16(3): 585–601.

Rabi, Muhammad. 1993. *Conflict Resolution and the Middle East Process*. Hamburg: Deutsches Orient-Institut.

—— 1994. *Conflict Resolution and Ethnicity*. Westport, CT: Praeger.

—— 1995. *U.S.-P.L.O. Dialogue: Secret Diplomacy and Conflict Resolution*. Gainesville, FL: University of Florida Press.

Radosevich, G.E. 1976. *International Conference on Global Water Law Systems*. Fort Collins, CO: Center for Economic Education and Department of Economics, Colorado State University.

—— 1977. Global Water Law Systems and Water Control. In *Water Needs for the Future – Political, Economic, Legal and Technological Issues in a National and International Framework*, ed. V.P. Nanda, 39–58. Boulder, CO: Westview Press.

—— 1995. The Mekong – A New Framework for Development and Management Under a Renewed Spirit of Cooperation. In *Asian Water Forum*. Asian Water Forum. Bangkok, Thailand: United Nations University, 30 January–1 February.

Rady, Essam. 1990. Water Management in Egypt. *Water International* 15 (1, March): 57–63.

Rady, Mohammed A. 1995. Satisfying National and International Water Demands. *Water International* 20 (1, March): 9.

Rahim, M. Afzalur. 1990. *Theory and Research in Conflict Management*. New York: Praeger.

Raiffa, Howard. 1982. *The Art and Science of Negotiation*. Cambridge, MA: Harvard University Press.

Ramjeawon, T. 1994. Water Resources Development on the Small Island of Mauritius. *Water Resources Development* 10(2): 143–56.

Rangarajan, L.N. 1985. *The Limitation of Conflict: A Theory of Bargaining and Negotiation*. New York: St Martin's Press.

Rangeley, Robert, Bocar M. Thiam, Randolph A. Andersen, and Colin A. Lyle. 1994. *International River Basin Organizations in Sub-Saharan Africa*. World Bank Technical Paper, Africa Technical Department Series, Vol. 250. Washington, DC: World Bank.

Rao, A. Ramachandra and T. Prasad. 1994. Water Resources Development of the Indo-Nepal Region. *Water Resources Development* 10(2): 157–74.

Raskin, P., E. Hansen, Z. Zhu, and D. Stavisky. 1992. Simulation of Water Supply and Demand in the Aral Sea Region. *Water International* 17(2): 55–67.

Rauscher, H.M. 1995. Natural Resource Decision Support: Theory and Practice. *AI Applications* 9(3): 1–2.

Rauschning, D. 1990. Indus Water Dispute. In *Encyclopedia of Public International Law 6*, ed. R. Bernhardt, 214–8. North Amsterdam: Holland.

Rausser, G.C., S.R. Johnson, and C. Willis. 1980. *Systems Methods in Natural Resource Economics*. Berkeley, CA: Giannini Foundation of Agricultural Economics, California Agricultural Experiment Station.

Raynal, Jose A. 1992. *Hydrology and Water Resources Education, Training, and Management: Proceedings of the International Symposium on Hydrology and Water Resources Education and Training: the Challenges to Meet at the Turn of the XXI Century, April 15–19, 1992, and Second North American Water Management Seminar, April 17, 1991, Chihuahua, Chih., Mexico*. Littleton, CO: Water Resources Publications.

Redclift, M. 1988. Economic Models and Environmental Values: A Discourse on Theory. In *Sustainable Environmental Management: Principles and Practice*, ed. R. Kerry Turner, 51–66. London, England: Belhaven Press.

—— 1991. The Multiple Dimensions of Sustainable Development. *Geography* 76, part 1 (330): 36–42.

Reeves, Randall R., Abdul Aleem Chaudhry, and Umeed Khalid. 1991. Competing for Water on the Indus Plain: Is There a Future for Pakistan's River Dolphins? *Environmental Conservation* 18 (Winter): 341–50.

Reguer, S. 1993. Controversial Waters: Exploitation of the Jordan River, 1950–80. *Middle Eastern Studies* 29(1): 53–90.

Remans, Wilfried. 1995. Water and War. *Humanitäres Völkerrecht* 8(1): 4–14.

Renner, Michael. 1989. *National Security: The Economic and Environmental Dimensions*. Worldwatch Paper, Vol. 89. Washington, DC: Worldwatch Institute.

Rhodes, Thomas C. and Paul N. Wilson. 1995. Sky Islands, Squirrels, and Scopes: The Political Economy of an Environmental Conflict. *Land Economics* 71 (1, February): 106–21.

Rice, Teresa and Lawrence J. MacDonnell. 1993. *Agriculture to Urban Water Transfers in Colorado: An Assessment of The Issues and Options*. Boulder, CO: Natural Resources Law Center, University of Colorado School of Law.

Rich, Vera. 1990. Central Europe II: The Battle of the Danube. *The World Today* 28(12): 217.

—— 1992. Swelling Chorus of Danube Blues. *New Scientist*, 26 September, 8.

—— 1993. The Murky Politics of the Danube. *The World Today* 49(8/9): 151–2.

Richards, Alan. 1993. Strengthening Markets to Build Peace: The General Case, Illustrated by the Example of Agriculture and Water. Conference on the Middle East Multilateral Conference. Los Angeles: University of California, Los Angeles, June.

Riebsame, W.E. et al. 1995. Complex River Basins. In *As climate Changes: International Impacts and Implications*, ed. K. Strzepek and J. Smith. Cambridge, MA: Cambridge University Press.

Rios Brehm, Monica and Jorge Quiroz. 1995. *The Market for Water Rights in Chile: Major Issues*. Washington, DC: World Bank.

Rizk, Edward. 1964. *The River Jordan*. Information Paper, Vol. 23. New York: Arab Information Center.

Robinson, Nicholas A. 1987. Marshalling Environmental Law to Resolve the Himalyaa-Ganges Problem. *Mountain Research and Development* 7: 305–15.

Rodda, J.C. 1995. *Facing Up to the Looming World Water Crisis*. Water Briefing, Financial Times Newsletters, Vol. 28.

Rodgers, A.B. and A.E. Utton. 1985. The Ixtapa Draft Agreement Relating to the Use of Transboundary Groundwaters. *Natural Resources Journal* 25 (3, July): 713–72.

Rodman, Margaret. 1989. *Deep Water: Development and Change in Pacific Village Fisheries.* Boulder, CO: Westview Press.

Rogers, Paul and Malcolm Dando. 1992. *A Violent Peace: Global Security After the Cold War.* London, UK: Brassey's.

Rogers, Peter. 1969. A Game Theory Approach to the Problems of International River Basins. *Water Resources Research* 5 (4, August): 749–60.

—— 1991. International River Basins: Pervasive Unidirectional Externalities. The Economics of Transnational Commons. Universita di Siena, Italy, 25–27 April.

—— 1992. *Comprehensive Water Resources Management: A Concept Paper.* Washington, DC: World Bank.

—— 1993. Integrated Urban Water Resources Management. *Natural Resources Forum* (February): 34–42.

—— 1993. A Model to Generate Pareto-Admissible Outcomes for International River Basin Negotiations. International Workshop on Economic Aspects of International Water Resources Utilization in the Mediterranean Basin. Milan, Italy: Fondazione ENI Enrico, 8–9, October.

—— 1993. The Value of Cooperation in Resolving International River Basin Disputes. *Natural Resources Forum* 17(2): 117–31.

—— 1994. Model to Generate Pareto-Admissible Outcomes for International River Basin Negotiations. *Fondazione Eni Enrico Mattei. Nota Di Lavoro* 48(94): 1–26.

——, R. Burden, and C. Lotti. 1978. Systems Analysis and Modeling Techniques Applied to Water Management. *Natural Resources Forum* 2: 349–58.

—— and Peter Lydon, eds. 1994. *Water in the Arab World: Perspectives and Prognoses.* Cambridge, MA: Harvard University Press.

Romer, T. and H. Rosenthal. 1978. Political Resource Allocation, Controlled Agendas and the Status Quo. *Public Choice* 33(4): 27–43.

Rosegrant, Mark W. 1995. Water Transfers in California: Potentials and Constraints. *Water International* 20 (2, June): 72–87.

Rosen, M. 1993. Conflict Within Irrigation Districts May Limit Water Transfer Gains. *California Agriculture* 46 (6, Nov/Dec): 4–7.

Rosenne, Shabtai. 1995. *The World Court: What it Is and How it Works.* Dordrecht, The Netherlands: Martinus Nijhoff.

Ross, Lee and Constance Stillinger. 1991. Barriers to Conflict Resolution. *Negotiation Journal* 7 (4, October): 389–404.

Rothman, Jay. 1991. Negotiation as Consolidation: Prenegotiation in the Israeli-Palestinian Conflict. *Jerusalem Journal of International Relations* 13(1).

—— 1995. Pre-Negotiation in Water Disputes: Where Culture is Core. *Cultural Survival Quarterly (U.S.)* 19 (3, Fall): 19–28.

Rowlands, Ian. 1991. Building International Regimes. *The Washington Quarterly* 14 (1, Winter): 99–118.

Rowley, Gwyn. 1993. Multinational and National Competition for Water in the Middle East: Towards the Deepening Crisis. *Journal of Environmental Management* 39: 187–97.

Rowse, J. 1988. Does an Exhaustible Resource Usually Have Many Near-Optimal Depletion Paths? *American Journal of Agricultural Economics* 70 (3, August): 646–53.

—— 1990. Using the Wrong Discount Rate to Allocate an Exhaustible Resource. *American Journal of Agricultural Economics* 72 (1, February): 121–30.

RuBino, Richard and Harvey Jacobs. 1990. *Mediation and Negotiation for Planning, Land Use Management, and Environmental Protection.* Chicago, IL: Council for Planning Librarians.

Russell, C. and J. Shogren, eds. 1993. *Theory, Modeling, and Experience in the Management of Nonpoint-Source Pollution.* Boston, MA: Kluwer Academic Publishers.

Russett, Bruce M. 1967. *International Regions and the International System: A Study in Political Ecology*. Chicago, IL: Rand McNally.

Ryans, Robert C. and United States Environmental Protection Agency. 1988. *Protection of River Basins, Lakes, and Estuaries: Fifteen Years of Cooperation Toward Solving Environmental Problems in the USSR and USA*. Bethesda, MD: Published for the U.S. Environmental Protection Agency by the American Fisheries Society.

Sabbagh, A. 1994. Conflict Over Water in the Middle East: From a Security and Strategic Point of View. *Studies in Environmental Science* 58: 505–14.

Sadler, B. 1990. The International Joint Commission. *Transboundary Resources Report* 4 (3, Winter): 1.

—— 1990. Sustainable Development and Water Resource Management. *Alternatives* 17(3): 14–24.

Saetevik, S. 1988. Environmental Cooperation Between the North Sea States: Success or Failure? *International Challenges* 8(4): 40–5.

Salacuse, Jeswald W. and Jeffrey Z. Rubin. 1990. Negotiating Processes At Work: Lessons from Four International Cases. *Negotiation Journal* 6 (4, October): 315–18.

Salem, Paul E. 1993. A Critique of Western Conflict Resolution from a Non-Western Perspective. *Negotiation Journal*, October, 361–9.

Salewicz, K.A. 1991. Management of Large International Rivers – Practical Experience from a Research Perspective. In *Hydrology for the Water Management of Large River Basins*, ed. F.H. Van de Ven et al. 57–69. International Association of Hydrological Sciences.

Saliba, Samir N. 1968. *The Jordan River Dispute*. The Hague: Martinus Nijhoff.

Salih, A.M. and A.A. Ali. 1992. Water Scarcity and Sustainable Development. *Nature and Resources* 28(1): 44–8.

Sanchez, R. 1993. Public Participation and the IBWC: Challenges and Options. *Natural Resources Journal* 33 (2, Spring): 283–313.

Sandstrom, Klas. 1995. *Forests and Water – Friends or Foes?* Linkoping Studies in Arts and Science, Vol. 120. Linkoping University.

Sarma, Sunile Sen. 1986. *Farakka: A Gordian Knot*. Calcutta: Ishika Publishing.

Sarris, A.H. 1983. *Food Security and Agricultural Production Strategies Under Risk in Egypt*. Berkeley, CA: Giannini Foundation of Agricultural Economics, California Agricultural Experiment Station.

Savage, Christopher. 1991. Middle East Water. *Asian Affairs* 22 (Fall): 3–10.

Schaake, J.C., R.M. Ragan, and E.J. Vanblargan. 1993. GIS Structure for the Nile River Forecast Project. In *Application of Geographic Information Systems in Hydrology and Water Resources Management*, ed. K. Kovar and H.P. Nachtnebel, 427–31. International Association of Hydrological Sciences.

Schelling, Thomas C. 1960. *The Strategy of Conflict*. Cambridge, MA: Harvard University Press.

Schmida, Leslie. 1983. *Keys to Control: Israel's Pursuit of Arab Water Resources*. Washington, DC: American Educational Trust.

Schmiedler, D. 1969. The Nucleolus of a Characteristic Function Game. *SIAM Journal on Applied Mathematics* 17: 1163–70.

Schmiesing, B. 1989. Theory of Marketing Cooperatives and Decision Making. In *Cooperatives in Agriculture*, ed. D. Cotia. Prentice Hall.

Schrijver, Nico. 1989. International Organization for Environmental Security. *Bulletin of Peace Proposals* 20(2): 115–22.

Schuemann, W. 1993. New Irrigation Schemes in Southeast Anatolia and in Northern Syria: More Competition and Conflict Over the Euphrates? *Quarterly Journal of International Agriculture* 32(3): 240–55.

Schulz, M. 1995. Turkey, Syria, and Iraq: A Hydropolitical Security Complex – The Case of Euphrates and Tigris. In *Hydropolitics*, ed. L. Ohlsson, 84–117. London, England: Zed Books.

Schwartz, Yehoshua and Aharon Zohar. 1991. *Water in the Middle East: Solutions to Water Problems in the Context of Arrangements Between Israel and the Arabs*. Tel Aviv: Jaffee Center for Strategic Studies.

Schwebel, Stephen M. 1981. *Third Report of the Law of Non-Navigational Uses of International Watercourses*. New York: United Nations.

Scudder, T. 1989. The African Experience with River Basin Development. *Natural Resources Forum* (May): 139–48.

Secretariat of the United Nations Commission for Europe. 1994. Protection and Use of Transboundary Watercourses and International Lakes in Europe. *Natural Resources Forum* 18(3): 171–80.

Sedjo, R.A. 1991. *Toward A Worldwide System of Tradeable Forest Protection and Management Obligations, ENR 91-16*. Washington, DC: Resources for the Future.

Sellers, Jackie. 1993. Information Needs for Water Resources Decision-Making. *Natural Resources Forum*, August, 228–34.

Serageldin, I. 1995. *Towards Sustainable Management of Water Resources*. Washington, DC: The World Bank, Directions in Development.

Sewell, W. 1966. *Comprehensive River Basin Planning: the Lower Mekong Experience*. Occasional Paper No. 2. Madison, Wisconsin: University of Wisconsin.

—— and A.E. Utton. 1986. Getting to Yes. *Natural Resources Journal* 26 (2, Spring): 201–5.

Shady, A., A. Adam, and K. Mohammed. 1994. The Nile 2002: the Vision Toward the Cooperation in the Nile Basin. *Water International* 19(2): 77–81.

Shah, R.B. 1994. Inter-state River Water Disputes: A Historical Review. *Water Resources Development* 10(2): 175–89.

Shahin, Mamdouh. 1985. *Hydrology of the Nile Basin*. New York: Elsevier.

—— 1986. Discussion of the paper entitled "Ethiopian Interests in the Division of the Nile River Waters." *Water International* 11(1): 16–20.

—— 1989. Review and Assessment of Water Resources in the Arab World. *Water International* 14.

Shapley, L.S. 1971. Cores of Convex Games. *International Journal of Game Theory* 1: 11–26.

Shaw, Brian R. 1996. When are Environmental Issues Security Issues? *Environmental Change and Security Project* (2, Spring): 39–44.

Shemesh, Moshe. 1988. *The Palestinian Entity 1959–1974: Arab Politics and the PLO*. Jerusalem: Frank Cass.

Sheriff, Muzufar. 1996. *Group Conflict and Cooperation: Their Social Psychology*. London, England: Routledge and Kegan Paul.

Sherk, G.W. 1989. Equitable Apportionment After Vermejo: The Demise of a Doctrine. *Natural Resources Journal* 29: 579–83.

Shrestha, Hari Man and Lekh Man Singh. 1995. The Ganges-Brahmaputra System: A Nepalese Perspective in the Context of Regional Cooperation. In *Asian Water Forum*. Asian Water Forum. United Nations University, 30 January–1 February.

Shubik, M. 1982. *Game Theory in the Social Sciences*. Cambridge, MA: MIT Press.

—— 1988. *A Game-Theoretic Approach to Political Economy*. Cambridge, MA: MIT Press.

Shue, Henry. 1992. The Unavoidability of Justice. In *The International Politics of the Environment: Actors, Interests, and Institutions*, ed. Andrew Hurrell and Benedict Kingsbury, 373–97. New York, NY: Oxford University Press.

Shuval, Hillel. 1980. *Water Quality Management Under Conditions of Scarcity*. New York: Academic Press.

—— 1992. Approaches to Resolving the Water Conflicts Between Israel and her Neighbors – a Regional Water-for-Peace Plan. *Water International* 17: 133–43.

—— 1994. Proposed Principles and Methodology for the Equitable Allocation of the Water Resources Shared by the Israelis, Palestinians, Jordanians, Lebanese, and Syrians. *Studies in Environmental Science* 58: 481–96.

Simms, R.A. 1989. Equitable Apportionment – Priorities and New Uses. *Natural Resources Journal* 29: 549–63.

Sinai, Anne and Allen Pollack. 1977. *The Hashemite Kingdom of Jordan and the West Bank: A Handbook.* New York: American Academic Association for Peace in the Middle East.

Siverson, Randolph M. and Paul F. Diehl. 1989. Arms Races, the Conflict Spiral, and the Onset of War. In *Handbook of War Studies*, ed. Manus I. Midlarsky, 195–218. Boston, MA: Unwin Hyman.

Skujins, J. 1991. *Semiarid Lands and Deserts: Soil Resource and Reclamation.* Marcel Dekker.

Smith, S.C. 1974. Economics and Economists in Water Resource Development. In *Man and Water*, ed. L.D. James, 81–101. Kentucky: The University of Kentucky Press.

Smith, Scot E. and Hussam M. Al-Rawahy. 1990. The Blue Nile: Potential for Conflict and Alternatives for Meeting Future Demands. *Water International* 15(4): 217–22.

Smith, V. Kerry. 1993. Nonmarket Valuation of Environmental Resources. *Land Economics* 69(1): 1–26.

Sofer, A. 1994. The Relevance of the Johnston Plan to the Reality of 1993 and Beyond. *Studies in Environmental Science* 58: 107–21.

Sohn, Louis B. 1987. Peaceful Settlement of Disputes and International Security. *Negotiation Journal* (2, April): 155–66.

Solanes, Miguel. 1987. The International Law Commission and Legal Principles Related to the Non-Navigational Uses of the Water of International Rivers. *Natural Resources Forum.*

—— 1992. Legal and Institutional Aspects of River Basin Development. *Water International* 17: 116–23.

Solbe, J.F., ed. 1986. *Effects of Land Use on Fresh Waters: Agriculture, Forestry, Mineral Exploitation, Urbanisation.* West Sussex, England: Ellis Horwood Limited.

Solem, Erik and Antony F. Scanlan. 1986. Oil and Natural Gas as Factors in Strategic Policy and Action: A Long-Term View. In *Global Resources and International Conflict: Environmental Factors in Strategic Policy and Action*, ed. Arthur H. Westing, 38–54. New York, NY: Oxford University Press.

Spalding, Mark J. 1995. Resolving International Environmental Disputes: Public Participation and the Right-to-Know. *Journal of Environment and Development* 4 (1, Winter): 141–54.

Special Issue on Himalaya Ganga: Contending with Complexity. 1994. *Water Nepal* 4 (1, September): 1994.

Spector, Lea and George Gruen. 1980. Waters of Controversy: Implications for the Arab-Israel Peace Process. *American Jewish Committee*, December.

Sprinz, Detlef. 1995. Regulating the International Environment: A Conceptual Model and Policy Implications. In *Prepared for the 1995 Annual Meeting of the American Political Science Association.* 1995 Annual Meeting of the American Political Science Association. Chicago, IL, 31 August–3 September.

Sprout, H. and M. Sprout. 1965. *The Ecological Perspective on Human Affairs, with Special Reference to International Politics.* Princeton, NJ: Princeton University Press.

Stahl, Michael. 1993. Land Degradation in East Africa. *Ambio* 22 (8, December): 505–8.

Stark, Robert C., David S. Bullock, and Wesley D. Seitz. 1994. Political Power Measures of Reservoir Interest Groups. In *Resolution of Water Quantity and Quality Conflicts*, ed. Ariel Dinar and Edna Loehman. New York, NY: Praeger.

Starkey, B.A. 1994. Negotiation Training Through Simulation: The ICON S International Negotiation Seminars. *Educational Exchange*, Spring, 6–11.

Starr, Harvey and Benjamin A. Most. 1976. The Substance and Study of Borders in International Relations Research. *International Studies Quarterly* 20: 581–620.

Starr, Joyce. 1989. *International Cooperation on Global Water Scarcity and Pollution*. Rome: Food and Agriculture Organization of the United Nations.

—— 1990. Water Politics in the Middle East. *Middle East Insight* 7(2/3): 64–70.

—— 1991. Water Wars. *Foreign Policy* (Spring): 17–36.

—— 1992. Middle East Water Security Framework: Historic Opportunity for Regional Stability and Sustainable Development. Council on Foreign Relations. New York, June.

—— and Daniel C. Stoll, eds. 1988. *The Politics of Scarcity: Water in the Middle East*. Boulder, CO: Westview Press.

—— and Daniel Stoll. 1987. *U.S. Foreign Policy on Water Resources in the Middle East*. Washington, DC: The Center for Strategic and International Studies.

Stauffer, Thomas R. 1982. The Price of Peace: The Spoils of War. *American-Arab Affairs*. (1, Summer): 43–54.

—— 1996. *Water and War in the Middle East: The Hydraulic Parameters of Conflict*. Information Paper, Vol. 5. Washington, DC: The Center for Policy Analysis on Palestine.

Stearns, Scott. 1991. Carter Initiative: Role of the International Negotiation Network in Conflict Resolution. *West Africa* (3872, 25 November–1 December): 1966.

Stein, Janice Gross. 1985. Structures, Strategies, and Tactics of Mediation: Kissinger and Carter in the Middle East. *Negotiation Journal* (October): 331–47.

—— 1988. International Negotiation: A Multidisciplinary Perspective. *Negotiation Journal* (July): 221–31.

—— and Raymond Tanter. 1967. *Rational Decision-Making: Israel's Security Choices, 1967*. Columbus, OH: Ohio State University Press.

Stein, Kenneth and Samuel Lewis. 1991. *Making Peace Among Arabs and Israelis: Lessons From Fifty Years of Negotiating Experience*. Washington, DC: United States Institute of Peace.

——, Samuel Lewis, and Sheryl J. Brown. 1991. *Making Peace among Arabs and Israelis*. Washington, DC: United States Institute of Peace.

Stevens, Georgiana G. 1965. *Jordan River Partition*. Livermore, CA: The Hoover Institution on War, Revolution, and Peace at Stanford University.

Stevens, Joe Bruce. 1966. A Study of Conflict in Natural Resource Use: Evaluation of Recreational Benefits as Related to Changes in Water Quality. thesis, 1–205. Corvallis: Oregon State University.

Stoett, Peter J. 1995. The International Whaling Commission: From Traditional Concerns to an Expanding Agenda. *Environmental Politics* 4 (1, Spring): 130–5.

Stone, Paula J. 1980. A Systems Approach to Water Resource Allocation in International River Basin Development. *Water Resources Research* 16 (1, February): 1–13.

Stork, Joe. 1983. Water and Israel's Occupation Strategy. *MERIP Reports* 116(13): 19–24.

Sun, Peter. 1994. *Multipurpose River Basin Development in China*. Economic Development Institute Seminar Series. Washington, DC: Economic Development Institute of the World Bank.

Susskind, Lawrence. 1994. *Environmental Diplomacy: Negotiating More Effective Global Agreements*. New York, NY: Oxford University Press.

—— and Jeffrey Cruikshank. 1987. *Breaking the Impasse: Consensual Approaches to Resolving Water Disputes*. New York: Basic Books, Inc.

——, Eric J. Dolin, and J.W. Breslin, eds. 1992. *International Environmental Treaty Making*. Cambridge, MA: The Program on Negotiation at Harvard Law School.

——— and A. Weinstein. 1980. Towards a Theory of Environmental Dispute Resolution. *Boston College Environmental Affairs Law Review* 9(2): 143–96.

Swain, Ashok. 1993. Conflicts over Water: The Ganges River Dispute. *Security Dialogue* 24 (December): 429–39.

Swift, Adam. 1993. *Global Political Ecology: The Crisis in Economy and Government.* Boulder, CO: Pluto Press.

Swingle, Paul. 1970. *The Structure of Conflict.* New York: Academic Press.

Symposium on International and Transboundary Water Resources Issues. 1990. *Proceedings of the Symposium on International and Transboundary Water Resources Issues.* Bethesda, MD: American Water Resources Association.

Szekely, A. 1992. General Principles and Planned Measures Provisions in the International Law Commission's Articles on the Non Navigational Uses of International Watercourses: A Mexican Point of View. *Colorado Journal of Environmental Law and Policy* 3(1): 93–101.

——— 1993. Emerging Boundary Environmental Challenges and Institutional issues: Mexico and the United States. *Natural Resources Journal* 33 (1, Winter): 33–58.

——— 1993. How to Accommodate an Uncertain Future into Institutional Responsiveness and Planning: The Case of Mexico and the United States. *Natural Resources Journal* 33 (Spring): 397–403.

Tacsan, Joaquin. 1992. *The Dynamics of International Law in Conflict Resolution.* Dordrecht: Martinus Nijhoff.

Tahovnen, Olli, Veijo Kaitala and Matti Pohjola. 1993. A Finnish-Soviet Acid Rain Game: Noncooperative Equilibria, Cost Efficiency, and Sulfur Agreements. *Journal of Environmental Economics and Management* 24: 87–100.

Tajfel, Henri and John C. Turner. 1986. The Social Identity Theory of Intergroup Behavior. In *Psychology of Intergroup Relations*, ed. William G. Austin and Stephen Worchel, 7–24. Chicago, IL: Nelson-Hall.

Talbot, A.R. 1983. *Settling Things: Six Case Studies in Environmental Mediation.* Washington, DC: The Conservation Foundation and Ford Foundation.

Tamburi, Alfred J. 1973. *A Bibliography and Literature Review of Groundwater Geology Studies in the Indus River Basin.* Fort Collins, CO: Colorado State University.

Tanner, Christopher, Gregory Myers, and Ramchand Oad. 1993. *Land Disputes and Ecological Degradation in an Irrigation Scheme: A Case Study of State Farm Divestiture in Chokwe, Mozambique.* LTC Research Paper, Vol. 111. Madison, WI: Land Tenure Center, University of Wisconsin-Madison.

Tanzi, Attila. 1997. Codifying the Minimum Standards of the Law of International Watercourses: Remarks on Part One and a Half. *Natural Resources Journal* 21 (2, May): 109–18.

Teclaff, L.A. 1967. *The River Basin in History and Law.* The Hague: Martinus Nijhoff.

——— 1991. The Checkered Development of International Water Law. *Natural Resources Journal* 31 (Winter): 45–73.

——— 1991. Fiat or Custom: the Checkered Development of International Water Law. *Natural Resources Journal* 31.

——— 1991. Treaty Practice Relating to Transboundary Flooding. *Natural Resources Journal* 31 (1, Winter): 109–22.

——— and E. Teclaff. 1994. Restoring River and Lake Basin Eco-systems. *Natural Resources Journal* 34 (Fall): 905–32.

——— and A.E. Utton, eds. 1974. *International Environmental Law.* New York: Praeger.

——— and Albert E. Utton, eds. 1981. *International Groundwater Law.* London: Oceana Publications.

Teerink, J.R. and M. Nakashima. 1993. *Water Allocation, Rights and Pricing – Examples from Japan and the United States*. Technical Paper, Vol. 198. Washington, DC: The World Bank.

Teerink, John R. 1978. *Mass Transfer of Water Over Long Distances for Regional Development and Its Effects on Human Environment*. New Delhi: International Commission on Irrigation and Drainage.

Tekeli, Shim. 1990. Turkey Seeks Reconciliation for the Water Issue Induced by the Southeastern Anatolia Project (GAP). *Water International* 15(4): 206–16.

Thacher, Peter S. 1992. The Role of the United Nations. In *The International Politics of the Environment: Actors, Interests, and Institutions*, ed. Andrew Hurrell and Benedict Kingsbury, 183–211. New York, NY: Oxford University Press.

Thomas, Caroline and Darryl A. Howlett. 1993. *Resource Politics: Freshwater and Regional Relations*. Philadelphia, PA: Open University Press.

Thorsell, J., ed. 1990. *Parks on the Borderline: Experience in Transfrontier Conservation*. Gland, Switzerland: IUCN.

Tobey, J.A. and G.V. Chomo. 1994. Resource Supplies and Changing World Agricultural Comparative Advantage. *Agricultural Economics: the Journal of the International Association of Agricultural Economists* 10 (3, May): 207–17.

Tolba, Mostafa K. 1988. Sustainable Water Development: Opportunities and Constraints. *Water International* 13(4): 189–92.

Tollison, R.D. and T.D. Willet. 1979. An Economic Theory of Mutually Advantageous Issue Linkages in International Negotiations. *International Organization* 33: 425–49.

Toman, M.A. 1982a. *The Benefits of International Energy Cooperation: A Theoretical Perspective*. Energy and National Security Series, Discussion Paper D-82H. Washington, DC: Resources for the Future.

—— 1982b. *The Dynamics of International Energy Policy Coordination*. Energy and National Security Series, Discussion Paper D-82J. Washington, DC: Resources for the Future.

—— 1982. *Energy Policy Alliances and Cooperative Games: A Survey of Issues and Solution Concepts*. Energy and National Security Series, Discussion Paper D-82K. Washington, DC: Resources for the Future.

—— 1986. "Depletion Effects" and Nonrenewable Resource Supply: A Diagrammatic Supply. *Land Economics* 62 (4, November): 341–52.

Touval, Saadia. 1987. Frameworks for Arab-Israeli Negotiations – What Difference do They Make? *Negotiation Journal* (January): 37–52.

Trendle, Giles. 1992. Whose Water is It? *The Middle East* (January): 18–20.

Trevin, J.O. and J. Day. 1990. Risk Perception in International River Basin Management: The Plata Basin Example. *Natural Resources Journal* 30 (1, Winter): 87–105.

Trolldalen, Jon Martin. 1992. *International Environmental Conflict Resolution: The Role of the United Nations*. Washington, DC: World Foundation for Environment and Development.

—— 1993. *A Role for the World Bank in International Environmental Conflict Resolution?* preliminary draft World Bank, June. Washington, DC: Pollution and Environmental Economics Division Environment Department, the World Bank.

—— 1995. Strategies and Institutional Aspects in Joint Management of Shared Aquifers. In *Joint Management of Shared Aquifers: The Second Workshop November 27–December 1, 1994*, ed. M. Haddad and E. Feitelson. Jerusalem, Israel: The Harry S. Truman Research Institute and the Palestine Consultancy Group.

—— 1997. Troubled Waters in the Middle East: the Process Towards the First Regional Water Declaration Between Jordan, the Palestinian Authority, and Israel. *Natural Resources Journal* 21 (2, May): 101–8.

Tungittplakorn, Waranoot. 1995. Highland-Lowland Conflict over Natural Resources: A

Case of Mae Soi, Chiang Mai, Thailand. *Society and Natural Resources* 8 (July–August): 279–88.

Turan, Ilter. 1993. Turkey and the Middle East: Problems and Solutions. *Water International* 18 (1, March): 23–9.

—— and Gun Kut. 1997. Political-ideological Constraints on Intra-basin Cooperation on Transboundary Waters. *Natural Resources Journal* 21 (2, May): 139–46.

Turfan, M. and Z. Bozkus. 1991. Hydropower Potential in Turkey. *Water Power and Dam Construction* 44 (12, December): 11–12.

Tweeten, Luther G. 1971. The Search for a Theory and Methodology of Research Resource Allocation. In *Resource Allocation in Agricultural Research*, ed. Walter L. Fishel, 25–61. Minneapolis, MN: University of Minnesota Press.

U.S. Army Corps of Engineers. 1990. *ADR Round Table: U.S. Army Corps of Engineers (South Atlantic Division, Corporate Contractors, and Law Firms)*. IWR Working Paper 90-ADR-WP-1. U.S. Army Corps of Engineers.

Ullman, Richard H. 1983. Redefining Security. *International Security* 8(1): 129–53.

UNDP. 1995. *The Aral in Crisis*. Tashkent: UNDP.

UNESCO, United Nations Environment Programme, International Institute for Applied Systems Analysis, International Association of Hydrological Sciences International Hydrological Programme, and Environmentally Sound Management of Inland Water. 1990. *The Impact of Large Water Projects on the Environment: Proceedings of an International Symposium Convened by UNESCO and UNEP and Organized in Cooperation with IIASA and the IAHS, 21–31 October 1986, UNESCO Headquarters, Paris*. Paris, France: UNESCO.

UNGA. 1973. *Co-operation in the Field of the Environment Concerning Natural Resources Shared by Two or More States*. United Nations General Assembly Resolution, Vol. 3129. New York, NY: International Legal Materials XIII.

United Nations. 1975. *Management of International Water Resources: Institutional and Legal Aspects*. New York, NY: United Nations Publications.

—— 1978. *Register of International Rivers*. New York: Pergamon Press.

—— 1983. *Experiences in the Development and Management of International River and Lake Basins*. New York: United Nations.

—— 1984. *Treaties Concerning the Utilization of Internationl Water Courses for Other Purposes than Navigation*. New York: United Nations.

United Nations Secreteriat. 1960. *Legislative Texts and Treaty Provisions Concerning the Utilization International Rivers for Other Purposes than Navigation*. New York: United Nations.

United States of America (Department of the Interior). 1964. *Report on Land and Water Development in the Indus Plain*. Washington, DC: US GPO.

United States of America (National Research Council). 1992. *Water Transfers in the West: Efficiency, Equity, and the Environment*. Washington, DC: National Academy Press.

Unver, Olcay. 1994. Innovations in Water Resources Development in the Southeastern Anatolia Project. In *Water as an Element of Cooperation and Development in the Middle East*, ed. Ali Ihsan Bagis. Ankara: Hacettepe University.

—— and Bruno Voron. 1993. Improvement of Canal Regulation Techniques: The Southeastern Anatolia Project – GAP. *Water International* 18 (3, September): 157–65, 166–77.

Upreti, B.C. and Nirala Pubci. 1993. *Politics of Himalayan Rivers*.

Urquhart, Brian. 1985. The Work of Peace. *Negotiation Journal*, January, 71–92.

Ury, William L. 1987. Strengthening International Mediation. *Negotiation Journal*, July, 225.

—— 1991. *Getting Past No: Negotiation With Difficult People*. New York: Bantam.

—— 1995. Conflict Resolution among the Bushmen: Lessons in Dispute Systems Design. *Negotiation Journal* 11(4).

—— and Richard Smoke. 1985. Anatomy of a Crisis. *Negotiation Journal*, January, 93.

——, Jeanne M. Brett, and Stephen B. Goldberg. 1988. Designing An Effective Dispute Resolution System. *Negotiation Journal*, October, 413.

——, Jeanne M. Brett, and Stephen B. Goldberg. 1988. *Getting Disputes Resolved; Designing Systems to Cut the Costs of Conflict*. London: Jossey-Bass Publishers.

Utton, A.E. 1970. *National Petroleum Policy: A Critical Review*. Albuquerque, NM: University of New Mexico Press.

—— 1982. Development of International Groundwater. *Natural Resources Journal* 22 (January): 95–119.

—— 1994. Water and the Arid Southwest: An International Region Under Stress. *Natural Resources Journal* 34 (Fall): 957–61.

—— and Daniel H. Henning, eds. 1973. *Environmental Policy: Concepts and International Implications*. New York: Praeger.

—— and L.A. Teclaff, eds. 1987. *Transboundary Resources Law*. Boulder, CO: Westview Press.

Valencia, M.J. 1986. Taming Troubled Waters: Joint Development of Oil and Mineral Resources in Overlapping Claim Areas. *San Diego Law Review* 23: 661.

Van der Keij, W., R.H. Dekker, H. Kersten, and J.A. De Wit, W. 1991. Water Management of the River Rhine: Past, Present, and Future. *European Water Pollution Control* 1(1): 9–18.

van de Klundert, Bram and Pieter Glasbergen. 1995. The Role of Mediation in the Process of Integrated Planning: Environmental Planning in the Area of the World's Biggest Harbour. In *Managing Environmental Disputes: Network Management as an Alternative*, ed. Pieter Glasbergen, 69–90. Boston, MA: Kluwer Academic Publishers.

Van de Ven, F.H., D. Gutknecht, D.P. Loucks, and K.A. Salewicz. 1991. *Hydrology for the Water Management of Large River Basins. Proceedings of an International Symposium, Vienna, August 1991*. International Association of Hydrological Sciences.

Venema, Henry David. 1995. Water Resources Planning for the Senegal River Basin. *Water International* 20 (2, June): 61–70.

Verghese, B.G. 1990. *Waters of Hope*. Oxford: Oxford University Press.

—— and R. Iyer, eds. 1993. *Harnessing the Eastern Himalayan Rivers: Regional Cooperation in South Asia*. Delhi, India: Konark Publishers.

——, R.R. Iyer, Q.K. Ahmad, B.B. Pradhan, and S.K. Malla. 1994. *Converting Water Into Wealth: Regional Cooperation in Harnessing the Eastern Himalayan Rivers*. Delhi, India: Konark Publishers.

—— 1995. Towards An Eastern Himalayan Rivers Concord. In *Asian Water Forum*. Asian Water Forum. United Nations University, 30 January–1 February.

Vermillion, Douglas L. and Carlos Garces-Restrepo. 1994. Transfer of Irrigation Management to Farmers in Colombia: Assessment of Process and Results. *Quarterly Journal of International Agriculture* 33 (October–December): 380–92.

Victoria, J.J. and W.W. Yeh, G. 1979. Systems Analysis and Decision Theory in International Water Resources Planning. *World Congress on Water Resources* 8: 3673–81.

Vidal-Hall, J. 1989. Wellsprings of Conflict. *South*.

Vinagradov, S. 1996. Transboundary Water Resoruces in the Former Soviet Union: Between Conflict and Cooperation. *Natural Resources Jounral* 36 (2, Spring): 393–415.

Vining, Joanne and Herbert W. Schroeder. 1989. Effects of Perceived Conflict, Resource Scarcity, and Information Bias on Emotions and Environmental Decisions. *Environmental Management* 13 (March–April): 199–206.

Vlachos, E., A. Webb, and I. Murphy. 1986. The Management of International River Basin Conflicts. Proceedings of the Workshop Held at the International Institute for Applied Systems Analysis. Laxenbourg, Austria, 22–25 September.

Vlachos, Evan. 1986. *The Management of International River Basin Conflicts*. Washington DC: George Washington University Press.

Votruba, Ladislav and Zdenek. Kos. 1988. *Analysis of Water Resource Systems*. New York, NY: Elsevier.

Wachtel, Boaz. 1992. The Peace Canal Project: A Multiple Conflict Resolution Perspective for the Middle East. Draft.

—— 1993. From Peace Pipelines to Peace Canals: The Search for a Solution to the Middle East Water Crisis. *Middle East Insight*, November/December, 25–9.

Wagner, R. Harrison. 1983. The Theory of Games and the Problem of International Cooperation. *American Political Science Review* 77 (June): 330–46.

Wakil, Mikhail. 1993. Analysis of Future Water Needs for Different Sectors in Syria. *Water International* 18 (1, March): 18–22.

Walker, Odell L. 1960. *Application of Game Theory Models to Decisions on Farm Practices and Resource Use*. Ames, IA: Agricultural and Home Economics Experiment Station, Iowa State University of Science and Technology.

Wallenstein, Peter. 1986. Food Crops as a Factor in Strategic Policy and Action. In *Global Resources and International Conflict: Environmental Factors in Strategic Policy and Action*, ed. Arthur Westing, 143–58. New York, NY: Oxford University Press.

Walters, Stanley D. 1970. *Water For Larsa*. New Haven, CT: Yale University Press.

Waterbury, John. 1979. *Hydropolitics of the Nile Valley*. Syracuse, NY: Syracuse University Press.

—— 1992. Three Rivers in Search of a Regime: The Jordan, the Euphrates and the Nile. William Stewart Tod Professor of Public and International Affairs, April.

—— 1993. Transboundary Water and the Challenge of International Cooperation in the Middle East. In *Symposium on Water in the Arab World*. Symposium on Water in the Arab World. Harvard University, 1–3 October.

—— 1997. Between Unilateralism and Comprehensive Accords: Modest Steps toward Cooperation in International River Basins. *International Journal of Water Resources Development* 13 (3, September): 279–90.

Waterman, Stanley. 1994. Spatial Separation and Boundaries. In *Political Boundaries and Coexistence: Proceedings of the IGU-Symposium, Basle/Switzerland, 24–27 May*, ed. Werner A. Galluser, 395–401. New York, NY: Peter Lang.

Wathern, P., ed. 1988. *Environmental Impact Assessment: Theory and Practice*. Unwim Hyman.

WCED (Expert Group on Environmental Law of the World Commission on Environment and Development). 1987. *Environmental Protection and Sustainable Development*. Graham and Trotman: Martinus Nijhoff Publishers.

Weiner, Myron. 1994. Security, Stability, and Migration. In *Conflict After the Cold War*, ed. Richard K. Betts, 394–412. New York, NY: MacMillan Publishing Company.

Wescoat, James L., Jr. 1984. Evaluation of Long-Term Change in Water Management Systems. *Transactions of the Internaitonal Commission on Irrigation and Drainage*. New Delhi.

—— 1990. Common Law, Common Property, and Common Enemy: Notes on the Political Geography of Water Resource Management for the Sundarbans Areas of Bangladesh. *Agriculture and Human Values* 7: 73–87.

—— 1991. Managing the Indus River Basin in Light of Climate Change. *Global Environmental Change*, December, 381–95.

—— 1992. Beyond the River Basin: The Changing Geography of International Water Problems and International Watercourse Law. *Colorado Journal of International Environmental Law and Policy* 3: 301–30.

—— 1996. Main Currents in Early Multilateral Water Treaties: A Historical-Geographic Perspective, 1648–1948. *Colorado Journal of International Environmental Law and Policy* 7(1): 39–74.

—— 1998. The Historical Geography of Indus Basin Management: A Long-Term Perspective, 1500–2000. In *The Indus River: Biodiversity, Resources, Humankind*, ed. Azra and Meadows. Karachi: Oxford University Press.

—— (forthcoming). Water Rights in South Asia and the United States: Comparative Perspectives, 1873–1996. In *Landed Property Rights and Global Change*, ed. John Richards.

—— Roger Smith, and David Schaad. 1992. Visits to the U.S. Bureau of Reclamation from South Asia and the Middle East, 1946–1990: An Indicator of Changing Internaitonal Programs and Politics. *Irrigation and Drainage Systems* 6: 55–67.

Wesley, James Paul. 1962. Frequency of Wars and Geographical Opportunity. *Journal of Conflict Resolution* 6: 387–9.

Westing, Arthur. 1986. Environmental Factors in Strategic Policy and Action: An Overview. In *Global Resources and International Conflict: Environmental Factors in Strategic Policy and Action*, ed. Arthur H. Westing, 3–20. New York, NY: Oxford University Press.

—— 1986. An Expanded Concept of International Security. In *Global Resources and International Conflict: Environmental Factors in Strategic Policy and Action*, ed. Arthur H. Westing, 183–200. New York, NY: Oxford University Press.

—— 1986. *Global Resources and International Conflict: Environmental Factors in Strategic Policy and Action*. New York, NY: Oxford University Press.

—— 1988. *Cultural Norms, War and the Environment*. Oxford: Oxford University Press.

—— 1989. The Environmental Component of Comprehensive Security. *Bulletin of Peace Proposals* 20(2): 129–34.

—— 1990. *Environmental Hazards of War: Releasing Dangerous Forces in an Industrialized World*. London, England: Sage Publications.

White, Gilbert F. 1963. The Mekong River Plan. *Scientific American* 208 (4, April).

—— 1974. Role of Geography in Water Resources Management. In *Man and Water*, ed. L.D. James, 102–21. Kentucky: The University of Kentucky Press.

—— 1977. *Environmental Effects of Complex River Development*. Boulder, CO: Westview Press.

White, Sally Blount and Margaret A. Neale. 1991. Reservation Prices, Resistance Points, and BATNAs: Determining the Parameters of Acceptable Negotiated Outcomes. *Negotiation Journal* 7 (4, October): 379–88.

White, T. Anderson and C. Ford Runge. 1994. Common Property and Collective Action: Lessons from Cooperative Watershed Management in Haiti. *Economic Development and Cultural Change* 43 (1, October): 1–41.

Whittington, D. 1986. Discussion of the Paper Entitled "Ethiopian Interests in the Division of the Nile River Waters." *Water International* 11(2): 88–9.

—— and Kingsley Haynes. 1985. Nile Water for Whom? Emerging Conflicts in Water Allocation for Agricultural Expansion in Egypt and Sudan. In *Agricultural Development in the Middle East*, ed. Peter Beaumont and K. McLachlan. New York: John Wiley and Sons.

—— and E. McClelland. 1992. Opportunities for Regional and International Cooperation in the Nile Basin. *Water International* 17(3): 144–54.

Wieriks, Koos and Anne Schulte-Wulwer-Leidig. 1997. Integrated Water Management for the Rhine River Basin: From Pollution Prevention to Ecosystem Improvement. *Natural Resources Journal* 21 (2, May): 147–56.

Wilkinson, K.P. 1974. *Diffusion of Technology and Political Information: What Theory Do We Have? Water Resource Planning*. New York, NY: A E Ext Department of Agricultural Economics, New York State College, Agricultural Contract College State University, Cornell University, 74(3): 100–8.

Williams, J.D. 1954. *The Compleat Strategyst, Being a Primer on the Theory of Games of Strategy*. New York: McGraw-Hill.

Wishart, David. 1985. The Political Economy of Conflict over Water Rights in the Jordan

Valley from 1890 to Present. diss. University of Illinois at Urbana-Champaign: University of Illinois at Urbana-Champaign.

—— 1989. An Economic Approach to Understanding Jordan Valley Water Disputes. *Middle East Review* 21 (4, Summer): 45–53.

—— 1990. The Breakdown of the Johnston Negotiations over the Jordan Waters. *Middle Eastern Studies* 26 (4, October): 536–46.

Witmer, T. Richard, ed. 1956. *Documents on the Use and Control of the Waters of Interstate and International Streams: Compacts, Treaties, and Adjudications*. Washington, DC: U.S. Department of the Interior.

Wolf, A.T. 1993. The Jordan Watershed: Past Attempts at Cooperation and Lessons for the Future. *Water International* 18 (1, March): 5–17.

—— 1993. Water-for-Peace in the Jordan River Watershed. *Natural Resources Journal* 33(3): 797–839.

—— 1994. Hydropolitical History of the Nile, Jordan, and Euphrates River Basins. In *International Waters of the Middle East: From Euphrates-Tigris to Nile*, ed. Asit K. Biswas, 5–43. Bombay, Delhi, Calcutta, Madras: Oxford University Press.

—— 1995. *Hydropolitics Along the Jordan River: The Impact of Scarce Resources on the Arab-Israeli Conflict*. Tokyo, Japan: United Nations University Press.

—— 1995. International Water Dispute Resolution: The Middle East Multilateral Working Group on Water Resources. *Water International* 20(3): 141–50.

—— 1996. Hydrostrategic Territory in the Jordan Basin: Water, War, and Arab-Israeli Peace Negotiations. Paper given at conference: Water: A Trigger for Conflict/A Reason for Cooperation. Bloomington, IN, 7–10 March.

—— 1997. International Water Conflict Resolution: Lessons from Comparative Analysis. *International Journal of Water Resources Development* 13 (3, September): 333–66.

—— 1999. Equitable Water Allocations: The Heart of Transboundary Water Conflicts. *Natural Resources Forum* (February).

—— and A. Dinar. 1994. Middle East Hydropolitics and Equity Measures for Water-Sharing Agreements. *The Journal of Social, Political, and Economic Studies* 19(1): 69–89.

—— and M. Murakami. 1995. Techno-political Decision Making for Water Resources Development: The Jordan River Watershed. *Water Resources Development* 11(2): 147–62.

—— and T. Ross. 1992. The Impact of Scarce Water Resources on the Arab-Israeli Conflict. *Natural Resources Journal* 32: 921–58.

Wolfe, Mary Ellen. 1992. The Milk River: Deferred Water Policy Transitions in an International Waterway. *Natural Resources Journal* 32 (1, Winter): 55–76.

Wood, W.B. 1992. Ecopolitics: Domestic and International Linkages. *Geographica* 27(1): 49–54.

World Bank. 1993. *Water Resources Management*. Washington, DC: World Bank.

World Bank/UNDP/UNEP. 1994. Aral Sea Program – Phase 1. Proposed Donors Meeting, 23–4.

World Water Council. 1995. *World Water Council Bulletin* 1 (1, December).

Worthington, P. 1986. Regulatory Remedies for Agricultural Pollution. In *Effects of Land Use on Fresh Waters: Agriculture, Forestry, Mineral Exploitation, Urbanisation*, ed. J.F. Solbe, 341–8. West Sussex, England: Ellis Horwood Limited.

Wrightsman, Lawrence S., Jr., John O'Connor, and Norma J. Baker. 1972. *Cooperation and Competition: Readings on Mixed-Motive Games*. Belmont, CA: Brooks/Cole Publishing Company.

Xepapadeas, Anastasios. 1994. Optimal Management of the International Commons. *Fondazione Eni Enrico Mattei. Nota Di Lavoro* 4(94): 1–36.

Yaron, Dan. 1994. An Approach to the Problem of Water Allocation to Israel and the Palestinian Entity. *Resource and Energy Economics* 16 (4, November): 271–86.

Yergin, Daniel. 1988. Energy Security in the 1990s. *Foreign Affairs* 67: 110–32.

—— and Martin Hillenbrand. 1982. *Global Insecurity: A Strategy for Energy and Economic Renewal*. Boston, MA: Houghton Mifflin Company.

Young, Gordon J., James C. Dooge, and John C. Rodda. 1994. *Global Water Resource Issues*. Cambridge, MA: Cambridge University Press.

Young, Ralph. 1991. The Economic Significance of Environmental Resources: A Review of the Evidence. *Review of Marketing and Agricultural Economics* 59 (3, December): 229–54.

Yuchtman-Yaar, Ephraim and Michael Inbar. 1986. Social Distance in the Israeli-Arab Conflict: A Resource-Dependency Analysis. *Comparative Political Studies* 19 (October): 283–316.

Zacklin, Ralph and Lucius Caflisch. 1981. *The Legal Regime of International Rivers*. The Hague: Martinus Nijhoff.

Zaman, M. 1982. The Ganges Basin and the Water Dispute. *Water Supply and Management* 6(4): 321–8.

—— 1983. The Ganges Basin Development: a Long-term Problem and Some Short-term Options. In *Water Resource Development*, ed. M. Zaman et al., 99–109. Dublin: Tycooly International Publishing.

—— 1983. Institutional and Legal Framework for Cooperation Between Bangladesh and India on Shared Water Resources. River Basin Development: Proceedings of the National Symposium on River Basin Development. Dacca, Bangladesh, 4–10 December.

Zarour, H. and J. Isaac. 1994. A Novel Approach to the Allocation of International Water Resources. *Studies in Environmental Science* 58: 389–98.

—— and J. Isaac. 1993. Nature's Apportionment and the Open Market: A Promising Solution to the Arab-Israeli Water Conflict. *Water International* 18(1): 40–53.

Zartman, I. William. 1978. *The Negotiation Process*. London: Sage Publications.

—— 1992. International Environmental Negotiation: Challenges for Analysis and Peace. *Negotiation Journal* 8 (2, April): 113–24.

—— and Maureen R. Berman. 1982. *The Practical Negotiator*. New Haven, CT: Yale University Press.

Zeitouni, N., N. Becker, and M. Shechter. 1994. Water Sharing Through Trade in Markets for Water Rights: An Illustrative Application to the Middle East. *Studies in Environmental Science* 58: 399–412.

Zohar, Aharon. 1991. Weighty Water: The Problem of Water in the Framework of a Settlement Between Israel and the Arab Countries. In *Winning the Peace*. Italy, December.

Zunes, Stephen. 1995. The Israeli-Jordanian Agreement: Peace or Pax Americana? *Middle East Policy* 3: 57–69.

Index